Pitman Research Notes in Mathematics Series

Titles in this series.

275 Green functions for second order parabolic integro-differential problems
M G Garroni and J F Menaldi

276 Riemann waves and their applications
M W Kalinowski

277 Banach C(K)-modules and operators preserving disjointness
Y A Abramovich, E L Arenson and A K Kitover

278 Limit algebras: an introduction to subalgebras of C*-algebras
S C Power

279 Abstract evolution equations, periodic problems and applications
D Daners and P Koch Medina

280 Emerging applications in free boundary problems
J Chadam and H Rasmussen

281 Free boundary problems involving solids
J Chadam and H Rasmussen

282 Free boundary problems in fluid flow with applications
J Chadam and H Rasmussen

283 Asymptotic problems in probability theory: stochastic models and diffusions on fractals
K D Elworthy and N Ikeda

284 Asymptotic problems in probability theory: Wiener functionals and asymptotics
K D Elworthy and N Ikeda

285 Dynamical systems
R Bamon, R Labarca, J Lewowicz and J Palis

286 Models of hysteresis
A Visintin

287 Moments in probability and approximation theory
G A Anastassiou

288 Mathematical aspects of penetrative convection
B Straughan

289 Ordinary and partial differential equations. Volume IV
B D Sleeman and R J Jarvis

290 *K*-theory for real *C**-algebras
H Schröder

291 Recent developments in theoretical fluid mechanics
G P Galdi and J Necas

292 Propagation of a curved shock and nonlinear ray theory
P Prasad

293 Non-classical elastic solids
M Ciarletta and D Ieşan

294 Multigrid methods
J Bramble

295 Entropy and partial differential equations
W A Day

296 Progress in partial differential equations: the Metz surveys 2
M Chipot

297 Nonstandard methods in the calculus of variations
C Tuckey

298 Barrelledness, Baire-like- and (LF)-spaces
M Kunzinger

299 Nonlinear partial differential equations and their applications. Collège de France Seminar. Volume XI
H Brezis and J L Lions

300 Introduction to operator theory
T Yoshino

301 Generalized fractional calculus and applications
V Kiryakova

302 Nonlinear partial differential equations and their applications. Collège de France Seminar Volume XII
H Brezis and J L Lions

303 Numerical analysis 1993
D F Griffiths and G A Watson

304 Topics in abstract differential equations
S Zaidman

305 Complex analysis and its applications
C C Yang, G C Wen, K Y Li and Y M Chiang

306 Computational methods for fluid-structure interaction
J M Crolet and R Ohayon

307 Random geometrically graph directed self-similar multifractals
L Olsen

308 Progress in theoretical and computational fluid mechanics
G P Galdi, J Málek and J Necas

309 Variational methods in Lorentzian geometry
A Masiello

310 Stochastic analysis on infinite dimensional spaces
H Kunita and H-H Kuo

311 Representations of Lie groups and quantum groups
V Baldoni and M Picardello

312 Common zeros of polynomials in several variables and higher dimensional quadrature
Y Xu

313 Extending modules
N V Dung, D van Huynh, P F Smith and R Wisbauer

314 Progress in partial differential equations: the Metz surveys 3
M Chipot, J Saint Jean Paulin and I Shafrir

315 Refined large deviation limit theorems
V Vinogradov

316 Topological vector spaces, algebras and related areas
A Lau and I Tweddle

317 Integral methods in science and engineering
C Constanda

318 A method for computing unsteady flows in porous media
R Raghavan and E Ozkan

319 Asymptotic theories for plates and shells
R P Gilbert and K Hackl

320 Nonlinear variational problems and partial differential equations
A Marino and M K V Murthy

319 Asymptotic theories for plates and shells
R P Gilbert and K Hackl

320 Nonlinear variational problems and partial differential equations
A Marino and M K V Murthy

321 Topics in abstract differential equations II
S Zaidman

322 Diffraction by wedges
B Budaev

L J Mason
St Peter's College and the Mathematical Institute, Oxford
L P Hughston
Merrill Lynch International, and King's College, London
P Z Kobak
Scuola Internazionale Superiore di Studi Avanzati, Trieste
(Editors)

Further advances in twistor theory

Volume II: Integrable systems, conformal geometry and gravitation

CRC Press
Taylor & Francis Group
Boca Raton London New York

CRC Press is an imprint of the
Taylor & Francis Group, an informa business
A CHAPMAN & HALL BOOK

CRC Press
Taylor & Francis Group
6000 Broken Sound Parkway NW, Suite 300
Boca Raton, FL 33487-2742

First published 1995 by Longman Group Limited

ISBN-13: 978-0-582-00465-8 (pbk)
ISBN-13: 978-0-429-33254-8 (ebk)

Visit the Taylor & Francis Web site at
http://www.taylorandfrancis.com

and the CRC Press Web site at
http://www.crcpress.com

British Library Cataloguing in Publication Data

A catalogue record for this book is
available from the British Library

Library of Congress Cataloging-in-Publication Data

A catalog record for this book is available

Contents

Editors and Contributors

Editors:

L.J.Mason, St. Peter's College and the Mathematical Institute, Oxford.
e-mail: lmason@maths.ox.ac.uk.

L.P.Hughston, Merrill Lynch International and Kings College, London.
e-mail: lane@lonnds.ml.com.

P.Z.Kobak, Scuola Internazionale Superiore di Studi Avanzati, Trieste.
e-mail: kobak@tsmi19.sissa.it.

Contributors:

A.Ashtekar, Center for Gravitational Physics and Geometry, Pennsylvania State University.
e-mail: ashtekar@phys.psu.edu. M.F.Atiyah, Trinity College, Cambridge.

T.N.Bailey, Department of Mathematics, University of Edinburgh. e-mail: tnb@maths.ed.ac.uk.

P.Baird, School of Mathematics, University of Leeds.

R.J.Baston, Paribas Capital Markets, London.

A.Dancer, Department of Applied Mathematics and Theoretical Physics, Cambridge.

A.J.Dougan, The Mathematical Institute, Oxford.

M.G.Eastwood, Department of Pure Mathematics, University of Adelaide.
e-mail: meastwoo@maths.adelaide.edu.au.

J.Fletcher, Fletcher & Partners, Salisbury, UK.

A.R.Gover, Department of Pure Mathematics, University of Adelaide.

T.R.Hurd, Department of Mathematics and Statistics, McMaster University, Ontario.
e-mail: Hurd@mcmaster.ca.

B.P.Jeffryes, Schlumberger Cambridge Research Ltd, Cambridge. e-mail: jeffryes@scr.slb.com.

R.Kelly, The Mathematical Institute, Oxford (but see note on p.257).

M.K.Murray, Department of Pure Mathematics, University of Adelaide.

R.Penrose, Wadham College and the Mathematical Institute, Oxford.

W.T.Shaw, Oxford Systems Solutions, Oxford.

I.A.B.Strachan, Department of Mathematics and Statistics, University of Newcastle.
e-mail: I.A.B.Strachan@newcastle.ac.uk. V.Thomas, The Mathematical Institute, Oxford.
e-mail: vthomas@maths.ox.ac.uk. K.P.Tod, St John's College and the Mathematical Institute, Oxford. e-mail: Tod@maths.ox.ac.uk.

J.C.Wood, School of Mathematics, University of Leeds. e-mail: pmt6jcw@gps.leeds.ac.uk.

N.M.J.Woodhouse, Wadham College and the Mathematical Institute, Oxford.
e-mail: nwoodh@maths.ox.ac.uk.

Preface

It was in 1976 that a group of us at the Mathematical Institute, Oxford, began to circulate *Twistor Newsletter*, an informal publication consisting of short articles written mostly by members of Roger Penrose's research group, relating to active work going on in twistor theory. This was around the time of the publication of the original non-linear graviton construction. It has been said that the art of doing mathematics consists in finding that special case which contains all the germs of generality. The non-linear graviton was just such a 'special case', and as a consequence interest in the theory increased significantly, both among physicists (especially relativists), but also among an increasing number of pure mathematicians. There was thus a small but steady demand from colleagues outside of Oxford and abroad for the informal communication of new ideas and the latest results, and *Twistor Newsletter* neatly fit the bill. It was no doubt an odd sort of journal, but it was successful in its own way, and after thirty-eight issues and eighteen years it continues to thrive.

By 1979 enough material had been stockpiled on this basis to warrant publication in a volume called *Advances in Twistor Theory*, edited by R.S.Ward and one of us (LPH). In it the collected articles were grouped into four broad (sometimes overlapping) categories: massless fields and sheaf cohomology, curved twistor spaces, twistors and elementary particles, and twistor diagrams. This proved to be a useful book, and so we were encouraged to gather together a further collection of subsequent *Twistor Newletter* articles to edit for publication under the general title *Further Advances in Twistor Theory*. Volume I of this new series (which appeared in 1990) was called *Applications of the Penrose Transform*, and here we present Volume II, which is called *Integrable Systems, Conformal Geometry, and Gravitation.*

The range of material in these volumes is extensive, and it has not been easy to find a simple way of logically grouping articles into volumes (let alone chapters!). The scheme we eventually settled on is as follows. Volume I contains material primarily concerned, in one way or another, with applications of sheaf cohomology on flat twistor space, and with the elucidation of the properties of a variety of related twistor contour integral formulae. There are also various serious attempts at applications to basic physics, as well as a host of mathematically motivated interesting generalizations to higher dimensional analogues. Some of the articles are speculative and exploratory, launching out in new directions; whereas others set about straightening out and clarifying (and in some cases greatly improving and generalizing) results already established or conjectured. All of this material refers in a generic way (whether it be loosely or rigorously) to the well known 'Penrose transform', and thus a characteristic feature of much of the analysis in Volume I is *linearity*. The contents include (1) an overview (with both mathematical background and the motivation for the ideas from basic physics), (2) concrete (i.e. contour integral) approaches to the Penrose transform, (3) abstract (i.e. cohomological) approaches to the Penrose transform, (4) twistor theory and elementary particle physics, (5) twistor diagram theory and scattering amplitude evaluation, and (6) sources and currents, relative cohomology, and non-Hausdorff twistor spaces.

The present work, Volume II, contains applications of *flat* twistor space to *non-linear* problems. (Methods that introduce 'deformed' twistor spaces will appear in volume III.) Flat twistor space can be applied when one studies non-linear structures lying either *in* or *on* twistor space. Alternatively, one can sometimes associate a flat twistor space in a natural way to some otherwise non-linear object. For example, local twistors are the flat space twistors associated to the conformally flat Minkowski space that approximates most closely a given conformal manifold to second order at some given point. Another example arises with 2-surface twistors, which form the solution space of the linear 2-surface twistor equation.

Chapter 1 contains articles on integrable or soluble non-linear equations. Here many of the non-linear constructions arise from the study of holomorphic vector bundles on twistor space, which give rise to solutions of the self-dual Yang-Mills equations by virtue of Ward's correspondence. This gives constructions for a wide variety of integrable equations. In fact it has emerged that most such equations arise as symmetry reductions of the self-dual Yang-Mills equations. Solutions of other non-linear equations can be shown to correspond to submanifolds of flat twistor space. Solutions of harmonic map equations and the equations for harmonic morphisms arise in this way.

Chapter 2 contains articles on conformal differential geometry. Here the linear twistor spaces involved are the spinors in six dimensions. These are applied to the study of generalizations of the Kerr theorem and Robinson's theorem in six dimensions. Cartan's conformal connection in four dimensions has structure group $SO(4,2)$. Its associated spin connection is the local twistor connection. This is used to study conformal invariants and invariant differential operators amongst other topics.

Chapter 3 contains articles on various aspects of general relativity. Flat space twistors are applied to cosmological models for which the underlying conformal structure is conformally flat. Space-times admitting solutions of the two-index *twistor equation* are also studied. In another range of applications, advantage is taken of the fact that the equations for space-times with two symmetries are equivalent to the self-dual Yang-Mills equations with two symmetries, so flat space twistor methods can be applied. We also include some 'exceptional' articles, viz., the Penrose-Ashtekar proof of the positive energy theorem and Thomas's study of the initial value problem for general relativity by exact sets and power series.

Finally, chapter 4 is concerned with the development of Penrose's *quasi-local mass* construction. This uses 2-surface twistors as essential ingredients in a definition of the energy momentum and angular momentum of the gravitational and matter fields threading through a space-like two-surface in space-time.

The chapters start with introductions that give some of the background to the material in each chapter, and a summary of the contents of each chapter. We hope that this introductory material, together with that in volumes O and I, will help make these volumes relatively self contained, even for the non-expert.

Our warm thanks to Roger Penrose, Florence Tsou, Debby Morgan, and all the contributors for their help in the production of this volume.

—L.J. Mason, L.P. Hughston, and P.Z. Kobak, November 1994.

A note on global structure and cross references. We refer to the original *Advances in Twistor Theory* (L.P. Hughston & R.S. Ward, editors, Pitman, 1979) as Volume O, and by §O.5.1 we mean article 1 in chapter 5 of that book. In the current *Further Advances in Twistor theory* series, the volume preceeding the present book is Volume I: *Applications of the Penrose Transform.* By §I.2.3 we mean article 3 of chapter 2 of that book. By §§II.4.5-8 we mean articles 5 to 8 of chapter 4 of the present work, Volume II: *Integrable Systems, Conformal Geometry, and Gravitation.*

Chapter 1

Integrable and soluble systems

§II.1.1 Introduction *by L.J.Mason*

Nonlinear differential equations. The most impressive applications of twistor theory arise in the analysis of nonlinear partial differential equations. There are now a number of twistor constructions where the solutions of certain nonlinear partial differential equations can be shown to be in one to one correspondence with deformations of twistor spaces or structures thereon. The utility of these constructions arises from the fact that the twistor descriptions are usually more tractable than the original equations. For local solutions the twistor description can usually be realized in terms of free functions, and in general it can be easier to pass from the twistor description to a solution of the original equations than to solve these nonlinear equations directly.

Probably the first construction in this general spirit is Weierstrass' construction of minimal surfaces in $\mathbb{R}^3$ in terms of a free holomorphic function. In today's parlance we would say that this free holomorphic function describes a holomorphic curve in minitwistor space, the holomorphic tangent bundle of the Riemann sphere. Modern applications started with Penrose's nonlinear graviton construction, (Penrose 1976), in which a correspondence is established between Ricci flat metrics with self-dual Weyl curvature and deformations of twistor space. R.S.Ward then found an analogous correspondence between solutions of the self-dual Yang-Mills equations and holomorphic vector bundles on regions in twistor space (Ward 1977).

In this chapter, articles §§II.1.2–11 are concerned with integrable systems and their twistor constructions. These are mostly based on Ward's correspondence (although there are some connections with the nonlinear graviton construction noted in §§II.1.7–8) and will be described in more detail later. There are also three articles on harmonic morphisms (§§II.1.12–14), which are Euclidean analogues of the Kerr theorem. The Kerr theorem states that a geodesic shear free congruence in Minkowski space corresponds to the intersection of a holomorphic hypersurface in twistor space with $\mathbb{PN}$. The basic idea of a harmonic morphism and its connections with the standard Kerr theorem are described in §§II.1.12–13. A detailed analysis of their properties, including connections with monopoles and spinor fields, is given in §II.1.14. Finally there is an article on harmonic maps (§II.1.15) which analyzes them using their complexification and an article on minimal spheres in (symmetric) Riemannian manifolds using generalizations of the Weierstrass construction (§II.1.16).

Twistor theory and integrable systems. The first few articles in this chapter are concerned with a circle of ideas originating with R.S.Ward concerning a possible 'universal' role for the self-dual Yang-Mills equations in the theory of integrable systems. Integrable and soluble systems are systems of nonlinear differential equations that, despite their nonlinearity, are still relatively tractable, so one can obtain large families of exact solutions and precise analytic information about general classes of solutions. What has emerged is that the twistor construction constitutes a paradigm for integrability, and leads to a substantial unification of the theory of completely integrable systems. This overview has been developed over a number of years by many workers including Ward (1985,1986), Hitchin (1987), Woodhouse & Mason (1988), Mason & Sparling (1989, 1992), and in various lectures by M.F.Atiyah. See also Witten (1992). There is a book in preparation on this topic by Mason & Woodhouse (1995).

I give in the following an outline of this new overview on integrable systems arising from twistor theory and self-duality. Most of the technical details are omitted but there are sufficient references to fill in the gaps. Further details of the basic Ward construction itself can be found in §I.4.0 (see also Ward 1977, Ward & Wells 1989).

The self-dual Yang-Mills equations. It is convenient, when discussing integrable systems, to work with 'low technology' versions of the Ward transform. The self-dual Yang Mills equations are defined on $\mathbb{R}^4$, with coordinates x^a, $a = 0, 1, 2, 3$, and metric $ds^2 = dx^0 \cdot dx^3 + dx^1 \cdot dx^2$. They are equations on a connection $D_a = \partial_a - A_a$ where $\partial_a = \partial/\partial x^a$ and $A_a = A_a(x^b) \in l$ where l is the Lie algebra of some fixed group G (usually a finite dimensional group of matrices). The connection A_a is defined modulo the gauge freedom:

$$A_a \longrightarrow g^{-1}A_a g - g^{-1}\partial_a g, \; g(x) \in G.$$

The self-dual Yang-Mills equations are

$$[D_0, D_2] = [D_1, D_3] = [D_0, D_3] + [D_1, D_2] = 0$$

which are the compatibility conditions $[D_0 + \lambda D_1, D_2 + \lambda D_3] = 0$ for the linear system of equations

$$(D_0 + \lambda D_1)\psi = (D_2 + \lambda D_3)\psi = 0 \qquad (1)$$

where $\lambda \in \mathbb{C}$ and ψ is a vector in some representation of l.

The Ward construction provides a one to one correspondence between gauge equivalence classes of solutions of the self-dual Yang-Mills equations and holomorphic vector bundles on regions in $\mathbb{PT}^*$. In 'low technology' language, one goes from a bundle on dual twistor space to a solution of the self-dual Yang-Mills equations as follows.

Choose local coordinates (λ, w_1, w_2) on $\mathbb{PT}^*$ so that the incidence relations are $w_1 = x_1 - \lambda x_0$ and $w_2 = x_3 - \lambda x_2$. One needs at least one more coordinate chart, $(\lambda', w_1', w_2') = (1/\lambda, w_1/\lambda, w_2/\lambda)$ for the region near $\lambda = \infty$ to cover a large enough region in twistor space (the region covered is the complement of the line corresponding to the point at infinity in space-time). Holomorphic vector bundles have a variety of different representations; the most basic are labelled 'Čech' and 'Dolbeault'. In the Čech description the bundle can be characterized (in this case) by a holomorphic $SL(N, \mathbb{C})$ valued 'patching' or 'clutching' function $P(\lambda, w_1, w_2)$ holomorphic away from $\lambda = 0$. The patching function is effectively a freely prescribable function but is subject to gauge freedom. The general local solution can be obtained from such a patching matrix.

When the bundle is described in terms of a patching function P, the solution $A_a(x^b)$ on space-time is reconstructed by differentiation of the solution $G_\pm$ of a parametrized Riemann-Hilbert problem:

$$G_+(x^a, \lambda) = G_-(x^a, \lambda)P(\lambda, x^1 - \lambda x^0, x^3 - \lambda x^2). \tag{2}$$

Here G_+ is nonsingular on $|\lambda| > 0$ including at the point $\lambda = \infty$, and G_- is nonsingular on $|\lambda| < \infty$. Note that the patching function P is only defined for $\infty > |\lambda| > 0$. Given a generic P with values in $SL(n, \mathbb{C})$ the Birkhoff factorization theorem tells us that, at least on the complement of a codimension–one subset of $\mathbb{M}$, there exist solutions to the above factorization problem. They are unique up to premultiplication by a matrix function of x^a alone by virtue of Liouville's theorem.

The self-dual Yang-Mills connection is reconstructed by attempting to find a connection A_a such that $G_\pm$ is a solution of the linear system

$$(D_0 + \lambda D_1)G_\pm = (D_2 + \lambda D_3)G_\pm = 0.$$

We find that we would have to have

$$A_0 + \lambda A_1 = \{(\partial_0 + \lambda\partial_1)G_\pm\}G_\pm^{-1},$$

with a similar formula for $A_2 + \lambda A_3$. It turns out that this equation is consistent and allows us to read off A_0 and A_1. The consistency follows because

$$\{(\partial_0 + \lambda\partial_1)G_-\}G_-^{-1} = \{(\partial_0 + \lambda\partial_1)G_+\}G_+^{-1}$$

which is a consequence of equation (2) and the fact that $(\partial_0 + \lambda\partial_1)P = 0$. One can read off A_0 and A_1 because the expression with G_- implies that the right hand side is holomorphic for $|\lambda| < \infty$ and the expression with G_+ has a simple pole at $\lambda = \infty$, so the whole expression must, by a simple extension of Liouville's theorem, be linear in λ. The corresponding connection must automatically satisfy the self-dual Yang-Mills equations as there are solutions, G_- and G_+ to equation (1).

See §I.4.0, Ward (1977), or Ward & Wells (1989) for a description in terms of patching functions. Alternatively, when the bundle is defined by a $\bar{\partial}$-operator, the connection A_a is reconstructed from the (parametrized) solution of a linear $\bar{\partial}$-equation on the sphere. See Atiyah, Hitchin & Singer (1978) for a $\bar{\partial}$-operator description appropriate to Euclidean signature; and Sparling in §I.4.2, for a $\bar{\partial}$-operator description appropriate to Lorentzian signature and in line with the brief description above and Chakravarty, Mason & Newman (1987) for a comparison.

These methods demonstrate the 'solubility' of the self-dual Yang-Mills equations since the reconstruction procedure requires us to solve only *linear* problems (which may nevertheless be hard to implement in practice). The nonlinear graviton construction (Penrose 1976) indicates that the self-dual Einstein equations are also integrable and Penrose's method also describes their solutions in terms of free data. The reconstruction procedure is, however, rather more subtle than the above.

The connection with integrable systems. The main link arises from the fact that virtually all completely integrable differential equations in one and two dimensions, and many integrable partial differential equations in three dimensions, are symmetry reductions of the self-dual Yang-Mills equations (SDYM). There are, however, exceptions. In particular, some systems in $2+1$ dimensions, such as the KP equations, are very likely *not* reductions of the self-dual Yang-Mills equations. Such equations still have a twistor-like correspondence (see Mason 1994 and its precursor in §II.1.9 for further discussion of the KP equations). Thus, the point is that *it is the existence of a twistor construction that is central to integrability.* Nevertheless, the self-dual Yang-Mills equations are sufficiently large to include most of the more famous integrable systems.

Integrable systems in one and two dimensions (and those in three dimensions that arise by reduction of the self-dual Yang-Mills equations) can therefore be classified as reductions of SDYM with reference to the various ingredients arising in the reduction. These ingredients are:

(a) The signature of the metric on $\mathbb{R}^4$, either (2,2) or (4,0)

(b) The subgroup of the conformal group (the symmetry group of SDYM) with $(4-n)$-dimensional orbits for a system in n dimensions

(c) The gauge group

(d) Certain constants of integration that arise in the reduction process (these typically only play a significant role when one or more of the symmetries is essentially null)

(e) A choice of gauge in which to represent the SDYM solution.

To illustrate the framework I present the case of two-dimensional translation subgroups. These reductions are classified partially by the signature of the metric restricted to the 2-plane spanned by the translations:

1. **Nondegenerate cases.**
   (a) $(+,+)$ signature leads to the equations

$$\partial_x(J^{-1}\partial_x J) + \partial_y(J^{-1}\partial_y J) = 0$$

where $J = J(x,y)$. J describes a harmonic map from $\mathbb{R}^2$ into:
(i) $G_{\mathbb{C}}/G$ for $(++++)$ signature,
(ii) G for $(++--)$ signature.

   (b) $(+,-)$ signature leads to

$$\partial_t(J^{-1}\partial_t J) - \partial_x(J^{-1}\partial_x J) = 0$$

where $J = J(x,t) \in G$. This is the Chiral model or 'wave map'.

   (c) We work with signature $(+,-)$ as above, but, for $G = SL(n)$, impose a $\mathbb{Z}_n$ rotational symmetry in the 2-plane spanned by the translations and require also that the generator of the $\mathbb{Z}_n$ symmetry acts on A_a at the axis by conjugation with

$$\begin{pmatrix} 1 & 0 & \cdots & 0 \\ 0 & \omega & \cdots & 0 \\ \vdots & \vdots & \ddots & \vdots \\ 0 & 0 & \cdots & \omega^{n-1} \end{pmatrix}$$

where $\omega^n = 1$. We then obtain the extended Toda lattice (and the sine-Gordon equations for $n = 2$). See Ward (1985, 1986) and Hitchin (1987) for details of these reductions.

2. **Partially degenerate case.** In this case the metric on the 2-plane spanned by translations has rank one, which requires that the signature for $\mathbb{R}^4$ must be $(++--)$. Choose coordinates so that the metric is

$$ds^2 = dx^2 - dy^2 + dv \cdot dt$$

and the symmetries are ∂_y and ∂_v, so that the latter is null. The reduced Lax pair (or linear system) can then be put in the form

$$(\partial_x - A_1 + \lambda\Phi)\psi = 0, \quad (\partial_t - A_2 + \lambda\partial_x)\psi = 0.$$

Two of the field equations that arise from the compatibility conditions lead to 'constants of integration'. These are:

(i) $\partial_x\Phi = 0$ which implies that Φ is a function of t only. However, the components of this l-valued (recall that l is the Lie algebra of the gauge group G) function of t persist into the final reduced equations, and we must actually demand that they be t-independent if we want the reduced equation to be independent of t. So we shall assume that $\partial_t\Phi = 0$ and regard Φ as a constant of integration.

(ii) $\partial_x A_1 = [\Phi, A_2]$ which implies that the image of A_1 in $l/[\Phi, l]$ is independent of x. The image of A_1 in $l/[\Phi, l]$ will also be assumed constant following the same considerations as above. Take a constant representative Ψ in l of the equivalence class determined by A_1 in $l/[\Phi, l]$.

Reductions of this form are classified by pairs (Φ, Ψ) modulo conjugation by the gauge group G together with the transformation $\Psi \mapsto \Psi + [a, \Phi], a \in l$. The two quantities can naturally be considered together as an element of the loop algebra $\hat{l}$ of l. Write

$$\Lambda = \Phi + \frac{1}{\lambda}\Psi \in \hat{l}.$$

Various examples are as follows:

(a) $G = SL(2)$, $\Lambda = \begin{pmatrix} 1 & 0 \\ 0 & -1 \end{pmatrix}$ leads to the non-linear Schrödinger equations (Mason & Sparling 1989, 1992).

(b) $G = SL(2)$, $\Lambda = \begin{pmatrix} 0 & 1/\lambda \\ 1 & 0 \end{pmatrix}$ leads to the Korteweg de Vries equations (Mason & Sparling 1989, 1992).

(c) G general, $\Lambda =$ Coxeter element of $\hat{l}$, leads to parts of the Drinfeld-Sokolov (1985) hierarchies (Mason & Singer 1994).

More general choices of Λ lead to generalizations of the Drinfeld & Sokolov hierarchies such as those discovered by Kacs & Wakimoto (1989). In the $SL(2)$ case, if the constants of integration are allowed to depend on t, one can still eliminate the functions and reduce to standard KdV and non-linear Schrödinger if one is allowed to perform a nonlinear Galilean coordinate transformation (Mason & Sparling 1992). This is explained and generalized in §II.1.10.

3. **Signature $(0,0)$, anti-self-dual.** There are two distinct types of 2-planes on which the metric restricts to zero, self-dual and anti-self-dual. (This can only happen in $(2,2)$ signature.) In the anti-self-dual case the Lax pair now reduces to:

$$(\partial_x - A_1 + \lambda\Phi_1)\psi = 0, \quad (\partial_x - A_2 + \lambda\Phi_2)\psi = 0.$$

(a) Use the gauge freedom to set $A_1 = A_2 = 0$. We then obtain the Wess-Zumino-Witten equations (Strachan 1992).

(b) Use $[\Phi_1, \Phi_2] = 0$ to put Φ_1 and Φ_2 both into simultaneous normal form under conjugation using the gauge freedom. There are two extreme normal forms:

(i) with Φ_1, Φ_2 diagonal, assumed constant and A_1 and A_2 assumed skew we get the N-wave equations (see §II.1.5 or Ablowitz & Clarkson 1992 and references therein).

(ii) Φ_1, Φ_2 nilpotent, assumed constant we obtain other parts of the Drinfeld-Sokolov hierarchies (Mason & Singer 1994).

4. **Signature $(0,0)$, self-dual.** There are no known applications.

Remarks. (1) Each equation in each hierarchy of all these equations can be obtained using higher order symmetries of the self-dual Yang-Mills equations. However to do so in a uniform manner requires the use the Bogomolny hierarchy and generalizations (see Mason & Sparling 1989, 1992, and Mason & Singer 1994).

(2) Non-translation symmetry groups generally lead to equations that are not translation invariant, although they may have some other residual symmetry such as the symmetry group of a hyperbolic metric (Atiyah 1987). See §II.1.10 for examples where the residual symmetry group is much larger and more interesting than one would have expected.

(3) The ordinary differential equations obtained by the Adler, Kostant & Symes procedure (see for example the description in the last chapter of Fadeev & Takhtajan 1987) are reductions of the Bogomolny hierarchy (Wong & Mason, preprint).

(4) The various Painlevé ordinary differential equations are reductions of the $SL(2)$ self-dual Yang-Mills equations with different choices of 3-dimensional abelian subgroups of the conformal group in 4-dimensions (Mason & Woodhouse, 1993).

Twistor theory and the theory of integrable systems. One can impose symmetries on the Ward construction for SDYM and encode the constants of integration to obtain a reduced twistor correspondence for solutions of each integrable equation. This provides a one to one correspondence between solutions of the equations and holomorphic vector bundles perhaps satisfying some additional conditions (associated with the constants of integration, the chosen reality structure and boundary conditions, etc.) on a reduced twistor space. These have been investigated by many authors.

A single translation symmetry in Euclidean space yields the *Bogomolny equations.* The quotient of twistor space by the corresponding symmetry on twistor is the holomorphic tangent bundle of $\mathbb{CP}^1$; this quotient space has become known as 'minitwistor space'. See Hitchin (1982, 1987) who characterizes the bundles corresponding to monopoles (solutions of the Bogomolny equations satisfying certain natural boundary conditions) in terms of *spectral curves* in minitwistor space. The Euclidean case with one rotational symmetry gives rise to the Bogomolny equations in hyperbolic three-space. The corresponding reduced twistor space is $\mathbb{CP}^1 \times \mathbb{CP}^1$ and is treated in Atiyah (1987); see also Jones & Tod (1985). This case is pursued further in §II.1.3 where the authors aim to link the spectral curves arising for these solutions with those arising in certain solutions of the Yang-Baxter equations for a family of integrable statistical mechanical models—the chiral Potts models. In §II.1.4 it is shown that the generic conformal symmetry leads to a reduced twistor space that is rather complicated. In particular, it is non-Hausdorff.

For two translations with signature $(1+0)$ see Mason & Sparling (1989, 1992) and with signature $(0+0)$ see §II.1.5 and Mason & Singer (1994). See Ward (1983), Woodhouse (1987), Woodhouse &

Mason (1989) and Woodhouse & Fletcher (1990) for the case of one rotation and one translation. This last case is of particular interest as this reduction effectively yields the full Einstein vacuum equations in the presence of two commuting Killing vectors. The articles §II.3.7–9 apply these ideas to the study of stationary axisymmetric space-times.

There is too much material for a systematic review, but we note the following general features that impinge on articles in this chapter:

(1) Riemann-Hilbert problems for obtaining solutions can always be obtained from the twistor correspondence via the presentation of bundles on twistor space using patching functions. The inverse scattering transform can be understood (perhaps rather naively in this context) as a particular normal form for the patching function on twistor space (Mason & Sparling 1992). In §II.1.9 it is proposed that a deeper geometric origin for the inverse scattering transform is available, and this is given for the full self-dual Yang-Mills equations (and the self-dual Einstein equations) in terms of a non-Hausdorff twistor correspondence in §II.1.11.

(2) In Mason & Sparling (1989, 1992) a very natural generalization of the Bogomolny equations was presented. This has a twistor correspondence using the complex line bundle of Chern class n on $\mathbb{CP}^1$. The motivation for this was that symmetry reductions of the hierarchy give rise to the standard hierarchies associated with the nonlinear Schrödinger and KdV equations. In §II.1.6 it is shown that other symmetry reductions of these equations give rise to a large class of systems in the literature based on higher order spectral problems.

(3) It is often possible to obtain a good description of the complete solution space via its description as a moduli space of of holomorphic vector bundles on some reduced twistor space. See for example Atiyah *et al.* (1978), Hitchin (1982), Atiyah (1987), Woodhouse & Mason (1988), and Mason & Singer (1994), amongst others. The moduli space for $SU(3)$ monopoles with non-maximal symmetry breaking is analyzed in §II.1.2. The construction in §II.1.11 gives an elegant parametrization of the solution space of global solutions of the self-dual Yang-Mills equations in $(2,2)$ signature paralleling the Atiyah *et al.* (1978) description in Euclidean sugnature.

(4) An important role is played by the self-dual Einstein equations. This is completely integrable in the twistor sense, but the twistor construction for it is qualitatively different. However, there are close connections with the self-dual Yang-Mills equations; the self-dual Einstein equations are a reduction of the self-dual Yang-Mills equations with various diffeomorphism groups as gauge group (Ashtekar, Jacobson & Smolin 1988, Mason & Newman 1989). In §II.1.7 this is used and extended to show that all integrable systems in two dimensions with 2×2 linear systems can be obtained as reductions of the self-dual Einstein equations (Mason 1990). This is extended further in §II.1.8 using the Moyal algebra.

(5) A striking feature of reductions of the self-dual Yang-Mills equations is that their symmetry

group is often larger than one would have expected from naive considerations (in particular, the conformal invariance of the standard 2 translation reduction from Euclidean space discovered by Hitchin). Old and new examples of this phenomena are explored in §II.1.10.

See §II.1.9 for a discussion of some open problems and avenues for further research.

References

Ablowitz, M.J. & Clarkson, P.A. (1992) *Solitons, Nonlinear evolution equations and inverse scattering*, L.M.S. Lecture note series, **149**, CUP.

Ashtekar, A. Jacobson, T. & Smolin, L. (1988) A new characterization of half-flat solutions to Einstein's equation, Comm. Math. Phys. **115**, 631-648.

Atiyah, M.F., Hitchin, N.J. & Singer, I.M. (1978) Self-duality in four-dimensional Riemanninian geometry, Proc. Roy. Soc. **A 362**, 425

Atiyah, M.F., Hitchin, N.J., Drinfeld, V.G. & Manin, Yu.I. (1978) Construction of Instantons, Phys. Lett. A, **65**, 185-187.

Atiyah, M.F. (1979) *Geometry of Yang-Mills fields*, Scuola Normale Superiore, Pisa.

Atiyah, M.F. (1987) Magnetic monopoles in hyperbolic spaces, *Vector bundles on algebraic varieties*, (M.F.Atiyah ed.), OUP.

Chakravarty, S., Ablowitz, M.J. & Clarkson, P.A. (1991) One dimensional reductions of self-dual Yang-Mills Fields and classical equations, in *Volume in honour of E.T.Newman's 60th birthday*, (A.Janis ed.).

Drinfeld, V.G. & Sokolov, V.V. (1985) Lie algebras and equations of Korteweg de Vries type, Jour. Sov. Math. **30**, 1975.

Fadeev, L. & Takhtajan, M. (1987) *Hamiltonian methods in the theory of solitons*, Springer.

Fletcher, J. & Woodhouse, N.M.J. (1990) Twistor characterization of stationary axisymmetric solutions of Einsteins equations, in *Twistors in mathematics and physics*, (T.N.Bailey & R,J,Baston eds.), LMS Lecture Notes Series **156**, CUP.

Hitchin (1982) Monopoles and Geodesics, Comm. Math. Phys., **83**, 589-602.

Hitchin (1987) Monopoles, minimal surfaces and algebraic curves, Seminaire de Mathematiques supérieures, NATO Advanced Study Institute, Montreal University Press.

Hughston, L.P. & Mason, L.J. (1988) *Class. Quant. Grav.*, **5**, p. 275.

Ivancovitch, J.S., Mason, L.J. & Newman, E.T. (1990) On the density of the Ward ansatze in the space of solutions of anti-self-dual Yang-Mills solutions, Comm. Math. Phys., **130**, 139-155.

Jones, P. & Tod, K.P. (1985) Minitwistor spaces and Einstein-Weyl spaces, Class. Quant. Grav. **2**, 565-77.

Kacs, V.G., Wakimoto, M. (1989) Exceptional hierarchies of soliton equations, in: *Proceedings of Symposia in Pure Mathematics*, **49**, 191, A.M.S.

Mason, L.J. (1992) Global solutions of the self-duality equations in signature (2,2), preprint.

Mason, L.J. (1994) Generalized twistor correspondences, d-bar problems and the KP equations, in *Proceedings of the Seale Hayne conference on twistor theory*, (S.A.Huggett, ed.), Marcel Dekker.

Mason, L. Chakravarty, S. & Newman, E.T. (1988) Backlünd transformations for the anti-self-dual Yang-Mills equations, J. Math. Phys. **29**, 4, 1005-1013.

Mason, L.J. & Newman, E.T. (1989) A connection between the Einstein and Yang-Mills equations, Comm. Math. Phys., **121**, 659-668.

Mason, L.J. & Singer, M.A. (1994) The twistor theory of equations of KdV type: part 1, to appear in Comm. Math. Phys.

Mason, L.J. & Sparling, G.A.J. (1989) Nonlinear Schrödinger and Korteweg de Vries are reductions of self-dual Yang-Mills, Phys. Lett. **A 137**(1,2), 29-33.

Mason, L.J. & Sparling, G.A.J. (1992) Twistor correspondences for the soliton hierarchies, J. Geom. Phys., **8**, 243-271.

Mason, L.J. & Woodhouse, N.M.J. (1992) Self-duality and the Painlevé transcendents, Nonlinearity.

Mason, L.J. & Woodhouse, N.M.J. (1993) Twistor theory and the Schlesinger equations', in *Proceedings of Nato ARW, Exeter* 1992, Kluwer.

Mason, L.J. & Woodhouse, N.M.J. (1995) Twistor theory, self-duality and integrability, OUP.

Newman, E.T. (1986) Gauge theories, the holonomy operator and the Riemann-Hilbert problem, J.Math. Phys. **27**, 2797.

Penrose, R. (1976) Nonlinear gravitons and curved twistor theory, Gen. Rel. Grav. **7**, 31-52.

Strachan, I.A.B. (1992) Phys. Lett., **282**, 63-66.

Ward, R.S. (1977) On self-dual gauge fields, Phys. Lett. **61A**, 81-2.

Ward, R.S. (1981) Ansatze for self-dual Yang-Mills fields, Comm. Math. Phys., **80**, 563-74.

Ward, R.S. (1983) Stationary axisymmetric space-times: a new approach, Gen. Rel. Grav. **15**, 105-9.

Ward, R.S. (1984) Completely solvable gauge field equations in dimension greater than four, Nucl. Phys. **B 236**, 381-396.

Ward, R.S. (1985) Integrable and solvable systems and relations among them, Phil. Trans. R. Soc. A **315**, 451-7.

Ward, R.S. (1986) Multidimensional integrable systems. in: *Field theory, quantum gravity and strings*, eds. H. de Vega and N. Sanchez, Lecture Notes in Physics Vol. 246, Springer, Berlin.

Ward, R.S. & Wells, R.O. (1989) *Twistor geometry and field theory*, CUP.

Ward, R.S. (1990) Integrable systems in twistor theory, in *Twistors in mathematics and physics*, (T.N.Bailey & R.J.Baston eds.), LMS Lecture Notes Series **156**, CUP.

Witten, E. (1992) J. Geom. Phys., **8**.

Wong, W.K. & Mason, L.J. (1992) The Adler, Kostant & Symes O.D.E.'s and the Bogomolny hierarchy, preprint.

Woodhouse, N.M.J. (1985) Real methods in twistor theory, Class. Quant. Grav. **2**, 257–291.

Woodhouse, N.M.J. (1987) Twistor description of the symmetries of Einstein's equations for stationary axisymmetric space-times, Class. Quant. Grav., **4**, 799–814.

Woodhouse, N.M.J. & Mason, L.J. (1988) The Geroch group and non-Hausdorff twistor spaces, Nonlinearity, **1**, 73–114.

§II.1.2 **Twistors and $SU(3)$ monopoles** *by A.Dancer* ($\mathbb{TN}$ 29, November 1989)

Hitchin (1983a) has shown that $SU(2)$-monopoles of charge k on $\mathbb{R}^3$ are equivalent to algebraic curves (*spectral curves*) of genus $(k-1)^2$, satisfying certain constraints, lying in the minitwistor space $T\mathbb{P}^1$. Now $T\mathbb{P}^1 = \{(\mathbf{u},\mathbf{v}) \mid \mathbf{u},\mathbf{v} \in \mathbb{R}^3; ||\mathbf{u}|| = 1, \mathbf{u}\cdot\mathbf{v} = 0\}$ so it may be identified with the space of oriented lines in $\mathbb{R}^3$. Also $T\mathbb{P}^1$ fibres over $\mathbb{P}^1$ and we may take coordinates (η,ζ) on $T\mathbb{P}^1$ where ζ is a coordinate on $\mathbb{P}^1$ and η is a fibre coordinate. There is a real structure on $T\mathbb{P}^1$; in terms of the above coordinates it is $\tau : (\eta,\zeta) \mapsto (-\overline{\eta}/\overline{\zeta}^2, -1/\overline{\zeta})$, but it is easier to think of it as just reversing the orientation of oriented lines in $\mathbb{R}^3$. We define line bundles L^t of degree 0 over $T\mathbb{P}^1$ by letting L^t be the bundle with transition function $\exp(t\eta/\zeta)$.

For each $SU(2)$-monopole there is just one associated spectral curve S in $T\mathbb{P}^1$. It satisfies:

(i) S is compact and has equation $\eta^k + a_1(\zeta)\eta^{k-1} + \ldots + a_k(\zeta) = 0$ where each a_i is a polynomial of degree $2i$.

(ii) L^2 is trivial over S; or equivalently (since deg $L^2 = 0$), $H^0(S, L^2) \neq 0$.

(iii) S is preserved by the real structure τ.

(iv) S has no multiple components.

(v) (nondegeneracy condition) $H^0(S, L^t(k-2)) = 0$ for $0 < t < 2$.

A parameter count gives the dimension of the moduli space of charge k $SU(2)$-monopoles as $4k-1$.

These results have been extended to the case of $SU(n)$-monopoles with symmetry broken to $U(1)\times\ldots\times U(1)$ by Murray (1983) who showed that such monopoles were generically determined by $n-1$ spectral curves (satisfying certain constraints) in minitwistor space.

We can also consider monopoles with nonmaximal symmetry breaking, i.e. symmetry broken to a nonabelian subgroup. In particular consider $SU(3)$ monopoles with symmetry broken to $U(2)$. As $SU(3)$ is the QCD gauge group such monopoles may be of particular physical interest.

We now have only one spectral curve (as opposed to two curves for $U(1)\times U(1)$ symmetry breaking). This curve satisfies conditions (i) and (iii) above; however the condition that there is a nontrivial element of $H^0(S, L^2)$ is replaced by the requirement that $H^0(S, L^3(p,p)) \neq 0$ (for the charge $2p$ monopole). Parameter counting, using results from algebraic geometry about the dimension of linear systems on algebraic curves, suggests that the charge $2p$ moduli space should have dimension $\leqslant 12p-1$; in fact the charge 2 moduli space should have dimension precisely 11 (or 8 once we fix the centre of the monopole in $\mathbb{R}^3$). This agrees with a result of Weinberg (1982) (Weinberg includes an S^1 phase to get 12 parameters).

Further investigations concerning nondegeneracy conditions suggest that the 4-dimensional space of $SU(2)$ charge 2 monopoles should arise as a boundary of a 5-dimensional quotient of the 8-

dimensional space of $SU(3)$ minimal symmetry breaking charge 2 monopoles. Now it is known (Hitchin 1983b) that $SU(2)$ charge k monopoles are equivalent to triples (T_1, T_2, T_3) of $k \times k$-matrix valued functions on [0,2] satisfying:

(1) $T_i^*(t) = -T_i(t)$

(2) $T_i(2-t) = -\overline{T}_i(t)$

(3) T_i is analytic on (0,2) with simple poles at $t = 0, 2$

(4) $\frac{dT_1}{dt} = [T_2, T_3]$ and cyclically (Nahm's Equations).

(5) The residues of the T_i at $t = 0, 2$ give an irreducible representation of $SU(2)$.

The pole at $t = 2$ corresponds to the bundle L^2 being trivial over the spectral curve S. Condition (2) reflects the quaternionic nature of $SU(2)(\cong Sp(1))$. In the $SU(3)$ case, therefore, we should drop these conditions. The resulting modified system of Nahm's equations may be solved (in the charge 2 case) explicitly using $SO(3)$ and $SU(2)$ symmetries and Jacobi elliptic functions. We obtain a 5-dimensional moduli space of centred $SU(3)$ charge 2 monopoles, having the moduli space of $SU(2)$ monopoles as a boundary. The moduli space of $SU(3)$ monopoles includes a spherically symmetric monopole and a 3-parameter family of axisymmetric monopoles; this agrees with results of Ward (1981) arrived at via twistor theory (Ward considers uncentred monopoles and so gets a 6-parameter family of axisymmetric solutions).

Comment (1993). Further material on Nahm moduli spaces and their relevance to $SU(3)$ monopoles may be found in Dancer (1992), Dancer (1994) and Dancer & Leese (1993).

References

Dancer, A.S. (to appear, 1994) Nahm's equations and hyperkähler geometry, Comm. Math. Phys.

Dancer, A.S. (1992) Nahm data and $SU(3)$ monopoles, Nonlinearity **5**, 1355–1373.

Dancer, A.S. and Leese, R.A. (1993) Dynamics of $SU(3)$ monopoles, Proc. Royal Soc. **A 440**, 421–430.

Hitchin, N.J. (1983a) Monopoles and Geodesics, Comm. Math. Phys. **83**, p. 579–602.

Hitchin, N.J. (1983b) On the Construction of Monopoles, Comm. Math. Phys. **89**, 145–190.

Murray, M.K. (1983) Monopoles and spectral curves for arbitrary Lie groups, Comm. Math. Phys. **90**, p. 263–271.

Ward, R.S. (1981) Magnetic Monopoles with gauge group $SU(3)$ broken to $U(2)$, Phys. lett. **B 107**, p. 281–284.

Weinberg, E.J. (1982) Fundamental and Composite Monopoles, in *Monopoles in Quantum Field Theory*, World Scientific, p. 153–154.

§II.1.3 Monopoles and Yang-Baxter equations *by M.F.Atiyah & M.K.Murray* ($\mathbb{TN}$ 30, June 1990)

It has been known for some time that the Yang-Baxter equations can be solved using elliptic curves. More recently it was discovered (Au-Yang *et al* 1987, Baxter *et al* 1988) that the Yang-Baxter equation for the N state chiral Potts model could be solved using special curves of genus $(N-1)^2$.

The curves that arise can be defined as follows. Let $k^2 + k'^2 = 1$ and then intersect the following two Fermat surfaces in $\mathbb{CP}_3$:

$$a^N + k'b^N = kd^N$$
$$k'a^N + b^N = kc^N.$$

This gives a curve $\hat{\Sigma}_N$ with a high degree of symmetry. In fact $\mathbb{Z}_N^4$ acts on $\mathbb{CP}_3$ by

$$(w_1, w_2, w_3, w_4)[a, b, c, d] = [w_1 a, w_2 b, w_3 c, w_4 d]$$

(where $w_i^N = 1$) and this fixes $\hat{\Sigma}_N$ (of course the diagonal (w, w, w, w) acts trivially so it is really a $\mathbb{Z}_N^3 = \mathbb{Z}_N^4/\Delta$ action). The quotient of $\hat{\Sigma}_N$ by a free action of a $\mathbb{Z}_N$ subgroup gives the curve Σ_N of genus $(N-1)^2$.

On a visit to Canberra in 1989 the first author conjectured that these curves should be related to the spectral curves of an $SU(2)$ monopole of charge N. These are also a special class of curves of genus $(N-1)^2$. We now understand how such a relationship exists for hyperbolic monopoles with Higgs field equal to zero. Consider the $\mathbb{C}^\times$ action on $\mathbb{CP}^3$ given by

$$\lambda : [a, b, c, d] \longmapsto [\lambda a, b, c, \lambda d].$$

If we remove the lines $C_1 = [0, b, c, 0]$ and $C_2 = [a, 0, 0, d]$ of fixed points this is a free action with quotient the quadric $Q = \mathbb{P}_1 \times \mathbb{P}_1$. In fact it realizes $\mathbb{CP}_3 - C_1 \cup C_2$ as the $\mathbb{C}^\times$ bundle of the line bundle $\mathcal{O}(1, -1)$ over the quadric. We shall call this bundle L.

The important fact is that $\hat{\Sigma}_N$ intersects the orbits of the $\mathbb{C}^\times$ action in the orbits of $\mathbb{Z}_N$ considered as a subgroup inside $\mathbb{C}^\times$. So projecting to Q divides $\hat{\Sigma}_N$ by $\mathbb{Z}_N$ to give the curve Σ_N in Q. It is easy to check that Σ_N is in the linear system $\mathcal{O}(N, N)$. In fact we can say more. If we factor the $\mathbb{C}^\times$ bundle $\mathbb{CP}_3 - C_1 \cup C_2$ by $\mathbb{Z}_N$ this gives the bundle $L^N \to Q$ and the curve $\hat{\Sigma}_N$ becomes a section over Σ_N. So Σ_N satisfies the constraint

$$L^N_{|\Sigma_N} \simeq \mathcal{O}.$$

In the theory of hyperbolic monopoles (Atiyah 1987) the monopole is determined by a spectral curve S_N in Q in the linear system $\mathcal{O}(N, N)$. This satisfies a constraint

$$L^{2p+N}_{|S_N} \simeq \mathcal{O}$$

where p is the norm of the Higgs field at infinity.

So this shows that the curve Σ_N is that for a monopole with zero Higgs field! Strictly speaking such monopoles are trivial so we have to interpret Σ_N as arising from some limit of monopoles. Work in progress suggests that this can be done via the rational map of the monopole. Finally notice that we can turn this discussion about and say that a curve Σ_N as above with the constraint $L^N/\Sigma_N \simeq 0$ is equivalent to a curve in $\mathbb{CP}_3$ with no constraint except invariance under $\mathbb{Z}_N \subset \mathbb{C}^\times$. Looked at from this point of view the curves $\hat{\Sigma}_N$ are special curves invariant under two more actions of $\mathbb{Z}_N$. The more general curves we have discussed here have moduli spaces of dimension $4N$ and it is hoped that this means that these curves and these associated solutions can be generalized.

Comment (1994). The interested reader is referred also to Atiyah (1991) where more details of these results appear.

References

Atiyah, M.F. (1987) Magnetic Monopoles in hyperbolic spaces, in *Proceedings of the Bombay Colloquium 1984 on vector Bundles on Algebraic Varieties* (Oxford University Press), p. 1–34.

Atiyah, M.F. (1991) Magnetic monopoles and the Yang-Baxter equation, in International Journal of Modern Physics A, Vol 6, No 16., 2761 - 2774.

Au-Yang, H., McCoy, B., Perk, J.H.H., Tange, S., and Yau, M.-L. (1987) Commuting transfer matrices in the chiral Potts models: Solutions of star-triangle equations with genus > 1, *Phys. Lett.* **A 123** (5), p. 219–223.

Baxter, R.J., Perk, J.H.H. and Au-Yang, H. (1988) New solutions of the star-triangle relations for the Chiral Potts model, *Phys. Lett.* **A 128** (3,4), p. 138–142.

§II.1.4 **A Non-Hausdorff Mini-twistor Space** *by K.P. Tod* ($\mathbb{TN}$ 30, June 1990)

This note is about another example of a non-Hausdorff complex manifold arising naturally in twistor theory. A mini-twistor space $\mathbb{K}$ is the 4-real-dimensional space of directed geodesics of a 3-real-dimensional Weyl space $\mathbb{W}$, which becomes a 2-complex-dimensional manifold if the Weyl space satisfies the Einstein-Weyl condition. Since it is defined as a space of geodesics, and geodesics can wind around in funny ways, a mini-twistor space is always liable to be non-Hausdorff. I will describe an example of a particularly simple Einstein-Weyl space where the mini-twistor space can be seen to be non-Hausdorff in a fairly tame way.

Recall first that a Weyl space $\mathbb{W}$ is a manifold with a symmetric connection D and a conformal metric $[g]$ which is preserved by D. Given a choice g_{ab} of representative metric, the compatibility between conformal metric and connection means that we can define D in terms of the metric connection and a 1-form ω_a. Under change-of-choice of representative metric we have

$$g_{ab} \to \Omega^2 g_{ab}\ ; \quad \omega_a \to \omega_a + 2\nabla_a \log \Omega \tag{1}$$

so that we can think of a Weyl space as the pair (g_{ab}, ω_a) subject to (1). For more details see e.g. Hitchin (1980), Jones & Tod (1985) and Pedersen & Tod (1993).

The connection D has a Riemann tensor and a Ricci tensor, but the Ricci tensor is not necessarily symmetric. The Einstein-Weyl condition on $\mathbb{W}$ is that the symmetrised Ricci tensor be proportional to the (conformal) metric. This can be written out as an equation on the Ricci tensor of the representative metric and the 1-form ω_a. In 3 dimensions the equation is

$$R_{ab} - \frac{1}{2}\nabla_{(a}\omega_{b)} - \frac{1}{4}\omega_a\omega_b = \Lambda g_{ab}, \quad \text{some } \Lambda. \tag{2}$$

This equation is, from its definition, conformally invariant and can be regarded as a conformally-invariant generalisation of the Einstein equations. Note that spaces conformal to Einstein spaces satisfy (2) since we can use (1) to eliminate ω_a. These examples can be recognised by the fact that ω_a is exact.

The example I want to consider comes about by conformally rescaling and making identifications on flat space. Take the metric and 1-form as

$$g = dr^2 + r^2(d\theta^2 + \sin^2\theta\, d\phi^2); \quad \omega = 0$$

and conformally rescale with $\Omega = \exp(-\chi)$, defining $\chi = \log r$:

$$g = d\chi^2 + d\theta^2 + \sin^2\theta\, d\phi^2; \quad \omega = -2d\chi$$

Now impose a periodicity in χ to obtain an Einstein-Weyl structure on $S^1 \times S^2$ (this example is given in Pedersen & Tod (1993); part of the interest of it is that this manifold has no Einstein metric). The periodictity in χ corresponds to identifying the radial coordinate r with λr for some λ with $0 < \lambda < 1$.

As I said at the beginning, the space of directed goedesics of a 3-dimensional Einstein-Weyl space $\mathbb{W}$ is a 2-dimensional complex manifold $\mathcal{H}$, the mini-twistor space of $\mathbb{W}$. For flat space, the mini-twistor space is the space of directed lines in $\mathbb{R}^3$ which can be thought of as pairs of 3-dimensional real vectors (a, b) where a is unit and b is orthogonal to a. Equivalently, this is $T\mathbb{P}_1$, the tangent

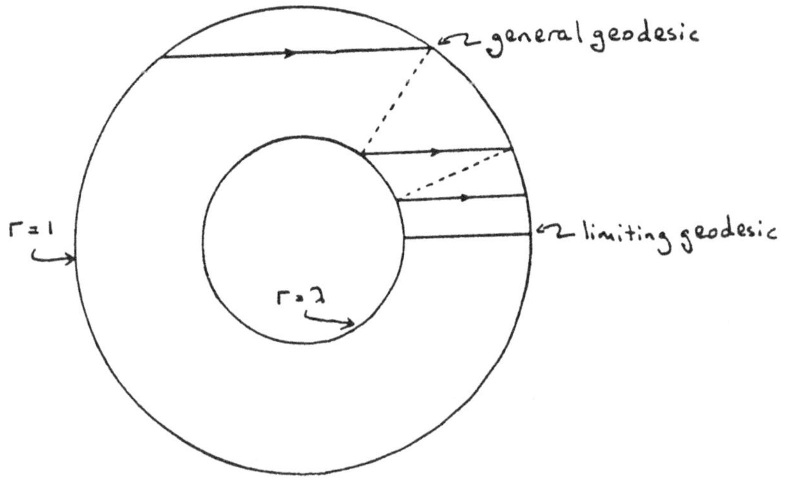

bundle of the complex projective line. For the example to be considered here we shall need to modify this a little.

A geodesic in the $S^1 \times S^2$ Einstein-Weyl structure is as shown in diag. 1. It is basically a straight line which, when it hits the outer sphere at $r = 1$, is brought back to the inner sphere at $r = \lambda$ making the same angle with the radius vector. This means that in the future, the geodesic tends to a limiting one which is radially outwards and closed, while in the past is tends to a limiting one which is radially inwards and closed. In particular, this means that there are 'shadows' in the space: given a point p, points on the diameter through p but on the other side cannot be reached by geodesics through p (I am grateful to Paul Gauduchon for the suggestion that there might be shadows in this example). We shall return to these shadows below.

To construct the mini-twistor space $\mathfrak{K}$, consider first the closed radial geodesics. These correspond to the zero-section of $T\mathbb{P}_1$, ie. to lines in $\mathbb{R}^3$ defined by pairs of the form $(a, 0)$, but there are two closed radial geodesics for each radial geodesic in flat-space so we need to double the zero-section. Next the non-radial geodesics: think of a line in flat-space as a pair (a, b). Then the process of bringing this back from the outer sphere to the inner sphere in the figure above corresponds to leaving a alone but rescaling b, ie. $b \to \lambda b$, with λ as before.

This is then the mini-twistor space: delete the zero-section from $T\mathbb{P}_1$; identify b with λb in the fibres; then put two copies of the zero-section back. It is non-Hausdorff at the radial geodesics, ie. at the doubled-up-zero-section, since any geodesic which is 'near to' a radially ingoing one is also 'near to' the continuation of it to the other side as a radially outgoing one.

A point p in the Einstein-Weyl space is represented by a holomorphic curve (a 'twistor line') in the mini-twistor space. The specification of this twistor line includes, at some stage, a choice of which of a pair of doubled-up points to take. Then any twistor line through the other of the relevant pair of doubled-up points in the mini-twistor space will correspond to a point of the Einstein-Weyl space in the 'shadow' of p.

A more complicated example of a non-Haudsorff mini-twistor space should be provided by the 'Berger sphere' Einstein-Weyl space, Jones & Tod (1985) and Pedersen & Tod (1993). This corresponds to a left-invariant metric on the 3-sphere. There is a special set of geodesics like the radial ones in the example above with the property that any other goedesic tends to one of them in the future and another in the past. The mini-twistor space seems to be a sort of deformed quadric with non-Hausdorff-ness along two generators of the same family. Henrik Pedersen and I have a description of it as a 'weighted projective space' but it is a little obscure.

Like my article II.1.13, the work for this was done during a most pleasant visit to Henrik Pedersen in Odense, and I gratefully acknowledge hospitality received.

References

Hitchin, N. (1980) in proceedings of *Twistor geometry and non-linear systems, Primorsko Bulgaria, 1980*, eds. Doebner and Palev (Springer Lecture Notes in Mathematics **970**).

Jones, P.E. & Tod, K.P. (1985) Class. Quant. Grav. **2**, pp. 565–577.

Pedersen, H. & Tod, K.P. (1993) Three-dimensional Einstein-Weyl geometry, Adv. in Math. **97**, pp. 74–109.

§II.1.5 **The 3-Wave Interaction from the Self-dual Yang Mills Equations** *by K.P.Tod*

(TN 33, November 1991)

There is a family of completely integrable systems called 'the n-wave interaction' (see eg Ablowitz & Segur 1981). According to current twistor dogma, these equations should be reductions of the self-dual Yang-Mills equations. While trying to do something different, I found a way of getting them by this route. I also found, though somewhat later, that Chakravarty & Ablowitz (1990) had a slightly different route with similar end-points.

The starting point is the self-dual Yang-Mills equations with 2 null symmetries. These are equivalent to the commutation relation

$$[D_1, D_2] = 0 \tag{1}$$

where

$$\begin{aligned} D_1 &= \partial_1 - A_1 + \zeta B_1 \\ D_2 &= \partial_2 - A_2 + \zeta B_2, \end{aligned} \tag{2}$$

the A_i and B_i are $n \times n$ complex matrices, functions of x^1 and x^2 only, and ζ is a complex constant.

Substituting (2) into (1) and equating separate powers of ζ to zero gives 3 equations. The $\mathcal{O}(\zeta^2)$ term is just

$$[B_1, B_2] = 0 \tag{3}$$

Mason & Singer (1991) (see also II.1.7) solve this by taking the B_i to be nilpotent and arrive at the n-th generalised KdV equation. The opposite extreme, which I shall take, is to suppose that the B_i are diagonalisable by the Yang-Mills gauge freedom, which is

$$\begin{aligned} B_i &\to G^{-1} B_i G; \\ A_i &\to G^{-1}(A_i G - \partial_i G). \end{aligned} \tag{4}$$

where G is an $n \times n$ complex matrix depending on x^1 and x^2. Now the $\mathcal{O}(\zeta)$ term in (1) is

$$\partial_1 B_2 - \partial_2 B_1 + A_2 B_1 + B_2 A_1 - A_1 B_2 - B_1 A_2 = 0 \tag{5}$$

The diagonal entries in (5) imply that there is a 'potential' for the B_i:

$$B_i = \partial_i C \tag{6}$$

while the off-diagonal entries imply that the off-diagonal entries of the A_i are proportional in a way that I shall write out explicitly below. Before that, we consider the $\mathcal{O}(1)$ term in (1) which is

$$\partial_2 A_1 - \partial_1 A_2 + A_1 A_2 - A_2 A_1 = 0 \tag{7}$$

The diagonal entries in (7) imply that the diagonal entries of the A_i have potentials in a way analogous to (6). A gauge transformation (4) with diagonal G preserves the diagonality of the B_i and can be chosen to remove the diagonal entries of the A_i.

To summarise the situation at this point in the argument:

(i) the B_i are diagonal and derived from a potential C as in (6);

(ii) the A_i are purely off-diagonal and can be expressed in terms of C and each other using (5);

(iii) finally (7) imposes some differential equations.

At what is essentially this point, Chakravarty & Ablowitz (1990) take the matrices B_i to be constant and arrive at the n-wave interaction. This is a specialisation in that the B's can't in general be made constant by a gauge transformation (4), but it leads to the same equations eventually as we shall see.

Now it is necessary to resort to taking components so for simplicity I will restrict to 3×3 matrices. Set

$$B_1 = \text{ diag } (\alpha, \beta, \gamma) = \partial_1 C \; ; \qquad B_2 = \text{ diag } (\lambda, \mu, \nu) = \partial_2 C \tag{8}$$

and

$$\begin{array}{lll} \alpha - \beta = \partial_1 P & \beta - \gamma = \partial_1 Q & \gamma - \alpha = \partial_1 R \\ \lambda - \mu = \partial_2 P & \mu - \nu = \partial_2 Q & \nu - \lambda = \partial_2 R \end{array} \tag{9}$$

so that

$$P + Q + R = 0. \tag{10}$$

We will eventually switch to using two of P, Q, R as independent variables.

With the choices (8) for the B_i, we can solve (5) for the A_i in terms of another off-diagonal matrix E. Set

$$A_1 = (a_{ij}), \quad A_2 = (b_{ij}), \quad E = (e_{ij}),$$

then (5) implies

$$a_{12} = (\alpha - \beta)e_{12} \; ; \qquad b_{12} = (\lambda - \mu)e_{12} \tag{11}$$

and the 5 equations obtained from this by the obvious permutations.

Finally, we substitute (11) into (7) to obtain differential equations on E. These differential equations can all be written with the aid of the Poisson bracket in (x^1, x^2). The typical one, from which the other 5 follow by permutations, is

$$\{e_{12}, R\} = e_{13}e_{32}\{P, Q\} \tag{12}$$

We can break the symmetry between P, Q, R by adopting P and Q as new independent coordinates. Write 'dot' and 'prime' for differentiation with respect to P and Q respectively and set

$$E = \begin{pmatrix} 0 & H & V \\ W & 0 & F \\ G & U & 0 \end{pmatrix}$$

then (12) becomes the system

$$\begin{array}{llll} F' = -VW & & U' = GH \\ \dot{G} = WU & & \dot{V} = -FH \\ \dot{H} - H' = -UV & & \dot{W} - W' = FG \end{array} \tag{13}$$

which is equivalent to the 3-wave interaction.

The further reduction 'dot = minus prime' leads, after some manipulating of constants, to the integrable Hamiltonian

$$h = p_1 p_2 q_3 + q_1 q_2 p_3$$

References

Ablowitz, M.J. & Segur, H. (1981) *Solitons and the Inverse Scattering Transform*, SIAM Philadelphia.

Chakravarty, S. & Ablowitz, M.J. (1990) On reductions of self-dual Yang-Mills equations, University of Colorado preprint **PAM 62**.

Mason, L.J. & Singer, M.A. (1994) The twistor theory of equations of KdV type, to appear in Comm. Math. Phys.

§II.1.6 **The Bogomolny Hierarchy and Higher Order Spectral Problems** *by I.A.B. Strachan* (TN 34, May 1992)

The starting point for the construction and solution of a wide range of integrable models is to write the equation as the integrability condition for the otherwise overdetermined linear system (where $\lambda \in \mathbb{CP}^1$ is the spectral parameter):

$$\begin{aligned} \partial_x s &= -U(\lambda).s\,, \\ \partial_t s &= -V(\lambda).s\,. \end{aligned} \tag{1}$$

The integrability condition for (1) is

$$\partial_x V - \partial_t U + [U, V] = 0\,, \tag{2}$$

and equating powers of λ (if U and V are polynomial in λ) yields the equation in question. Many of those systems which are known to have a twistorial description (such as the KdV, mKdV, NLS, SG and N-wave equations) arise from a so-called first order spectral problem, with

$$\begin{aligned} U &= \lambda A + Q(x,t)\,, \\ V &= \sum_i \lambda^i A_i(x,t)\,. \end{aligned}$$

In this article the matrices will be taken to be $sl(2,\mathbb{C})$-valued, with $A = \begin{pmatrix} \kappa & 0 \\ 0 & -\kappa \end{pmatrix}$ and $Q = \begin{pmatrix} 0 & p \\ q & 0 \end{pmatrix}$, i.e. $A \in \boldsymbol{h}$ and $Q \in \boldsymbol{k}$, where $\boldsymbol{h}$ is the Cartan subalgebra and $\boldsymbol{k}$ is the complement. In the terminology of Mason & Sparling, the fields are of type β; type α fields will not be considered here.

A higher order spectral problem is one for which U and V are general polynomial functions, namely:

$$U = \lambda^p A + \lambda^{p-1} Q_1 + \ldots + Q_p \,,$$
$$V = \lambda^n V_0 + \lambda^{n-1} V_1 + \ldots + V_n \,.$$

The simplest example ($p = 2\,, n = 4$ and $Q_2 = 0$) results in the derivative non-linear Schrödinger (or DNLS) equation. The purpose of this article is two-fold: firstly to show how such systems are nothing more than a reduction of the Bogomolny hierarchy introduced in Mason & Sparling (1992) and secondly to generalise these systems to $(2+1)$-dimensions while retaining their integrability.

The following method to generate the matrices $Q_1, \ldots, Q_p, V_0, \ldots, V_n$ for these higher order problems is due to Crumey (1992). Let

$$u = \lambda^p . A \,,$$
$$v = \lambda^n . A \,.$$

These trivially satisfy (2). However, this equation is gauge invariant, so if $\omega(x,t)$ is a λ-dependent gauge transformation (often called a 'dressing transformation'), defined by $\omega = \exp \sum_{i=1}^{\infty} \omega_i \lambda^{-i}$ with $\omega_i \in sl(2,\mathbb{C})$, then U and V, defined by

$$U = \omega u \omega^{-1} - \omega_x \omega^{-1} \,,$$
$$V = \omega v \omega^{-1} - \omega_t \omega^{-1} \,,$$

will also satisfy (2). Assuming that ω is chosen so that U and V involve only non-negative powers of λ yields, on projecting onto positive (including the λ^0 term) and negative powers of λ, the equations

$$U = (\lambda^p \omega A \omega^{-1})_+ \,, \qquad V = (\lambda^n \omega A \omega^{-1})_+ \,,$$
$$\omega_x \omega^{-1} = (\lambda^p \omega A \omega^{-1})_- \,, \qquad \omega_t \omega^{-1} = (\lambda^p \omega A \omega^{-1})_- \,.$$

These simplify further by decomposing ω as $\omega = h.k$, where $h = \sum_{i=1}^{\infty} h_i(x,t)\lambda^{-i}$, $h_i(x,t) \in \boldsymbol{h}$ and $k = \sum_{i=1}^{\infty} k_i(x,t)\lambda^{-i}$, $k_i(x,t) \in \boldsymbol{k}$. One then has

$$U = (\lambda^p k A k^{-1})_+ \,, \qquad V = (\lambda^n k A k^{-1})_+ \,.$$

Let A_{n-i} denote the coefficient of λ^{-i} in the expansion of kAk^{-1} (the reason for this skew choice will become apparent later), i.e.

$$A_{n-i} = \sum_{r=1}^{i} \frac{1}{r!} \sum_{(\{s_j\}:\sum s_j = i)} [k_{s_1}, [k_{s_2}, \ldots, [k_{s_r}, A] \ldots]] \,.$$

From this procedure one obtains the general form of the functions U and V. The matrices $k_1, \ldots, k_p$ are matrix valued fields. The integrable equation itself (which connects the time evolution of these fields with their spacial derivatives), together with the remaining matrices, may be found using the above equations, or equivalently, equation (2).

Having found the general form of U and V it remains to show how these are contained within the Bogonolny hierarchy. Assuming $m \equiv n - p \geqslant 0$, the matrix V may be written in the form

$$V = \lambda^m.U + \sum_{i=0}^{m-1} \lambda^i A_i \,,$$

and hence the original system (1) may be rewritten as

$$\begin{aligned} \partial_x s &= -\Big\{\sum_{i=0}^{p} \lambda^i A_{m+i}\Big\} s \,, \\ \partial_t s - \lambda^m \partial_x s &= -\Big\{\sum_{i=0}^{m-1} \lambda^i A_i\Big\} s \,. \end{aligned} \tag{3}$$

Recall from Mason & Sparling (1992) that given the minitwistor space $\mathcal{O}(n)$, the line bundle over the Riemann sphere of Chern class $n \geqslant 1$, the Ward construction gives rise to the linear system

$$\big\{[\partial_{z_i} + A_i] - \lambda[\partial_{z_{i+1}} + B_{i+1}]\big\} s = 0 \,, \qquad i = 0, \ldots, n-1 \,,$$

where A_i and B_{i+1} are $sl(2,\mathbb{C})$-valued gauge potentials. With the symmetry generated by ∂_{z_n}, together with $B_i = 0, i = 1, \ldots,, n-1, B_n \equiv A_{n+1}, A_0 = A$, relabelling $z_0 = t, z_m = x$, and eliminating the other variables results in (3) :

$$\left.\begin{aligned} [\partial_t - \lambda\partial_{z_1}]s &= -A.s \\ &\vdots \\ [\partial_{z_{m-1}} - \lambda\partial_x]s &= -A_{m-1}.s \end{aligned}\right\} \Rightarrow [\partial_t - \lambda^m \partial_x]s = -\Big\{\sum_{i=0}^{m-1} \lambda^i A_i\Big\}.s \,,$$

$$\left.\begin{aligned} [\partial_x - \lambda\partial_{z_{m+1}}]s &= -A_m.s \\ &\vdots \\ [\partial_{z_{n-1}} - \lambda\partial_{z_n}]s &= -[A_{n-1} + \lambda A_n].s \end{aligned}\right\} \Rightarrow \partial_x s = -\Big\{\sum_{i=0}^{p} \lambda^i A_{m+i}\Big\}.s \,.$$

Thus these higher order spectral problems may all be embedded within the Bogomolny hierarchy. Solutions of the simplest example, that of the DNLS equation, correspond to bundles over the space $\mathcal{O}(4)$ with certain symmetries.

These systems have an elegant generalisation to (2+1)-dimensions, Strachan (1993). By replacing the term $\lambda^m\partial_x$ in (3) by $\lambda^m\partial_y$ one naturally obtains examples of $(2+1)$-dimensional integrable systems. Thus the DNLS equation has the following generalisation:

$$\begin{aligned} i\partial_t\psi &= \partial_{xy}\psi + 2i\partial_x[V.\psi] \,, \\ \partial_x V &= \partial_y|\psi|^2 \,. \end{aligned}$$

These may be given a twistorial description by introducing a weighted twistor space defined by $\mathbb{T}_{m,p} = \{(Z_0, Z_1, Z_2, Z_3)\}/\sim$, where Z_0, Z_1 are coordinates on the Riemann sphere, $Z_2, Z_3 \in \mathbb{C}$, and $\sim$ is the equivalence relation

$$(Z_0, Z_1, Z_2, Z_3) \sim (\mu Z_0, \mu Z_1, \mu^m Z_2, \mu^p Z_3) \,, \quad \forall \mu \in \mathbb{CP}^1 \,.$$

Reimposing the symmetry $\partial_x = \partial_y$ corresponds to factoring out by a non-vanishing holomorphic vector field on $\mathbb{T}_{m,p}$ to recover $\mathcal{O}(m+p)$, exactly analogous to the construction of the minitwistor space $\mathcal{O}(2)$ from standard twistor space.

References

Mason, L.J. & Sparling, G.A.J. (1992) Twistor Correspondences for the Soliton Hierarchies, Journal of Geometry and Physics, **8**, 243-271.

Crumey, A. (1992) Integrable Hierarchies, Homogeneous Spaces and Kac-Moody Algebras, Leeds preprint.

Strachan, I.A.B. (1993) Some Integrable models in $(2+1)$-dimensions and their twistor description, J. Math. Phys. **34**, 243-259.

§II.1.7 **H-Space: a universal integrable system?** *by L.J.Mason* ($\mathbb{TN}$ 30, June 1990)

The following speculations have not been fulfilled yet (and may never) but I feel that the concrete aspects of the ideas are of interest and the various relations involved are intriguing.

Motivation. There is a large forest of integrable systems. R.S.Ward, amongst others, has pointed out that many, if not indeed most integrable systems are reductions of the self-dual Yang-Mills equations. This observation isn't just a question of bookkeeping, it gives a substantial insight into the theory underlying these equations, as the inverse scattering transform for these systems can be understood as a symmetry reduction of the Ward construction for solutions of the self-dual Yang-Mills equations. (See for example Mason & Sparling 1989 and 1992; the symmetry reduction can, however, be somewhat nontrivial, see in particular Woodhouse & Mason 1988 in which non-Hausdorff Riemann surfaces play an essential role).

Two gaps in the story are as follows. Firstly there appear to be genuine difficulties in incorporating the KP and Davey-Stewartson equations. There is little difficulty in incorporating integrable systems into some kind of twistor framework if the inverse scattering transform is realised by means of the solution of a Riemann-Hilbert problem. However the inverse scattering problem for the KP equations is more subtle and requires the solution of a 'non-local Riemann-Hilbert problem'. This gap is particularly irritating in view of the theoretical importance that the KP equations have acquired with their relations to the theory of Riemann surfaces and infinite dimensional grassmanians and so on. The second gap is that there appears to be little role for the self-dual vacuum equations and its twistor construction, Penrose's nonlinear graviton construction—this, it should be pointed out, is not based on the solution of a Riemann-Hilbert problem either. However I should like to state the following conjecture:

CONJECTURE. *The KP and Davey-Stewartson equations are reductions of the self-dual Einstein equations.*

The circumstantial evidence is as follows. (The self-duality equations are taken to be concerned with space-times with metric of signature (2,2).)

LEMMA 1. *KP can be obtained in the limit as $n \to \infty$ of the $SL(n)$ self-dual Yang-Mills equations reduced by two orthogonal null translations. (This extends the results of Mason & Sparling 1989).*

LEMMA 2. *(Hoppe, J.) The Lie algebra of the area preserving diffeomorphism group of a surface $\Sigma^2, S\,Diff(\Sigma^2)$ can be approximated arbitrarily closely by that of $SL(n)$ as $n \to \infty$.*

LEMMA 3. *The self-dual Einstein equations are equivalent to the self-dual Yang-Mills equations reduced by two orthogonal null translations with gauge group $S\,Diff(\Sigma^2)$. (This extends the results of Mason & Newman 1989).* □

REMARK. If it were the case that $SL(n)$ were a subgroup of $SL(\infty) = S\,Diff(\Sigma^2)$ then these results would imply that all 2-dimensional integrable models are obtainable as reductions from the self-dual Yang-Mills equations (at least by translations). Hence the title of this note and the question mark. However, my current opinion is that $SL(n)$ is only a subgroup of $S\,Diff(\Sigma^2)$ for $n = 2$. This still yields a reasonable class of integrable systems and certainly the more famous ones such as the KdV, nonlinear Schrödinger and sine-Gordon equations.

Proof of lemma 1. I shall use the presentation of the KP hierarchy due to Gelfand and Dickii. See for instance Segal & Wilson (1985) in the proceedings of the I.H.E.S for a description of this approach. The equations of the KP hierarchy are the consistency conditions for the existence of a solution ψ; to the following system of linear partial differential equations

$$(\partial_{t_2} - (Q^2)_+)\psi = 0, \quad (\partial_{t_3} - (Q^3)_+)\psi = 0, \ \ldots, \ (\partial_{t_r} - (Q^r)_+)\psi = 0, \ \ldots$$

where $(Q^r)_+$ is an rth order O.D.E. in the x variable, $Q_r = (\partial_x)^r + ru(\partial_x)^{r-2} + \ldots + w_r$ and $u(x, t_2, t_3, \ldots)$ is the subject of the KP hierarchy equation and w_r is some function which will be determined in terms of u by the equations. The notation is intended to indicate that the ordinary differential operators $(Q^r)_+$ are the differential operator part of the pseudo-differential operator Q

raised to the rth power where $Q = \partial_x + u(\partial_x)^{-1} +$ (lower order) and where $(\partial_x)^{-1}$, is a formal pseudo-differential operator defined by the relation

$$(\partial_x)^{-1}f = f(\partial_x)^{-1} + \sum_{i=1}^{\infty}(-\partial_x)^i f(\partial_x)^{-i-1}.$$

The original KP equation is the equation on $u(x, t_1, t_2)$ that follows from the consistency conditions for $(\partial_{t_2} - (Q^2)_+)\psi = 0$ and $(\partial_{t_3} - (Q^3)_+)\psi = 0$ alone. The evolution in the higher time variables are symmetries of the basic equations (and each other). If one imposes invariance in the nth time variable t_n, then the reduced system is referred to as the nth generalized KdV hierarchy ($n = 2$ gives the standard KdV hierarchy and $n = 3$ the Boussinesq).

The idea is that the operators $(Q^r)_+$ can be thought of as infinite dimensional matrices acting on $L^2(\mathbb{R})$ where x is a coordinate on $\mathbb{R}$. One can approximate this by $n \times n$ matrices by imposing a symmetry in the nth time variable since (setting $\partial_{t_n}\psi = \lambda\psi$) we have $(Q^n)_+\psi = \lambda\psi$ and we consider only ψ lying in the n-dimensional solution space of this equation, represented, say, by ψ and its first $(n-1)$-derivatives with respect to x. With this reduction we have:

$(\partial_{t_2} - (Q^2)_+)\psi = 0$ reduces to

$$\left\{\partial_{t_2} - \begin{pmatrix} 2u & 0 & 1 & 0 & \dots & 0 \\ \cdot & 2u & 0 & 1 & \ddots & \vdots \\ \cdot & \cdot & 2u & 0 & \ddots & 0 \\ \cdot & \cdot & \cdot & \ddots & \ddots & 1 \\ \cdot & \cdot & \cdot & \ddots & \ddots & 0 \\ \cdot & \cdot & \cdot & \cdot & \cdot & (2-n)u \end{pmatrix} + \lambda \begin{pmatrix} 0 & 0 & 0 & \dots & 0 \\ \vdots & \ddots & 0 & \dots & 0 \\ 0 & \ddots & \ddots & \ddots & \vdots \\ 1 & 0 & \ddots & \ddots & 0 \\ 0 & 1 & 0 & \dots & 0 \end{pmatrix}\right\} \underline{\psi} = 0$$

and $(\partial_{t_3} - (Q^3)_+)\psi = 0$ reduces to

$$\left\{\partial_{t_2} - \begin{pmatrix} u_x & u & 0 & 1 & 0 & \dots & 0 \\ \cdot & u_x & u & \ddots & \ddots & \ddots & \vdots \\ \cdot & \cdot & \cdot & \ddots & 0 & \ddots & 0 \\ \cdot & \cdot & \cdot & \cdot & \ddots & 0 & 1 \\ \cdot & \cdot & \cdot & \cdot & \ddots & u & 0 \\ \cdot & \cdot & \cdot & \cdot & \ddots & \ddots & u \\ \cdot & \cdot & \cdot & \cdot & \ddots & \cdot & \cdot \end{pmatrix} + \lambda \begin{pmatrix} 0 & 0 & 0 & \dots & 0 \\ \vdots & & \dots & \ddots & \vdots \\ 1 & \ddots & 0 & \dots & 0 \\ 0 & 1 & \ddots & \dots & 0 \\ 0 & 0 & 1 & \dots & 0 \end{pmatrix}\right\} \underline{\psi} = 0$$

where 0_r is the $r \times r$ zero matrix. This matrix linear system is linear in the spectral parameter λ and can be seen to be the linear system of a reduction of SDYM with 2 null orthogonal translation symmetries. □

NOTE. A large gap in the above discussion is that the linear system is shown to be contained within the SDYM linear systems, but I have not characterised those SDYM solutions with the 2 orthogonal null symmetries that give rise to the nth KdV system.

Proof of Lemma 2. These ideas are standard. One presents the Lie algebra of the area preserving diffeomorphisms of a torus by using the area form as a symplectic form and representing vector fields corresponding to elements of *Lie S Diff*(Σ^2) by their Hamiltonians. Let θ_1 and θ_2 be angular coordinates on the torus such that the area form is $d\theta_1 \wedge d\theta_2$, then a basis for the Hamiltonians is $H_{\mathbf{A}} = \exp\{2\pi i(A_1\theta_1 + A_2\theta_2)\}$ where $\mathbf{A} = (\mathbf{A_1}, \mathbf{A_2}) \in \mathbb{Z} \times \mathbb{Z}$. The Lie bracket is the Poisson bracket:

$$\{H_{\mathbf{A}}, H_{\mathbf{B}}\} = (\mathbf{A} \wedge \mathbf{B})\mathbf{H_{A+B}} \qquad \text{where } (\mathbf{A} \wedge \mathbf{B}) = \mathbf{A_1B_2} - \mathbf{A_2B_1}.$$

For $SL(N)$ we use a basis for the Lie algebra constructed using a pair of matrices U, V satisfying the quantum plane relations: $UV = \xi VU$ where $\xi^N = 1$. An explicit representation has U diagonal with powers of ξ down the diagonal $U_{ij} = \xi^i \delta_{ij}$ and V a shift matrix $V_{ij} = \delta_{i(j+1 \bmod N)}$.

A basis for the Lie algebra of $SL(N)$ is then furnished by

$$T_{\mathbf{A}} = N\xi^{\frac{A_1A_2}{2}} U^{A_1} V^{A_2}.$$

The commutators are then given by:

$$[T_{\mathbf{A}}, T_{\mathbf{B}}] = N \sin \frac{2\pi \mathbf{A} \wedge \mathbf{B}}{\mathbf{N}} T_{\mathbf{A+B}} \xrightarrow[N \to \infty]{} (\mathbf{A} \wedge \mathbf{B})\mathbf{T_{A+B}}$$

which gives the same commutation relations as above for $H_{\mathbf{A}}$ in the limit as $N \to \infty$. □

Proof of lemma 3. This is a corollorary of the results in Mason & Newman (1989). In that paper it was shown that if you take the algebraic relations obtained by imposing four translational symmetries on the self-dual Yang-Mills equations and take the gauge group to be the group of volume preserving diffeomorphisms of some 4-manifold then, roughly speaking, one obtains the self-dual vacuum equations. Lemma 3 can be reformulated so as to be a special case of this. The self-dual Yang-Mills equations on $\mathbb{R}^4$ with metric $ds^2 = du\, dy + dv\, dx$ (signature 2,2) are the integrability conditions on connection components (A_u, A_v, A_x, A_y) in the Lie algebra of the gauge group for the the linear system

$$\{\partial_u + A_u + \lambda(\partial_x + A_x)\}\psi = 0$$
$$\{\partial_v + A_v + \lambda(\partial_y + A_y)\}\psi = 0.$$

When G is *S Diff*(Σ^2) the connection components are all vector fields on Σ^2 (depending also on the coordinates on $\mathbb{R}^4$). Impose two translational symmetries on the $\mathbb{R}^4$ so that the connection components depend only on the coordinates on the quotient, $\mathbb{R}^2$. The linear system then reduces to the system $\{V_u + \lambda V_x\}\psi = 0 = \{V_v + \lambda V_y\}\psi$ where the V's are vector fields on $\mathbb{R}^2 \times \Sigma^2$. These vector fields preserve the natural volume form on $\mathbb{R}^2 \times \Sigma^2$ and so determine elements of the Lie algebra of the volume preserving diffeomorphism group. The linear system is precisely that for the self-dual Yang-Mills equations with 4 translational symmetries and gauge group the volume preserving diffeomorphisms of $\mathbb{R}^2 \times \Sigma^2$.

Concretely introduce coordinates (p,q) on Σ^2 so that the area form is the symplectic form $dp\wedge dq$, and suppose the symmetries to be in the x and y directions so that the variables depend only on the coordinates (u,v) on $\mathbb{R}^2$. Represent the vector fields A_x on Σ^2 by their Hamiltonians denoted h_x etc.. The field equations are

$$[\partial_u + A_u + \lambda A_x, \partial_v + A_v + \lambda A_y] = 0$$

The first implication of this is that $\lambda^2[A_x, A_y] = 0$ so that we can choose coordinates on Σ^2 so that $A_x = \partial_q$ and $A_y = \partial_p$. The term proportional to λ implies $\partial_q h_v = \partial_p h_u$, so that $h_v = \partial_p g$ and $h_u = \partial_q g$ for some $g \equiv g(u,v,q,p)$. The final equation yields in terms of g

$$\partial_u\partial_p g - \partial_v\partial_q g + (\partial_p^2 g)(\partial_q^2 g) - (\partial_q\partial_q g)^2 = 0$$

which is Plebanski's second heavenly equation. □

Thanks to G.A.J. Sparling and E.T. Newman for discussions.

References

Mason, L.J. & Newman, E.T. (1989) A connection between the Einstein and Yang-Mills equations, *Comm. Math. Phys.*, **121**, 659-668.

Mason L.J. & Sparling, G.A.J. (1989) Non-linear Schrodinger and KdV are reductions of the self-dual Yang-Mills equations, *Phys. Lett. A.*, **137**, 29-33.

Mason L.J. & Sparling, G.A.J. (1992) Twistor correspondences for the soliton hierarchies, J. Geom. Phys. **8**, 243-271.

Ward, R.S. (1985) Integrable and solvable systems and relations among them, *Phil. Trans. R. Soc.* **A 315**, p. 451.

Woodhouse, N.M.J. & Mason, L.J. (1988) The Geroch group and non-Hausdorff Riemann surfaces, Nonlinearity, **1**, 73-114.

§II.1.8 **Integrable Systems and Curved Twistor Spaces** *by I.A.B.Strachan* (TN 35, December 1992)

One of the ways in which the self-dual Einstein equations may be understood is as a two dimensional chiral model with the gauge fields taking values in the Lie algebra $sdiff(\Sigma^2)$ of volume preserving diffeomorphisms of the 2-surface Σ^2 (Q. Han Park 1990). Moreover, since $sl(2,\mathbb{C})$ is a subalgebra of $sdiff(\Sigma^2)$, solutions of certain integrable systems associated with $sl(2,\mathbb{C})$ may be encoded within the geometry of the nonlinear graviton (Mason §II.1.7). This description breaks down for higher rank algebras, which are not subalgebras of $sdiff(\Sigma^2)$. However, by generalising the algebras such a description may be achieved. Another reason for studying integrable systems with infinite dimensional gauge groups is that the equations often simplify , and in some cases even linearise, (Ward 1992).

Let $\{\ ,\ \}$ be a generalised Poisson bracket acting on some manifold $\mathcal{N}$, satisfying the conditions:

- $\{f,g\} = -\{g,f\}$ (antisymmetry)
- $\{f,gh\} = \{f,g\}h + \{f,h\}g$ (derivation)
- $\{f,\{g,h\}\} + \text{cyclic} = 0$ (Jacobi identity)

With respect to a basis x^i , $i = 1,\ldots,\dim\mathcal{N}$, one may take

$$\{f,g\} = \sum_{i,j} G^{ij}(x)\frac{\partial f}{\partial x^i}\frac{\partial g}{\partial x^j}, \tag{1}$$

where $G^{ij}(x)$ is constrained by the equations

$$G^{ij} + G^{ji} = 0,$$
$$\sum_{l=1}^{\dim\mathcal{N}} G^{li}\frac{\partial G^{kj}}{\partial x^l} + G^{lj}\frac{\partial G^{ik}}{\partial x^l} + G^{lk}\frac{\partial G^{ji}}{\partial x^l} = 0. \tag{2}$$

Such generalised Poisson structures were first studied by Sophus Lie.

Given such a structure one may define an associated Lie algebra Ham of Hamiltonian vector fields. Let $L_f \in Ham$, where

$$L_f = \sum_{i,j} G^{ij}(x)\frac{\partial f}{\partial x^i}\frac{\partial}{\partial x^j}.$$

The Lie bracket for the algebra may be defined in two different, but equivalent, ways:

- Regard L_f and L_g as differential operators, and define the Lie bracket for the algebra by $[L_f, L_g] = L_f L_g - L_g L_f$,
- Regard L_f and L_g as vector fields on $\mathcal{N}$ and define the Lie bracket for the algebra be the Lie bracket of vector fields $[L_f, L_g]_{Lie}$.

In both cases $[L_f, L_g] = L_{\{f,g\}}$. The fact that this forms a Lie algebra follows trivially from (1) and (2) . The idea now is to study the self-dual Yang-Mills equations with gauge potentials taking values in this infinite dimensional Lie algebra.

Let $y^{AA'}$ be spinor coordinates for $\mathbb{C}^4$ (or perhaps $\mathbb{R}^{2+2}$ etc. depending on a choice of reality condition). The self-dual Yang-Mills equations are the compatibility condition for the otherwise overdetermined linear system:

$$\mathcal{L}_A\Psi = \pi^{A'}\left\{\frac{\partial}{\partial y^{AA'}} + A_{AA'}\right\}\Psi, \quad A, A' = 0,1, \quad \pi^{A'} \in \mathbb{CP}^1. \tag{3}$$

The $A_{AA'}(y)$ are Lie algebra valued functions known as gauge potentials. In what follows it will be assumed that these take values in the Lie algebra Ham constructed above. Thus the $A_{AA'}$'s are represented by vector fields $A_{AA'} \leftrightarrow L_{f_{AA'}}$, where the functions $f_{AA'}$ depend on both the coordinates on $\mathbb{C}^4$ and on $\mathcal{N}$.

With this, the linear operators $\mathcal{L}_A$ are now vector fields on $\mathbb{C}^4 \times \mathcal{N}$,

$$\mathcal{L}_A = \pi^{A'} \left\{ \frac{\partial}{\partial y^{AA'}} + \sum_{i,j} G^{ij}(x) \frac{\partial f_{AA'}}{\partial x^i} \frac{\partial}{\partial x^j} \right\}. \tag{4}$$

Owing to the equivalent definition of the Lie bracket, the self-duality equations are a special case of the (Frobenius) integrability conditions for the distribution (4), i.e. $[\mathcal{L}_0, \mathcal{L}_1]_{Lie} = 0$. The integral surfaces of this distribution may be regarded as curved twistor surfaces, and the space of such surfaces as a curved twistor space, fibred over the Riemann sphere.

The converse construction involves studying an appropriate Riemann-Hilbert problem for the infinite dimensional group. Similiar ideas have been applied to the $SU(\infty)$-Toda equations in Takasaki & Takebe (1991) which also developes the notion of a τ-function for this system and its associated hierarchy.

As mentioned at the beginning of this article, Mason in §II.1.7 shows that one could give a curved twistor space construction to certain integrable systems associated with $sl(2,\mathbb{C})$ by embedding it in the algebra $sdiff(\Sigma^2)$. The same is true for any finite dimensional Lie algebra g. Let the structure constants for the Lie algebra g, with respect to some basis e^i, $i = 1, \ldots, \dim g$ be $c^{ij}{}_k$, so $[e^i, e^j] = \sum_k c^{ij}{}_k e^k$. From this one may define a generalised Poisson bracket by setting

$$G^{ij}(x) = \sum_k c^{ij}{}_k x^k$$

(the conditions (2) are automatically satisfied due to the properties of the structure functions), and let the associated infinite dimensional Lie algebra of Hamiltonian vector field be denoted by $\tilde{g}$. The original Lie algebra is now a subalgebra of $\tilde{g}$, since

$$[L_{x^i}, L_{x^j}] = \sum_k c^{ij}{}_k L_{x^k}\,.$$

Thus any solution to the self-dual Yang-Mills equations with a finite dimensional algebra may be encoded within the structure of a curved twistor space by first embedding g in $\tilde{g}$.

Another approach is to use a deformation of $sdiff(\Sigma^2)$ known as the Moyal algebra, Strachan (1992), in which higher order derivatives are present. This leads to some interesting results, but a direct geometrical interpretation of the results is absent.

References

Q. Han Park, (1990) Phys. Lett. **B238**, 287–290.

Mason, L.J. (1990) Twistor Newsletter **30**, 14–17 and §II.1.7.

Ward, R.S. (1992) J. Geom. Phys. **8**, 317–326.

Takasaki, K. & Takebe, T. (1991) SDiff(2) Toda equations Hierarchy, Tau Function and Symmetries, Lett. Math. Phys. **23**, 205-214.

Strachan, I.A.B. (1992) Phys. Lett. **B282**, 63–66.

§II.1.9 **Twistor theory and integrability** *by L.J.Mason* (𝕋ℕ 33, November 1991)

This note consists mostly of certain speculative and conjectural comments that I gave or would have liked to give in my 5 minute contribution to the special twistor workshop to celebrate the 60th birthday of the founder of many of these ideas.

In this note I wish to make the point that the recently established links between twistor theory and integrable systems should lead to new techniques and results in twistor theory as well as unification and hopefully new results in the theory of integrable systems.

In various articles it has emerged that, with a small number of exceptions, most integrable systems are symmetry reductions of the self-dual Yang-Mills equations, the most notable exception being the KP hierarchy. Furthermore, much of the theory and structure of these equations can be understood in a reasonably direct way as various features of the symmetry reductions of the Ward correspondence for the self-dual Yang-Mills equations. See Ward (1986), Mason & Sparling (1992) and references therein.

As far as the equations are concerned, it appears that we can classify most integrable systems as reductions from self-dual Yang-Mills in 4-dimensions by choice of:

a) a gauge group,

b) a symmetry group (with a possible discrete component),

c) a normal form for the gauge and the various constants of integration that arise in the reduced equation.

For example, the Drinfeld Sokolov systems can all be understood in this way as can various other large classes of integrable systems.

The standard theory of the equations consists of such constructions as the inverse scattering transform and realizations of the partial differential equations as flows on grassmanians. These can be understood as various ansatze and normal forms for the patching data of the holomorphic vector bundles on twistor space with the appropriate symmetry properties that arise from the corresponding symmetry reductions of the Ward transform for the self-dual Yang-Mills equations.

However, these ideas from the theory of integrable systems are in many cases refinements of twistor ideas, and in others are completely new in twistor theory. There is therefore the possibility of methods from the theory of integrable systems being used to solve problems in twistor theory. The following conjectures and connections include examples where twistor theory may benefit from this interaction.

1) Inverse scattering. The inverse scattering transform provides a parametrization of the solution space of integrable partial differential equations. The parameters can be used to build patching data for the Ward bundle on twistor space directly. For example for the attractive nonlinear Schrodinger equation we get the solution space identified with: $Map(S^1 \mapsto D) \times \amalg_{k=1}^{\infty} S^k\{\mathbb{C}^* \times D\}$ where D is the unit disc in the complex plane $\mathbb{C}$, $\mathbb{C}^*$ are the non-zero complex numbers, $\amalg$ is the disjoint union and S^k is the symmetrized cartesian product. The first factor are solutions that one would expect from linearizing the equations (for which they are the Fourier transform) but the second factor are the soliton solutions which do not have a linear analogue.

One would expect this pattern to be generic for solutions of the self-dual Yang-Mills equations in indefinite signature. So one would expect for example that on the compactified 4-dimensional Minkowski space with signature (2,2) that the solution space of the $SU(n)$ self-dual Yang-Mills equations is a Cartesian product of maps from $\mathbb{RP}^3$ to unit determinant Hermitean $n \times n$ matrices with a soliton type sector, which would presumably be the (2,2) analogues of instantons. It is perhaps worth mentioning that the first factor can be understood as a nonlinear generalization of the Radon transform. A similar picture should hold for the symmetry reductions to equations in $2+1$ dimensions and other $1+1$ dimensional systems.

2) The inverse scattering transform in 2+1 dimensions such as for the KP hierarchy has features that distinguish it clearly from existing twistor correspondences so that there seems little real hope of incorporating it into the above framework. Nevertheless, it is a natural generalization of the framework for the KdV equations and leads to a coherent inverse scattering transform based on a non-local Riemann-Hilbert transform. One may hope, then, that the transform can be articulated geometrically so that it leads to some new category of twistor constructions.

It is perhaps worth remarking that the pseudo-differential operators that play such a prominent role in the KP equations also arose naturally in one of Penrose's earlier discussions of the googly problem—the patching operation was given by a pseudo-differential operator that can be represented by integration against a kernel just as in the KP inverse scattering problem.

Another point is that the inverse scattering transform does work for many other field equations in higher dimensions but is no longer implementable by linear procedures and hence does not lead to practical solution generation methods. It may nevertheless lead to a workable framework for understanding general relativity using spin 3/2 fields and Penrose's elemental states based on asymptotic twistors (see Penrose's article in TN 10 to appear also in volume III of this work).

3) It is possible to use the Ward correspondence to understand the connections between the KdV type equations and the 2-dimensional quantum field theory of free Fermions, developed by the Japanese school and described in Segal & Wilson (1985) and Witten (1988). Solutions (at least those that are reflectionless) of the KdV equations are given by amplitudes associated to flows acting on certain special vectors in the free Fermion Fock space. The link is that the free Fermions are the holomorphic sections of the Ward bundle on twistor space restricted to a complex projective line, and the quantum field theoretic amplitude in question is the 2-point function that gives rise to the Greens function for the $\bar{\partial}$-operator. Finding the Greens function is equivalent to trivializing the vector bundle on the line which is the key step in obtaining the self-dual Yang-Mills field in terms of the bundle.

One may ask the question then of whether its possible to realize other more complicated twistor constructions such as the nonlinear graviton construction as a more complicated, perhaps interacting 2-dimensional quantum field theory. In particular this might explain the remarkable link discovered by Ooguri & Vafa between $N = 2$ string theory and the self-dual Einstein equations.

4) There is much scope for using ideas from the quantum inverse scattering transform to understand how to use twistor methods in the context of integrable quantum field theory. In particular the Russian school's introduction of the R-matrix to describe the Poisson bracket structure should pass over directly to give the Poisson bracket relations for the twistor patching data. Other workers have managed to show that the inverse scattering transform survives quantization so that one can hope to quantize on twistor space and then transform the results to obtain a quantum field theory on space-time. The existing theory is still in need of further insights that twistor theory may be able to provide.

5) Witten (1989) has attempted a unification of the theory of integrable statistical mechanical models using Chern-Simons quantum field theory. This produces R-matrices that are sufficient for understanding knot polynomials. Unfortunately it does not provide the dependence of the R-matrices on the spectral parameter that is so crucial to integrability. So it is not possible to regard this as a satisfactory understanding of integrable statistical mechanics. One may conjecture that by studying a quantum field theory of self-dual Yang-Mills reduced to 3-dimensions this gap would be remedied.

It is perhaps also worth drawing attention to the Atiyah-Murray conjecture also in this context (see their article §II.1.3).

Author's comment (1994): There has already been significant progress on some of the above projects. This will appear in a forthcoming book by the author and N.M.J.Woodhouse called *Twistor theory and integrability* to be published by Oxford University Press. More immediate developments are as follows.

1) The first has been realized by observing that the appropriate twistor space for the compactified Minkowski space with signature $(2,2)$ is non-Hausdorff being two copies of $\mathbb{CP}^3$ glued together over some small open neighbourhood of $\mathbb{RP}^3$. These leads to a direct analogue of the Inverse scattering transform for self-dual Yang-Mills in the form alluded to above. See §II.1.11 for further details.

2) The second has led to a new type of twistor construction in which the the $\bar{\partial}$-operator of the Ward bundle on twistor space is generalized so as to be replaced by a more general linear differential operator, a Dirac operator on the sphere depending on parameters, with non-trivial index. This leads to a new construction for the KP equations based on the 'd-bar' approach due to Fokas & Ablowitz and Zakharov & Manakov (see Ablowitz, M. & Clarkson, P.A. 1991). For further details see Mason (1994).

3) There has also been progress in clarifying the connections between Grassmanians and 2-d quantum field theory and twistor theory. This is to appear in Mason & Singer (1994). Part II is still in preparation.

References

Ablowitz, M. & Clarkson, P.A. (1991) *Solitons, Nonlinear Evolution Equations and Inverse Scattering*, LMS lecture note series 149, CUP.

Mason, L.J. (1994) Generalised twistor correspondences, d-bar problems and the KP equations, in proceedings of the Seale Hayne conference on Twistor Theory, ed. S.Huggett, Marcel Dekker.

Mason, L.J. & Singer, M.A. (1994) The twistor theory of equations of KdV type part I, Comm. Math. Phys.

Mason, L.J. & Sparling, G.A.J. (1989) Korteweg de Vries and nonlinear Schrodinger are reductions of the self-dual Yang-Mills equations, Phys. Lett. A, **137**, #1,2, 29-33.

Mason, L.J. & Sparling (1992) Twistor correspondences for the soliton hierarchies, J.Geom. & Physics, **8**, 243-271.

Segal, G. & Wilson, G. (1985) Loop groups and equations of KdV type, Publ. I.H.E.S. **65**, 5-65.

Ward, R.S. (1986) Multi-dimensional integrable systems, in *Field Theory, Quantum Gravity and Strings*, eds. H.J. de Vega and N. Sanchez, Lecture Notes in Physics **246**, Springer, Berlin.

Witten, E. (1988) Quantum field theory and grassmanians, Comm. Math. Phys., **113**, 529-600.

Witten, E. (1989) Nucl. Phys. B.

§II.1.10 **On the symmetries of the reduced self-dual Yang-Mills equations**
by L.J.Mason (TN 34, May 1992)

Introduction. One of the remarkable features of reductions of the self-dual Yang-Mills equations to systems in two dimensions is that the symmetry group of the reduced equations (in the context of space-time symmetries) is much larger than one might have expected, often being infinite dimensional. A priori, one would expect the symmetry group of the reduced equations to be just the projection of those conformal symmetries in 4–dimensions that normalize the invariance group that one is reducing by. In Hitchin (1987) it was observed that the reductions of the self-dual Yang-Mills equations on Euclidean 4-dimensional space by two translations are actually conformally invariant in the infinite dimensional sense in the residual 2-dimensional space (a priori one would only expect the equations to be invariant under the 2-dimensional Euclidean group plus scalings). In Mason & Sparling (1992) it was observed that reductions of self-dual Yang-Mills by 2 translations spanning a 2-plane on which the metric has rank one also has an infinite dimensional symmetry group at least when the gauge group is $SL(2)$—nonlinear analogues of the Galilean group in so called $(1+0)$–dimensions as opposed to just the linear Galilean group in 2-dimensions..

The purpose of this note is to clarify the geometry underlying this result and state it independently of the gauge group. I also discuss two other examples of this phenomena, one being the reduction by symmetries spanning a totally null ASD 2-plane where the symmetry group is the whole diffeomorphism group (rather than just $GL(2)$), and the other being the reduction by two rotations (or a rotation and a translation) in which the symmetry group is the hyperbolic group in 2-dimensions ($SL(2,\mathbb{R})$).

An important corollary of Hitchin's result is that it makes it possible to transfer the equations to a general Riemann surface where they considerably enrich the theory of holomorphic vector bundles. The above results give alternate ways of transferring different reductions of the self-dual Yang-Mills equations to 2-dimensional surfaces endowed with different geometric structures.

The Yang-Mills Higgs equation on a Riemann surface. I first give a brief review of Hitchin's equations. We use $(z, w, \tilde{z}, \tilde{w})$ as coordinates on $\mathbb{R}^4$ that are independent and real for signature $(2,2)$ or complex with $\tilde{z} = \bar{z}$ etc. for Euclidean signature. We start with the Lax pair formulation of the self-dual Yang-Mills equations.

The self-dual Yang-Mills equations are the compatibility conditions for the pair of operators:

$$L_0 = D_z - \lambda D_{\tilde{w}}, \quad L_1 = D_w + \lambda D_{\tilde{z}}.$$

where $\lambda \in \mathbb{C}$ is an auxiliary complex parameter and D_z is the covariant derivative of some Yang-Mills connection in the direction $\partial/\partial z$.

For Hitchin's equations we start in Euclidean signature and impose symmetries in the $\partial/\partial w$ and $\partial/\partial \bar{w}$ directions. In an invariant gauge (i.e. one in which the gauge potentials are independent of $(w, \bar{w})$, $D_w = \partial/\partial w + \bar{\Phi}'$ and cc. and we can throw away the derivatives with respect to $(w, \bar{w})$ to leave the pair of operators (with a little rearrangement):

$$L_0 = D_z - \lambda\Phi', \quad L_1 = D_{\bar{z}} + \frac{1}{\lambda}\bar{\Phi}'.$$

We can make this more geometric by multiplying L_0 by dz and L_1 by $d\bar{z}$ and defining $\Phi = \Phi' dz$. We then obtain the one-form valued operator:

$$L = dz \otimes L_0 + d\bar{z} \otimes L_1 = D - \lambda\Phi + \frac{1}{\lambda}\bar{\Phi}.$$

The Yang-Mills Higgs equation on a Riemann surface are the consistency conditions for these operators:

$$D^2 = \Phi \wedge \bar{\Phi}, \quad D\Phi = 0, \quad D\bar{\Phi} = 0.$$

Where D is the covariant exterior derivative. These equations are invariant under the conformal group in two dimensions as they only require a bundle with connection and a complex structure to define Φ and $\bar{\Phi}$. One solution will be transformed to another if $z \mapsto z'(z)$ and Φ and D pull back.

Alternatively, these equations depend only on the $*$-operator on 1-forms on the quotient space of the symmetries. The data consists of a connection D on a bundle, E and a section $\Gamma = \Phi + \bar{\Phi}$ of $\Omega^1 \otimes End(E)$. The operator L is

$$L = D + \left(-\lambda\frac{1 - i*}{2} + \frac{1 + i*}{2\lambda}\right)\Gamma.$$

Thus the field equations arising from the consistency conditions of this operator are invariant under the diffeomorphisms preserving $*$, i.e. the conformal transformations in 2-dimensions.

The Galilean analogue. If, in $(2,2)$ signature, we impose one null symmetry along $\partial/\partial\tilde{w}$ and one non-null symmetry along $\partial/\partial z - \partial/\partial\tilde{z}$ we obtain the Lax pair:

$$L_0 = D_x - \lambda\Phi, \quad L_1 = D_w + \lambda(D_x + \Psi)$$

where $x = (z + \tilde{z})$ and we have reorganized the covariant derivative in the x direction to include part of the Higgs field associated to the symmetry in the $\partial_z - \partial_{\tilde{z}}$ direction. We can again perform the above trick, multiplying L_0 by dx and L_1 by dw to and adding together to obtain

$$L = dxL_0 + dwL_1 = D + \lambda(\Gamma + dwD_x)$$

where $\Gamma = -\Phi dx + \Psi dw$. To write this more geometrically, we introduce a degenerate $*$–operator that can be thought of as a map from 1–forms to 1–forms:

$$* = dw\frac{\partial}{\partial x}, \quad \alpha \mapsto \alpha(\frac{\partial}{\partial x})dw.$$

The operator L then becomes:

$$L = D + \lambda(*D + \Gamma).$$

The field equations arising from the consistency equations for this system are:

$$D^2 = 0, \quad D\Gamma = 0, \quad D * \Gamma + \Gamma \wedge \Gamma = 0$$

where D above is acting as the covariant exterior derivative so that the equations are all 2-form equations. Geometrically these equations determine a flat connection D on a bundle E, together with a section Γ of $\Omega^1 \otimes End(E)$.

It is clear, now, that the field equations arising from the consistency conditions for this operator will be invariant under diffeomorphisms of $\mathbb{R}^2$ preserving the degenerate $*$–operator, $dw \otimes \partial/\partial x$. These are the nonlinear Galilean transformations referred to previously:

$$(w, x) \mapsto (w', x') = (h(w), (\partial_w h(w))x + g(w))$$

where $h(w)$ and $g(w)$ are free functions except that $\partial_w h \neq 0$.

These equations embed the nonlinear Schrodinger and KdV equations and most of their generalizations to higher rank gauge groups (the Drinfeld Sokolov hierarchies etc.) into a Galilean invariant system. At least in the $SL(2)$ case, this coordinate freedom is completely fixed by the reduction to KdV and NLS.

The totally null case. In the case where the symmetries span an anti self-dual null 2–plane we obtain the linear system

$$L = D + \lambda\Gamma$$

where again D is a flat connection on a bundle E and Γ is again a section of $\Omega^1 \otimes End(E)$. The field equations are now:

$$D^2 = 0, \quad D\Gamma = 0, \quad \Gamma \wedge \Gamma = 0.$$

These equations are now invariant under the full 2-dimensional diffeomorphism group (preserving the 'zero' $*$–operator).

These equations are therefore 'topological' and indeed are another way of writing the Wess-Zumino-Witten equations (Strachan 1992). Their reductions include the n–wave equations and those parts of the Drinfeld-Sokolov hierarchies not obtainable from the Galilean reductions. These further reductions require that their exists coordinates and a gauge in which the components of Γ are constant. In the $SL(3)$ case one can fix the coordinate freedom by using these additional conditions.

Stationary axisymmetric systems. In Fletcher & Woodhouse (1990) it was observed that the reduction of SDYM by 2 rotations gave the same field equations as the reduction by a translation and a rotation. This fact alone endows the 2 rotation reduction with one unexpected symmetry; the residual translation symmetry. However, more is true. These equations are invariant under $SL(2,\mathbb{R})$; the group of motions of the residual space preserving a hyperbolic metric. While this was in some sense clear from the reduced twistor correspondence in Woodhouse & Mason (1987), it was difficult to see on space-time.

To see this we impose a rotational invariance with respect to θ in the $w = y\exp(i\theta)$ plane and set $x = z + \bar{z}$ and impose a symmetry in the $\partial_z - \partial_{\bar{z}}$ direction. We obtain the linear system:

$$D_x - iA + \lambda e^{i\theta}(D_y + \frac{i}{y}(\partial_\theta - B)),\ e^{-i\theta}(D_y - \frac{i}{y}(\partial_\theta - B)) - \lambda(D_x + iA).$$

We cannot just throw away the ∂_θ as there is explicit dependence on θ in the operators. This is connected with the fact that the Lie derivative of a spinor and hence λ along ∂_θ is not zero. To work independently of θ and to avoid derivatives with respect to the 'spectral parameter' we must use, instead of λ the parameter

$$\gamma = \frac{ye^{i\theta}\lambda}{2} + x - \frac{y}{2e^{i\theta}\lambda}$$

as this is the simplest function on the spin bundle that is both invariant and constant along the twistor distribution.

If we introduce the complex coordinate $\xi = x + iy$, a bit of massage yields the following form for the linear system:

$$2D_\xi + i\sqrt{\frac{\gamma - \bar{\xi}}{\gamma - \xi}}(A + \frac{i}{y}B),\quad 2D_{\bar{\xi}} + i\sqrt{\frac{\gamma - \xi}{\gamma - \bar{\xi}}}(A - \frac{i}{y}B).$$

In order to bring out the invariance properties of this system, we can first of all multiply the first operator by $d\xi$ and the second by $d\bar{\xi}$ and add them together. Then introduce homogeneous coordinates $\gamma_A = (\gamma_0, \gamma_1)$ with $\gamma = \gamma_1/\gamma_0$ and similarly for ξ_A. Define $\bar{\xi}_A$ to be the componentwise complex conjugate of ξ_A and denote the skew product $\gamma_1\xi_0 - \xi_1\gamma_0 = \gamma \cdot \xi$. The linear system then reduces, after some further massage, to:

$$2D + i\sqrt{\frac{\gamma \cdot \bar{\xi}}{i\xi \cdot \bar{\xi}\gamma \cdot \xi}}\Phi\xi \cdot d\xi + i\sqrt{\frac{\gamma \cdot \xi}{i\xi \cdot \bar{\xi}\gamma \cdot \bar{\xi}}}\bar{\Phi}\bar{\xi} \cdot d\bar{\xi}$$

where we have put $\Phi = (\sqrt{y}A + iB/\sqrt{y})/\xi_0$.

It can now be seen that the linear system is invariant under $SL(2,\mathbb{R})$; the Mobius transformations on ξ_A preserving the reality structure $\xi_A \mapsto \bar{\xi}_A$ and hence the hyperbolic metric $\xi \cdot d\xi \odot \bar{\xi} \cdot d\bar{\xi}/(i\xi \cdot \bar{\xi})^2$.

The integrability conditions are equations for a connection D on a bundle E and a section $\Phi \in \Gamma(\mathcal{O}(-1) \otimes E)$ that is a dual spinor valued section of End(E).

The field equations are

$$D^2 = [\Phi, \bar{\Phi}]\frac{\xi \cdot d\xi \wedge \bar{\xi} \cdot d\bar{\xi}}{\xi \cdot \bar{\xi}}, \quad \eth\Phi = \frac{\bar{\Phi}}{2\xi \cdot \bar{\xi}}, \quad \eth\bar{\Phi} = -\frac{\Phi}{2\xi \cdot \bar{\xi}}$$

where $\eth$ and $\bar{\eth}$ here denote the 'eth' operator and its complex conjugate, the $(0,1)$ and $(1,0)$ parts of the covariant derivative respectively.

So the 'Higgs fields' Φ and $\bar{\Phi}$ together constitute a Dirac field and satisfy the background coupled massive Dirac equation. Their commutator provides the curvature of the connection.

Remarks. Just as in Hitchin's case, one might hope to be able to transfer the other equations above to a Riemann surface also.

In the hyperbolic case, the equations can clearly be transferred for $g > 2$ using the unique metric on the Riemann surface with curvature -1.

For the Galilean analogue, instead of endowing the Riemann surface with a complex structure, one might endow it with a measured foliation which corresponds to a limit of a complex structure; indeed the space of measured foliations modulo certain equivalence relations is the Thurston boundary for Teichmuller space. It turns out that this is *not* the same concept as the degenerate $*$-operator introduced above. They both determine a foliation of the Riemann surface, but the degenerate $*$-operator has an affine structure on the leaves, but no structure transverse to the leaves, whereas the measured foliation has a measure transverse to the leaves, but no structure on the leaves. Nevertheless, one might hope that one could prove an equivalence between the two, modulo diffeomorphisms in the global context as a kind of uniformization result. Even if this is feasible, it is still perhaps not clear that one can obtain a good existence theory for solutions of this equation as the linearized analogues of these equations have $*\Gamma$ covariant constant along the leaves of the foliation, a condition that will have no solutions when the leaves are dense.

Further analysis is required for the other cases. The totally null reduction will presumably not give rise to any difficulty as the equations are underdetermined anyway. This leaves the Hyperbolic case for which more analysis is required.

Thanks to Jorgen Andersen for conversations.

References

Fletcher, J. & Woodhouse, N.M.J. (1990) Twistor characterization of stationary axisymmetric solutions of Einsteins equations, in *Twistors in Mathematics and Physics*, eds. Bailey, T.N. & Baston, L.M.S. Lecture Notes Series 156, CUP.

Hitchin, N.J. (1987) The self-duality equations on a Riemann surface, Proc. L.M.S.

Mason, L.J. & Sparling, G.A.J. (1992) Twistor theory of the soliton hierarchies, J.Geom. Phys., **8**, 243-271.

Mason, L.J. & Singer, M.A. (1994) The twistor theory of equations of Korteweg de Vries type part I, to appear in Comm. Math. Phys.

Strachan, I.A.B. (1992) Phys. Lett. **B282**, 63-66.

§II.1.11 Global solutions of the self-duality equations in split signature

by L.J. Mason ($\mathbb{TN}$ 35, December 1992)

1. Introduction. The purpose of this note is to give a parametrization of the space of global solutions of the self duality equations in signature (2,2). This is obtained by examining the appropriate globalization of twistor correspondences in signature (2,2). Despite some unusual features of the global geometry, it leads to a particularly simple and appealing parametrization of the solution space of the self-duality equations. In particular the treatment of the Ward construction for global solutions of the self-dual Yang-Mills equations in split signature leads to the non-linear graviton construction for anti-self-dual metrics of signature (2,2) on $S^2 \times S^2$ which specializes to give the twistor description of the examples of conformally anti-self-dual metrics described by K.P.Tod in his article which will appear in volume III of this work. Part of the motivation for this construction arose from a desire to understand the globalization of other aspects of the Penrose transform in signature $(2,2)$ and thereby give a twistor description of the Radon transform and the inverse scattering transform, see also §II.1.9 for further discussion.

When one considers boundary conditions for solutions of conformally invariant equations on $\mathbb{R}^4$ in signature $(2,2)$, the first condition one might try would be to require that solutions should extend to the conformal compactification of $\mathbb{R}^4$, $S^2 \times S^2/\mathbb{Z}_2$ as the equations are conformally invariant. We will see, however, that these boundary conditions eliminate all solutions in linear theory. Furthermore, the α-planes in $S^2 \times S^2$ are not simply connected and have fundamental group $\mathbb{Z}_2$ so that one cannot immediately apply the Penrose transform. These problems are rectified if we go to the double cover, $S^2 \times S^2$ (where one can ask the question as to which fields are invariant under the $\mathbb{Z}_2$ action).

2. The global geometry. Affine $(2,2)$ Minkowski space $\mathbb{M}_0$ is $\mathbb{R}^4$ with metric of signature $(2,2)$. The conformal compactification $\mathbb{M}$, obtained by adjoining a 'light cone at infinity' to $\mathbb{M}_0$, is the projective quadric in $\mathbb{RP}^5$ given by the zero set of a quadratic form Q in $\mathbb{R}^6$ of signature $(3,3)$. The conformal structure is determined by asserting that the light cones of $\mathbb{M}$ are the intersections of $\mathbb{M}$ with the tangent planes of points of $\mathbb{M}$.

To see that the topology of $\mathbb{M}$ is $S^2 \times S^2/\mathbb{Z}_2$ diagonalize Q using a pair of Euclidean 3-vectors $\underline{w}$ and $\underline{y}$ as coordinates on $\mathbb{R}^6$ such that $Q = \underline{w} \cdot \underline{w} - \underline{y} \cdot \underline{y}$. Set the scale by requiring $\underline{w} \cdot \underline{w} = 1$ so that $\underline{y} \cdot \underline{y} = 1$ also on $Q = 0$; this yields $S^2 \times S^2$ in $\mathbb{R}^6$. However, in $\mathbb{RP}^5$, $(\underline{w}, \underline{y}) \sim (-\underline{w}, -\underline{y})$ so the topology of $Q = 0$ in $\mathbb{RP}^5$ is $S^2 \times S^2/\mathbb{Z}_2$.

The conformal structure is simply realized on the double cover $\widetilde{\mathbb{M}} = S^2 \times S^2$ by taking the pullback of the round sphere metric $d\Omega^2$ on each factor and taking the difference

$$ds^2 = p_1^* d\Omega^2 - p_2^* d\Omega^2$$

where p_1, p_2 are the projections onto the first and second factors respectively. This is $\mathbb{Z}_2$ invariant and so descends to $\mathbb{M}$.

3. The global correspondence. We will be interested in the cases where the original region on which the fields are defined are $\mathbb{M}$ and its double cover $\widetilde{\mathbb{M}}$. We will now consider the correspondence for these cases (or deformations thereof for ASD metrics).

As, usual, in order to define the twistor space of a space-time U, we shall assume all fields are analytic and consider fields on a small complexification of U which will also, by an abuse of notation, be denoted by the U. We then define its twistor space $\mathbb{PT}(U)$ to be the space of connected components of totally null self-dual 2-planes (α-planes) in U. For $Z \in \mathbb{PT}(U)$ we will denote the corresponding α-plane in U by $\hat{Z}$.

Twistor space, $\mathbb{PT}(U)$, can be equivalently defined as the quotient of the projective spin bundle $\mathbb{PS}(U)$ over U by the twistor distribution spanned by $\pi^{A'}\nabla_{AA'}$. By definition, points of $\mathbb{PT}(U)$ correspond to α-planes in U. Conversely, points $x \in U$ correspond to $\mathbb{CP}^1$'s denoted L_x in $\mathbb{PT}(U)$. Points of L_x are the α-planes through x.

3.1. The correspondence for $\mathbb{M}$. Just as compactified complexified Minkowski space $\mathbb{CM}$ is the space of complex lines in $\mathbb{CP}^3$ via the complex Klein correspondence, $\mathbb{M}$ is the space of real lines in $\mathbb{RP}^3$ via the real Klein correspondence. In the context of the complex correspondence, point of $\mathbb{M}$ are complex lines in $\mathbb{CP}^3$ that intersect $\mathbb{RP}^3$ in a real line. Alternatively, they are complex lines in $\mathbb{CP}^3$ that are mapped into themselves by the complex conjugation $Z^\alpha \to \bar{Z}^\beta$ given by standard complex conjugation, component by component.

According to the definition above we have $\mathbb{PT}(\mathbb{M}) = \mathbb{CP}^3$. Given $Z \in \mathbb{CP}^3$, then if $Z = \bar{Z}$, $Z \in \mathbb{RP}^3$ and any real line in $\mathbb{RP}^3$ through Z corresponds to a point in $\mathbb{M}$ on $\hat{Z}$. In this case, $\hat{Z}$ intersects $\mathbb{M}$ in an $\mathbb{RP}^2$. If $Z \neq \bar{Z}$ then the complex line through Z and $\bar{Z}$ is real and corresponds to a point of $\mathbb{M}$. In fact the complex α-plane $\hat{Z}$ intersects $\mathbb{M}$ in the unique point corresponding to this line.

3.2. Linear theory. The linear problem was completely solved in the case of the wave equation by Fritz-John using the X-ray transform. In twistor notation, the general solution of the hyperbolic wave equation on $\mathbb{R}^4$ satisfying appropriate boundary conditions can be obtained from the integral formula

$$\phi(x^{AA'}) = \oint f(x^{AA'}\pi_{A'}, \pi_{A'})\pi^{A'}d\pi_{A'}.$$

Here f is a freely specifiable smooth section of $\mathcal{O}(-2)$ on $\mathbb{RP}^3$. That ϕ is a solution of the ultrahyperbolic wave equation follows by differentiation under the integral sign.

One might naively think that the function ϕ is naturally a function on the space of lines in $\mathbb{RP}^3$, $\mathbb{M}$. However, ϕ is defined by integrating f along lines and in order to perform the integration, one needs to have an orientation of the line. This means that ϕ is actually defined on the space of *oriented* lines in $\mathbb{RP}^3$. This is $\widetilde{\mathbb{M}}$ the double cover of $\mathbb{M}$. Clearly ϕ changes sign under reversal of orientation of the line and so does not descend to $\mathbb{M}$. Indeed, we will see that there are no solutions of the conformally invariant wave equation on $\mathbb{M}$.

Remark. Actually, there is a possible confusion here owing to the Grgin phenomena—the real point is that these solutions are anti-Grgin. Solutions of the wave equation are sections of $\mathcal{O}[-1]$, the inverse conformal weight bundle. Given just the metric $ds^2 = p_1^*d\Omega^2 - p_2^*d\Omega^2$ on $S^2 \times S^2/\mathbb{Z}_2$ there are two possible choices for $\mathcal{O}[-1]$, the trivial bundle or the Mobius bundle. The correct choice as far as the twistor correspondence is concerned is the Mobius bundle as this is just the restriction of the tautological bundle from $\mathbb{RP}^5$. The solutions above are actually sections of the trivial bundle which is wrong as far as the twistor correspondence is concerned. So one can simply write them down as even *functions* on $S^2 \times S^2$ but as odd sections of $\mathcal{O}[-1]$.

3.3. The correspondence for $\widetilde{\mathbb{M}}$. We must therefore study twistor correspondences for $\widetilde{\mathbb{M}}$. We shall abuse notation and denote also by $\widetilde{\mathbb{M}}$ a small complex thickening of $\widetilde{\mathbb{M}}$. We have:

Lemma 3.1. $\mathbb{PT}(\widetilde{\mathbb{M}})$ is the (non-Hausdorff) space obtained by gluing together two copies of $\mathbb{CP}^3$, denoted $\mathbb{CP}^3_+$ and $\mathbb{CP}^3_-$, together along some small thickening of $\mathbb{RP}^3$ using the identity map.

Remark. The non-Hausdorff points are those on the boundary of the glued down region—each point in this boundary has a partner on the other copy of $\mathbb{CP}^3$ with which it would be identified if the closure of the neighbourhood of $\mathbb{RP}^3$ were glued down. Any open set of one such point intersects any neighbourhood of its partner on the other sheet.

Proof. Points in the complement of (the thickening of) $\mathbb{RP}^3$ in $\mathbb{CP}^3$ correspond to α-planes that intersect $\mathbb{M}$ in a topologically trivial region and these are necessarily covered by two components in the double cover $\widetilde{\mathbb{M}}$. Whereas, points in the (thickening of) $\mathbb{RP}^3$ correspond to α-planes in (the small thickening of) $\mathbb{M}$ with topology $\mathbb{RP}^2$ so that when one takes the double cover the α-plane has topology S^2.

Thus $\mathbb{PT}(\widetilde{\mathbb{M}})$ double covers $\mathbb{CP}^3$ over the complement of the thickening of $\mathbb{RP}^3$ and the double covering is glued together over the thickening of $\mathbb{RP}^3$.

We reconstruct $\widetilde{\mathbb{M}}$ as the space of complex lines in $\mathbb{PT}(\widetilde{\mathbb{M}})$ that are cut into two pieces by $\mathbb{RP}^3$ with one piece lying in $\mathbb{CP}^3_+$'s and the other in $\mathbb{CP}^3_-$. This yields the space of oriented lines in $\mathbb{RP}^3$; the line is given by the intersection of complex line with $\mathbb{RP}^3$ and the orientation is determined by multiplying the arrow from the intersection with $\mathbb{CP}^3_-$ to that with $\mathbb{CP}^3_+$ by i, thereby rotating it by 90 degrees.

The non-Hausdorffness arises as twistor space is a quotient of the spin bundle of (the slightly complexified) $\widetilde{\mathbb{M}}$ and as one deforms a leaf of the foliation, it can break into two disconnected leaves.

3.4. Complex conjugation. Complex conjugation on the small thickening of $\widetilde{\mathbb{M}}$ sends α-planes to α-planes and hence leads to a conjugation on $\mathbb{PT}(\widetilde{\mathbb{M}})$. This covers the standard complex conjugation on $\mathbb{CP}^3$ that fixes $\mathbb{RP}^3$ and is lifted to $\mathbb{PT}(\widetilde{\mathbb{M}})$ by requiring that it interchange $\mathbb{CP}^3_+$ and $\mathbb{CP}^3_-$. Thus $[Z] \in \mathbb{CP}^3_-$ goes to $[\bar{Z}] \in \mathbb{CP}^3_+$ and the real lines of the conjugation are those described above that correspond to points of $\widetilde{\mathbb{M}}$.

4. The X-ray transform. We can now understand the X-ray transform in this context. Solutions of the wave equation on $\widetilde{\mathbb{M}}$ correspond to elements of $H^1(\mathbb{PT}(\widetilde{\mathbb{M}}), \mathcal{O}(-2))$. These can be studied by means of the Meyer-Vietoris sequence using the covering of $\mathbb{PT}(\widetilde{\mathbb{M}})$ by $\mathbb{CP}^3_+$ and $\mathbb{CP}^3_-$. Using the fact that $H^1(\mathbb{CP}^3, \mathcal{O}(-2)) = 0 = H^0(\mathbb{CP}^3, \mathcal{O}(-2))$ we find

$$H^1(\mathbb{PT}(\widetilde{\mathbb{M}})) = H^0(\mathbb{CP}^3_+ \cap \mathbb{CP}^3_-, \mathcal{O}(-2)) = H^0(\mathbb{RP}^3, \mathcal{O}(-2))$$

and the formula for the Penrose transform using these representatives is precisely the X-ray transform.

5. ASD Yang-Mills fields and the inverse scattering transform. For anti-self-dual Yang-Mills fields on $\widetilde{\mathbb{M}}$, we can obtain an analogous parametrization of the general solution.

Theorem 5.1 There is a 1-1 map from ASD Yang-Mills connections on a bundle F with structure group $G = \mathrm{SL}(n, \mathbb{R})$ or $\mathrm{SU}(n)$ and second Chern class $c_2(F) = k$ on $\widetilde{\mathbb{M}}$, and pairs consisting of

(a) a holomorphic vector bundle E on $\mathbb{CP}^3$ with vanishing first and third Chern classes, and $c_2(E) = k/2$, and

(b) for $G = \mathrm{SL}(n, \mathbb{R})$, a nondegenerate map $P : E|_{\mathbb{RP}^3} \to \bar{E}|_{\mathbb{RP}^3}$ with $\bar{P} = P^{-1}$ or, for $G = \mathrm{SU}(n)$, $P : E|_{\mathbb{RP}^3} \to \bar{E}^*|_{\mathbb{RP}^3}$ with $P = \bar{P}^t$

Remark. Here $\bar{E}$ denotes the conjugate holomorphic vector bundle defined by $E|_Z = \overline{E|_{\bar{Z}}}$. For the bundle F on space-time, $\bar{F} = F$ when $G = \mathrm{SL}(n, \mathbb{R})$ or $\bar{F} = F^*$ when $G = \mathrm{SU}(n)$.

Proof. The Ward transform gives a 1-1 map from ASDYM fields on $\widetilde{\mathbb{M}}$ to bundles on $\mathbb{PT}(\widetilde{\mathbb{M}})$. Given a bundle on $\mathbb{PT}(\widetilde{\mathbb{M}})$, we can restrict it to $\mathbb{CP}^3_+$ to obtain the bundle E on $\mathbb{CP}^3$. The restriction to $\mathbb{CP}^3_-$ will be a different bundle, but the complex conjugation shows that it must be $\bar{E}$ for $G = \mathrm{SL}(n, \mathbb{R})$ or $\bar{E}^*$ for $\mathrm{SU}(n)$ by complex conjugation.

The rest of the data of the bundle on $\mathbb{PT}(\widetilde{\mathbb{M}})$ is encoded in the patching P over $\mathbb{CP}_- \cap \mathbb{CP}_+$ as given in part (b). For compatibility with the complex conjugation, P must satisfy the conditions stated.

Note that this implies that k is even. There are two extreme examples of the above that serve to illustrate the general case.

1) When $k = 0$, the bundle E is necessarily trivial, and all the information is contained in the matrix function satisfying $P = \bar{P}^{-1}$ or $P = \bar{P}^t$ on $\mathbb{RP}^3$.
2) When P extends over all $\mathbb{CP}^3$, all the information is contained in a bundle E on $\mathbb{CP}^3$ satisfying $\bar{E} = E$ or $\bar{E} = E^*$ depending on G.

Remark. The connection with the inverse scattering transform is effectively by paradigm. In Mason & Sparling 1992 it was shown that one could realize the inverse scattering transform as a coordinate realization of the Ward transform, but it was not clear how the normal form for the patching function came about. The patching function could be factorized into an 'algebreo geometric' part which parametrized the 'solitonic' degrees of freedom of the solutions and have no analogue in linearized theory and the 'scattering data' consisting of C^∞ functions of the real twistor coordinates parametrizing the radiative/dispersive modes of the solution—these reduce to the Fourier transform in linear theory. In this four dimensional analogue we see that the role of the algebro-geometric part is played by the bundle E on $\mathbb{CP}^3$ that has no linearized analogue and the role of the scattering data is played by the map P that generalizes the Radon transform from linearized theory.

A direct connection with the inverse scattering transform requires the examination of symmetry reductions of this framework. This is not straightforward as it will change the boundary conditions and hence the details of the global structure.

6. Deformations of $\mathbb{PT}(\widetilde{\mathbb{M}})$. The nonlinear graviton construction implies that (small) ASD deformations of the conformal structure on $S^2 \times S^2$ correspond to (small) deformations of $\mathbb{PT}(\widetilde{\mathbb{M}})$. Since $\mathbb{CP}^3$ is rigid, the only deformable part is the gluing along $\mathbb{RP}^3$. In order to guarantee that the reality structure is preserved, the gluing map P from some open set in $\mathbb{CP}^3_+$ to one in $\mathbb{CP}^3_-$ must be compatible with the conjugation that sends $Z \in \mathbb{CP}^3_+$ to $\bar{Z} \in \mathbb{CP}^3_-$. This yields the condition that $P^{-1} = \bar{P}$ where $\bar{P}$ is the conjugate map. This can be arranged as follows:

Take a small analytic deformation ρ of the standard embedding of $\mathbb{RP}^3$ into $\mathbb{CP}^3$ so that ρ has a small analytic extension to a neighbourhood U of $\mathbb{RP}^3$ in $\mathbb{CP}^3$. Then we also have the conjugate embedding $\bar{\rho}$ of U into $\mathbb{CP}^3$ which is the complexification of the complex conjugate embedding (it is also holomorphic). The deformed gluing from $\mathbb{CP}^3_+$ to $\mathbb{CP}^3_-$ is then done with the map $P = \bar{\rho} \circ \rho^{-1}$.

The complex conjugation map of the deformed glued down twistor space can then be defined by sending the point $Z \in \mathbb{CP}^3_+$ to the point $\bar{Z} \in \mathbb{CP}^3_-$. This conjugation clearly fixes the image of $\mathbb{RP}^3$ and is antiholomorphic and acts globally.

The space-time with deformed ASD conformal structure is then reconstructed by constructing the complex lines in the deformed space that are divided into two parts by the glued down region and are half in $\mathbb{CP}^3_+$ and half in $\mathbb{CP}^3_-$.

Thus we have a $1-1$ map from ASD deformations of the conformal structure on $S^2 \times S^2$ and such real gluing maps as above. These can be thought of as the space of (analytically) embedded $\mathbb{RP}^3$'s in $\mathbb{CP}^3$ or $map\{\mathbb{RP}^3 \to \mathbb{CP}^3\}/Diff\{\mathbb{RP}^3\}$ as a diffeomorphism of $\mathbb{RP}^3$ does not affect the final P.

Examples. The examples that K.P.Tod writes down in his article in volume III are obtained from a split signature analogue of LeBrun's hyperbolic Gibbons-Hawking ansatz, LeBrun (1991). The basic idea is to take a global holomorphic vector field on $\mathbb{CP}^3$ that is real on $\mathbb{RP}^3$ and to drag the standard gluing some fixed amount along the imaginary part of the vector field. This is a global version of the construction of Jones & Tod (1985).

Use use 2×2 matrices as homogeneous coordinates on twistor space with columns (λ_A, μ_A). The real slice $\mathbb{RP}^3$ sits inside as PSU(2) with λ_A the SU(2) complex conjugate of μ_A. The vector field $V = i\lambda_A\partial/\partial\lambda_A - i\mu_a\partial/\partial\mu_A$ corresponds to right multiplication by diagonal SU(2) matrices. The quotient by the complexified vector field is the quadric Q with coordinates $([\lambda_A], [\mu_a])$ and real slice S^2. On $\widetilde{\mathbb{M}}$ the symmetry can be represented so that it rotates just one of the S^2 factors and leaves the other invariant.

Choose a real analytic function on S^2, $f([\lambda_A], [\mu_A])$ defined for $\lambda_A = \hat{\mu}_A$ and continue it to some small thickening of the real slice. Then we identify $(\exp(-f)\lambda_a, \exp(f)\mu_A)$ in $\mathbb{CP}^3_+$ with $(\exp(f)\lambda_A, \exp(-f)\mu_A)$ in $\mathbb{CP}^3_-$ where $[\lambda_A]$ and $[\mu_A]$ are close to being conjugate. Thus, if we take out the lines $\lambda_A = 0$ and $\mu_A = 0$, the twistor space $\mathbb{P}\mathcal{T}$ is a complex line bundle over the space $\tilde{Q}$ obtained by gluing one copy of the quadric Q_+ to another Q_- over some thickening of the real slice.

The space-time metric is given in standard form in Jones & Tod (1985) as

$$ds^2 = d\Sigma_3^2 + (d\phi + \omega)^2/V^2$$

where here $d\Sigma_3^2$ is the Einstein-Weyl space corresponding to the quadric which is just Lorentzian hyperbolic space and ω and V are the parts of an ASD Maxwell field that are respectively orthogonal and tangent to the symmetry direction and thus satisfy $d\omega = *_3 dV$ where $*_3$ is the hodge dual with respect to $d\Sigma_3^2$.

It is a straightforward exercise to show that the metrics in Tod's article can be put into this form after a conformal rescaling (the main non trivial part is to show that the 3-metric $d\theta^2/\sin^2\theta - 4d\zeta d\bar{\zeta}/\sin^2\theta(1+|\zeta|^2)^2$ is the Lorentzian hyperbolic metric).

Discussion. It is possible to write down metrics that are Ricci flat with conformal structures that extend over $S^2 \times S^2$ but whose null infinity cuts the space in half. This uses the same construction but with a translation symmetry generated by $\pi^A\partial/\partial\omega^A$ where the real slice is now given by real values for the components of (ω^A, π^A). The gluing identifies $\omega^A - if\pi^A$ on $\mathbb{CP}^3_+$ with $\omega^A + if\pi^A$ on $\mathbb{CP}^3_-$ where $f := f(\omega^A\pi_A, \pi_A)$ has homogeneity zero and is rapidly decreasing as $\omega^A\pi_A/(\pi_0^2+\pi_1^2) \to \infty$.

References

Fadeev, L. & Takhtajan, L. (1987) *Hamiltonian methods in the theory of solitons*, Springer, Berlin.

Jones, P. and Tod, K.P. (1985) Minitwistor spaces and Einstein Weyl spaces, Class. & Quant. Grav. **2** 565-577.

LeBrun, C.R. (1991) Explicit self-dual metrics on $\mathbb{CP}^2\#\ldots\#\mathbb{CP}^2$, J. Differential Geometry, **34** (1991) 223-253.

Mason, L.J. & Sparling, G.A.J. (1992) Twistor correspondences for the soliton hierarchies, J. Geom. Phys. **8** 243-271.

Ward, R.S. (1977) On self-dual gauge fields, Phys. Lett. **A 61**, 81-82

Ward (1985) Integrable and solvable systems and relations among them, Phil. Trans. Roy. Soc. Lond. **A315**, 451-457.

§II.1.12 Harmonic morphisms and mini-twistor space *by K.P. Tod* (TN 29, November 1989)

A harmonic morphism is a map $\varphi : M \to N$ of Riemannian manifolds M, N with the following property: $f : N \to \mathbb{R}$ is harmonic iff $\varphi \circ f : M \to \mathbb{R}$ is. As a concrete example, take M to be

$\mathbb{R}^3$ with coordinates x, y, z and N to be $\mathbb{R}^2$ with coordinates u, v. The map φ is defined by giving $u(x, y, z)$, $v(x, y, x)$ satisfying

$$\nabla^2 u = \nabla^2 v = 0 = \nabla u \cdot \nabla v; \qquad |\nabla u|^2 = |\nabla v|^2 \tag{1}$$

In this case Baird & Wood (1988) find that φ is locally defined by a holomorphic curve in $T\mathbb{P}_1$, the tangent bundle of the complex projective line. They go on to use this fact to classify globally defined harmonic morphisms in this case, and also in the cases $S^3 \rightarrow$ surface and $H^3 \rightarrow$ surface.

Since $T\mathbb{P}_1$ is the mini-twistor space of $\mathbb{R}^3$ it is natural to wonder what, if anything, is the relation to twistor theory of this property of φ. In the case when dim $M = 3$, dim $N = 2$, the inverse images of points of N give curves in M. One purpose of this note is to observe that *the defining property of harmonic morphisms is equivalent to the condition that this congruence of curves be a geodesic and shear-free congruence.*

Now $T\mathbb{P}_1$ is ths space of geodesics of the flat metric on $\mathbb{R}^3$ and so a congruence of geodesics corresponds to a 2-real parameter surface in $T\mathbb{P}_1$. As one might anticipate from the Kerr theorem, there is a mini-Kerr theorem that this surface is a holomorphic curve iff the congruence is shear-free. In particular, this leads to an explicit formula for such congruences: if the generator is

$$L = \frac{1-\alpha\bar{\alpha}}{1+\alpha\bar{\alpha}}\frac{\partial}{\partial z} + \frac{\alpha+\bar{\alpha}}{1+\alpha\bar{\alpha}}\frac{\partial}{\partial x} - \frac{\mathrm{i}(\alpha-\bar{\alpha})}{1+\alpha\bar{\alpha}}\frac{\partial}{\partial y}$$

then $\alpha(x, y, z)$ is given implicitly by

$$f(x(1-\alpha^2) + iy(1+\alpha^2) - 2\alpha z, \alpha) = 0 \quad \text{or in spinors} \quad F(x^{AB}\alpha_A\alpha_B, \alpha_C) = 0$$

for arbitrary holomorphic f or holomorphic and homogeneous F (a formula similar to this is in Baird & Wood (1988)).

As Baird & Wood remark, to find solutions of (1) was set as a problem by Jacobi. This now falls into the class of non-linear differential-geometric problems solvable by twistor theory (see II.1.13 for more on this subject).

I am grateful to John Wood and Paul Baird for telling me about harmonic morphisms.

References

Baird, P. & Wood, J. (1988) Math. Ann. **280**, p. 579–603.

Baird, P. & Wood, J. (1987) Ann. Inst. Fourier, Grenoble **37**, p. 135–173.

Baird, P. & Wood, J. (1989) Harmonic morphisms and conformal foliations by geodesics of three-dimensional space-forms, University of Melbourne Department of Mathematics Research Report **2**.

§II.1.13 **More on harmonic morphisms** *by K.P.Tod* (TN 30, June 1990)

I wish to make some observations which might make my Twistor Newsletter article II.1.12, clearer, and to describe what I think is a new way of looking at the Kerr theorem appropriate to Riemannian twistor theory (though in this last connection see Hughston & Mason 1988).

Recall that the generator of a geodesic shear-free congruence is

$$L = \frac{1-\alpha\bar{\alpha}}{1+\alpha\bar{\alpha}}\frac{\partial}{\partial z} + \frac{\alpha+\bar{\alpha}}{1+\alpha\bar{\alpha}}\frac{\partial}{\partial x} - i\frac{(\alpha-\bar{\alpha})}{1+\alpha\bar{\alpha}}\frac{\partial}{\partial y} \tag{1}$$

where $\alpha(x, y, z)$ is given implicitly by

$$f(x(1-\alpha^2) + iy(1+\alpha^2) - 2\alpha z, \alpha) = 0. \tag{2}$$

In the interest of clarity, I should have added in II.1.12 that:

(a) If $\alpha = u + iv$ then

$$\nabla^2 u = \nabla^2 v = 0 = \nabla u \cdot \nabla v; \qquad |\nabla u|^2 = |\nabla v|^2$$

ie. the complex function $\alpha(x, y, z)$ obtained from (2) defines the harmonic morphism of II.1.12.

(b) $\alpha(x, y, z)$ is constant along L as given by (1).

(c) For a given constant value of α, the real and imaginary parts of (2) each define a plane and L is then tangent to the line of intersection of the two planes.

There is an 'elementary' interpretation of (2) as follows: a line in $\mathbb{R}^3$ is given by an equation of the form

$$r \wedge a = b \tag{3}$$

where a is a unit vector and b is orthogonal to a. Parametrise the unit vectors with the complex number α according to (1), then (3) can be written

$$x(1-\alpha^2) + iy(1+\alpha^2) - 2\alpha z = \beta \tag{4}$$

where β parametrises the Argand plane orthogonal to a. Now (2) is equivalent to taking β to be an arbitrary holomorphic function of α, ie. the 'intercept' is an arbitrary holomorphic function of the 'direction', both understood as complex variables. When this function is linear, the congruence is a 3-dimensional picture of the Kerr congruence and is sketched in Baird & Wood (1988).

'Harmonic morphisms' is also the answer to the question 'what do you get from the Kerr theorem in Riemannian twistor theory (when there are no shear-free geodesic congruences)?' To see this, take a homogeneous holomorphic twistor function $F(Z^{\alpha})$ and intersect the zero-set of F with a line in twistor space which is 'real' in the sense appropriate to Riemannian twistor theory. This gives

$$F(a + b\zeta, -\bar{b} + \bar{a}\zeta, 1, \zeta) = 0 \tag{5}$$

writing $(1, \zeta)$ for the π-spinor (rather than $(1, \alpha)$ which I used at the beginning). Here a and b are complex coordinates on $\mathbb{R}^4$ and the metric is

$$ds^2 = da\, d\bar{a} + db\, d\bar{b}. \tag{6}$$

Solving (5) gives a function $\zeta(a, b, \bar{a}, \bar{b})$ with

$$\frac{\partial \zeta}{\partial \bar{a}} + \zeta \frac{\partial \zeta}{\partial \bar{b}} = 0; \qquad \frac{\partial \zeta}{\partial b} - \zeta \frac{\partial \zeta}{\partial a} = 0 \tag{7}$$

from which it follows that ζ has vanishing Laplacian and null gradient in the metric (6):

$$\frac{\partial^2 \zeta}{\partial a \partial \bar{a}} + \frac{\partial^2 \zeta}{\partial b \partial \bar{b}} = 0; \qquad \frac{\partial \zeta}{\partial a} \cdot \frac{\partial \zeta}{\partial \bar{a}} + \frac{\partial \zeta}{\partial b} \cdot \frac{\partial \zeta}{\partial \bar{b}} = 0 \tag{8}$$

If we think of ζ as the stereographic coordinate on the sphere, then (8) is easily seen to be the conditions for (5) to define a harmonic morphism from $\mathbb{R}^4$ to $\mathbb{S}^2$. However, (7) is stronger in that it implies that, as well as defining a harmonic morphism, ζ is constant on flat 2-planes. Note that if ζ satisfies (8) then so does any holomorphic function of ζ. In this sence, ζ defines a family of holomorphically-related harmonic morphisms constant on flat 2-planes, one of which satisfies (7).

For the converse suppose that η satisfies (8) and is constant on flat 2-planes. Define ζ by

$$\frac{\partial \eta}{\partial \bar{a}} + \zeta \frac{\partial \eta}{\partial \bar{b}} = 0 \quad \text{so that also} \quad \frac{\partial \eta}{\partial b} - \zeta \frac{\partial \eta}{\partial a} = 0$$

then it follows from the conditions on η that ζ is a holomorphic function of η and so in turn satisfies (7) and (8).

To summarise, a family of holomorphically-related harmonic morphisms from $\mathbb{R}^4$ to $\mathbb{S}^2$ which are constant on 2-planes defines and is defined by a holmorphic function in twistor space. This can be called 'the Riemannian Kerr theorem'.

This view of the Kerr theorem arose in discussions with Henrik Pedersen.

References

Baird, P. & Wood, J. (1988) Math. Ann. **280**, p. 579–603.

Hughston, L.P. and Mason, L.J. (1988) A generalized Kerr-Robinson theorem, Class. Quant. Grav. **5**, p. 275.

§II.1.14 Monopoles, harmonic morphisms and spinor fields *by P.Baird and J.C. Wood*
(TN 31, October 1990)

Introduction. In this paper we draw attention to the connection between static monopoles, harmonic morphisms and spinor fields defined on 3- and 4-spaces.

Hitchin has shown how every static monopole of charge k may be constructed canonically from an algebraic curve by means of the Atiyah-Ward Ansatz $\mathcal{A}_k$ (Hitchin 1982). This algebraic curve is a real curve (that is, invariant under a real structure τ) of degree k in the complex surface TS^2, satisfying certain other technical conditions. In an apparently independent line of ideas, we studied harmonic morphisms from domains in Euclidean 3-space (see below), obtaining their classification, again in terms of algebraic curves in TS^2 (Baird & Wood 1988). Thus, remarkably, these solutions to two quite different equations, one relating the gradient of the Higgs field to the curvature of an SU_2-bundle, the other a second order elliptic equation in a complex valued function with a first order conformality condition, are classified in terms of the same objects. A direct relationship between solutions remains obscure. However, here we are able to describe harmonic morphisms on $\mathbb{R}^3$ and $\mathbb{R}^4$ (also on 4-dimensional Minkowski space M^4) in terms of spinor fields satisfying some simple first order equations. Furthermore, we write down all solutions to these equations explicitly. Indirectly, therefore, we express a static monopole in terms of a (multiple-valued) spinor field on $\mathbb{R}^3$. This description may give a more illuminating picture of monopoles. For instance, in examples such as the axially symmetric solutions of Prasad & Rossi, (see section 1) the region of physical interest corresponds to singularities of the spinor field. Note that harmonic morphisms on $\mathbb{R}^3$ may also be viewed as harmonic morphisms on $\mathbb{R}^4$ invariant under translation in some particular direction.

Now let φ be a harmonic morphism from a domain in $\mathbb{R}^4$ to a surface. At regular points of φ we are able to interpret the spinor equations in terms of holomorphicity properties of sections γ^1, γ^2 of twistor bundles. This ties in with the description of the second author for submersive harmonic morphisms from a Riemannian 4-manifold to a surface (Wood 1986). Unlike the $\mathbb{R}^3$ case (Baird & Wood 1988), the sections γ^1, γ^2 may have 'unremovable' singularities at critical points of φ. However, the spinor formulation has the great advantage that the spinor solutions can be chosen so that they extend continuously over critical points.

Another connection between harmonic morphisms and physical fields is the following: A submersive harmonic morphism from a Riemannian m-dimensional manifold M to a surface is locally equivalent to an $(m-2)$-dimensional conformal foliation of M by minimal submanifolds. In the case

$m = 3$, we can remove the restriction 'submersive' and the foliation is by geodesics. Such conformal foliations are the Riemannian analogue of the well-known shear-free, null geodesic congruences, much studied by relativists in connection with zero rest mass fields (see Baird & Wood 1991, 1992 and II.1.12, II.1.13).

Throughout, we use spinors as described in the Appendix of Penrose & Rindler (1986). This enables us to consider spinors defined on a vector space with metric of arbitrary signature.

1. Harmonic morphisms from $\mathbb{R}^3$ in terms of spinors. Let $M \subset \mathbb{R}^m$ be an open subset and $\varphi : M \to \mathbb{C}$ a smooth map. Then φ is said to be *horizontally conformal* if

$$\sum_a \left(\frac{\partial \varphi}{\partial x^a}\right)^2 = 0, \tag{1.1}$$

and *harmonic* if

$$\sum_a \frac{\partial^2 \varphi}{(\partial x^a)^2} = 0. \tag{1.2}$$

Here $(x^1, \ldots, x^m)$ are standard coordinates on $\mathbb{R}^m$ and summation is over $a = 1, \ldots, m$. The map φ is called a *harmonic morphism* if both (1.1) and (1.2) hold. There are other equivalent definitions, for example, harmonic morphisms are characterized as mappings which pull back local harmonic functions to local harmonic functions (see Fuglede 1978, Ishihara 1979). Equivalently they are mappings which send Brownian paths to Brownian paths (see Bernard, Campbell & Davie 1979). The two latter definitions then generalize the notion to mappings between arbitrary Riemannian manifolds, the equations (1.1), (1.2) becoming more complicated in this case (see Fuglede 1978, Ishihara 1979). Note that in the case $m = 2$ the only solutions to (1.1), (1.2) are $\pm$-holomorphic (by which we mean holomorphic or antiholomorphic) functions $\varphi(x^1 \pm \mathrm{i}x^2)$; in the case $m \geqslant 3$ there are many more, see below. For all m the equations are invariant under conformal transformations in the range, that is, if $\varphi : M \to V \subset \mathbb{C}$ is a solution, then so is the composition $\rho \circ \varphi : M \to \mathbb{C}$ where $\rho : V \to \mathbb{C}$ is any weakly conformal (equivalently $\pm$-holomorphic) map. In particular we may consider harmonic morphisms $\varphi : M \to N$ with values in a *Riemann surface* N; these will satisfy (1.1) and (1.2) with respect to a local complex coordinate on N.

We consider $\mathbb{R}^3$ with its standard Euclidean metric. Vectors x^a may be expressed in terms of spinors by the correspondence

$$x^a = (x^1, x^2, x^3) \longleftrightarrow x^{AB} = \frac{1}{\sqrt{2}} \begin{pmatrix} x^2 + \mathrm{i}x^3 & -x^1 \\ -x^1 & -x^2 + \mathrm{i}x^3 \end{pmatrix}.$$

Writing $\partial_a = \partial/\partial x^a$, the spinor covariant derivatives D_{AB} are given by

$$D_{00} = (\partial_2 - \mathrm{i}\partial_3)/\sqrt{2}$$
$$D_{01} = -\partial_1/\sqrt{2}$$
$$D_{10} = -\partial_1/\sqrt{2}$$
$$D_{11} = (-\partial_2 - \mathrm{i}\partial_3)/\sqrt{2}$$

(see also Sommers 1980 equation (14)).

Let $M \subset \mathbb{R}^3$ be an open subset and $\varphi : M \to \mathbb{C}$ a smooth mapping. Then φ is horizontally conformal if and only if

$$\det D_{AB}\,\varphi = 0$$

(that is $\nabla\varphi$ is a *spatial null vector field*), and this holds if and only if

$$D_{AB}\,\varphi = \xi_A \eta_B.$$

But the symmetry of $D_{AB}\,\varphi$ then implies that $\eta_B = \lambda\xi_B$ for some scalar function λ. So we deduce that φ is horizontally conformal if and only if

$$D_{AB}\,\varphi = \xi_A \xi_B \tag{1.3}$$

for some spinor field ξ_A defined over M.

If φ is horizontally conformal, so that (1.3) holds, then φ is harmonic, and so is a harmonic morphism, if and only if

$$D_{AB}\xi^A\xi^B = 0. \tag{1.4}$$

Conversely, as in Sommers (1980), given a spacial null vector field $v^a \leftrightarrow \mu^A\mu^B$, then curl v^a is given by $-\mathrm{i}\sqrt{2}D^{C(A}\mu^{B)}\mu_C$. Combining this with equation (1.4), we obtain

THEOREM 1.1. *Let $M \subset \mathbb{R}^3$ be a simply connected open subset. There is a correspondence between harmonic morphisms $\varphi : M \to \mathbb{C}$ and spinor fields ξ^A on M satisfying the spinor equation*

$$D_{AB}\xi^A\xi^C = 0. \tag{1.5}$$

REMARKS 1.2.

(i) This correspondence is one-to-one if we identify two harmonic morphisms $\varphi, \tilde{\varphi}$ whenever $\tilde{\varphi} = \varphi + c$, for some complex number c,and define the spinor field up to sign only.

(ii) The spinor field ξ^A extends continuously over critical points of φ. For if φ is a harmonic morphism, then at a critical point (i.e. a point where the differential of φ has rank less than 2) the derivative collapses completely (Fuglede 1978), that is $D_{AB}\,\varphi \equiv 0$. Setting $||D_{AB}\,\varphi||^2$ to be the Hilbert-Schmidt norm of the derivative (the sum of the squares of the moduli of its components), then $||D_{AB}\,\varphi||^2 = |\xi_A|^4$, where $|\xi_A|^2 = |\xi_0|^2 + |\xi_1|^2$. Since φ is smooth, it follows that ξ_A extends with the value 0 at a critical point.

(iii) A spinor field $\psi^{AB} = \xi^A\xi^B$ satisfying equation (1.5) may be interpreted as a null, source free, time independent solution to Maxwell's equations. This is clear by expressing $\nabla\varphi = E + iB$

in real and imaginary parts. Then horizontal conformality implies $E \cdot B = 0$ *and* $|E| = |B|$ *(nullity),* curl $E =$ curl $B = 0$ *is automatic and harmonicity gives* div $E =$ div $B = 0$.

(iv) *In Baird & Wood (1988) the notion of a generalized harmonic morphism was defined. This corresponds to a holomorphic curve in the mini-twistor space* TS^2, *the space of oriented lines in* $\mathbb{R}^3$, *with its natural complex structure. Such a curve determines an oriented line field defined on* $\mathbb{R}^3$. *Locally such a line field corresponds to a harmonic morphism. Globally it should be thought of as corresponding to a multiple-valued harmonic morphism, where envelope points (points where nearby lines become infinitesimally close) correspond to branch points of the harmonic morphism. In fact such a multiple-valued harmonic morphism* $z = z(x), x \in \mathbb{R}^m$, *is determined by an equation of the form*

$$P(x, z) = 0, \tag{1.6}$$

where P *is holomorphic in* z *and a harmonic morphism in* x, *that is*

$$\sum_a \left(\frac{\partial P}{\partial x^a}\right)^2 = 0 \quad \text{and} \quad \sum_a \frac{\partial^2 P}{(\partial x^a)^2} = 0.$$

In general for each $x \in \mathbb{R}^m$ this may have several roots. Branch points of the resulting multivalued function $z = z(x)$ correspond to points (x, z) where $\partial P/\partial z = 0$. Frequently P is polynomial in z in which case they correspond to values x for which two roots z of (1.6) coincide. We may think of such multiple-valued harmonic morphisms as a natural generalization to higher dimensions of the (multiple-valued) analytic functions of Riemann surface theory, see Baird (1987, 1990) for some examples and Gudmundsson & Wood (1993) for a general theory.

Given a harmonic morphism $\varphi : M \to \mathbb{C}, M$ open in $\mathbb{R}^3$, we can associate a Gauss map $\gamma : M \to S^2$, given by $\gamma(x) =$ unit positive tangent to the fibre of φ through x (see Baird 1987, Baird & Wood 1988). In fact γ extends smoothly across critical points (Baird & Wood 1992). Then it is easily checked that in the chart given by stereographic projection $S^2 \to \mathbb{C} \cup \infty$, γ is represented by ξ^0/ξ^1. The equation (1.5) now has the simple interpretation of (i) minimality of the fibres $\varphi =$ const., and (ii) horizontal $\pm$-holomorphicity of the Gauss map (Baird 1987, Wood 1986).

Harmonic morphisms $\varphi : M \to \mathbb{C}$ from open subsets of Euclidean space $\mathbb{R}^3$ have been completely classified in Baird & Wood (1988). Indeed the fibres are (parts of) straight lines which form a smooth foliation $\mathcal{F}$ of M and, locally, up to composition with a weakly conformal map on an open subset of $\mathbb{C}, \varphi$ is given implicitly by an equation

$$\alpha_a(\varphi(x))x^a = 1 \tag{1.7}$$

where

$$\alpha = \frac{1}{2h}\left(1 - g^2,\ i(1+g^2), -2g\right)$$

and g, h are meromorphic functions on a certain Riemann surface N. (Indeed, because of the horizontal conformality condition (1.2), the foliation $\mathcal{F}$ has a transverse conformal structure and N is a local leaf space.)

Note that writing $z = \varphi(x)$, equation (1.7) is of the form (1.6). The solution to (1.5) corresponding to φ is given at $x \in M$ by

$$\xi_A = \frac{1}{\sqrt{2^{1/2}\alpha'.x}}\left(\frac{1}{\sqrt{h}}, \frac{g}{\sqrt{h}}\right), \tag{1.8}$$

where g, h are evaluated at $\pi(x)$, π being the natural projection onto the leaf space N. By a result in Baird & Wood (1988), the only harmonic morphisms defined *globally* on $\mathbb{R}^3$ with values in $\mathbb{C}$ (or, more generally in a Riemann surface) are given by an orthogonal projection followed by a weakly conformal map of $\mathbb{C}$. In this case after appropriate choice of coordinates, $N \approx \mathbb{C}$, g is constant and $h(z) = z$, and $[\xi_A]$, the *projectivized* spinor field, is constant.

Note that, by the Implicit Function Theorem, (1.7) has smooth solutions φ away from points where $\alpha_a'(\varphi(x))x^a = 0$. Thinking of the equation $\alpha_a(z)x^a = 1$ $(z \in N)$ as defining a family of lines (parametrized by $z \in N$) these points are the *envelope points* of this family where nearby lines get infinitesimally close (see Baird & Wood 1988, Gudmundsson & Wood 1993). At such points the spinor field ξ_A has a singularity.

The simplest monopoles are given by Hitchin (1982, 1983):

(i) (Monopole of charge $k = 1$). $g(z) = z, h(z) = z$. This corresponds to a two-valued harmonic morphism known as radial projection, see Baird (1987), Baird & Wood (1988), Gudmundsson & Wood (1991). The fibres consist of lines through the point $(0, 0, -1)$, with both orientations. The envelope of the fibre is the single point $(0, 0, -1)$.

(ii) (The Prasad-Rossi monopole of charge $k = 2$). $g(z) = z, h(z) = \sqrt{-z^2}$. This also corresponds to a rotationally symmetric two-valued harmonic morphism. One branch of this is called the *disk* example (Bernard, Campbell & Davie 1979, Baird 1987, Baird & Wood 1988). In this case the envelope is the circle

$$C : x^3 = 0, (x^1)^2 + (x^2)^2 = 1.$$

Lines twist through the interior of this circle and the corresponding tangent line field is discontinuous as we cross the (x^1, x^2)-plane outside the circle C. (There is also a version of this example where the line field is discontinuous as we cross the (x^1, x^2)-plane *inside* the circle C, see Gudmundsson & Wood (1993).)

2. Harmonic morphisms from $\mathbb{R}^4$ in terms of spinors. We consider $\mathbb{R}^4$ with its standard Euclidean metric. Vectors x^a may now be expressed in terms of spinors by the correspondence

$$x^a = (x^0, x^1, x^2, x^3) \quad \leftrightarrow \quad x^{AA'} = \frac{1}{\sqrt{2}} \begin{pmatrix} \mathrm{i}x^0 + x^1 & x^2 + \mathrm{i}x^3 \\ x^2 - \mathrm{i}x^3 & \mathrm{i}x^0 - x^1 \end{pmatrix}.$$

Let $M \subset \mathbb{R}^4$ be an open subset. Then, as for $\mathbb{R}^3$, a map $\varphi : M \to \mathbb{C}$ is horizontally conformal if and only if the gradient $\nabla\varphi$ is a complex null field and this holds if and only if

$$\nabla_{AA'}\varphi = \xi_A \eta_{A'} \tag{2.1}$$

for some spinor fields $\xi_A, \eta_{A'}$ defined on M, where the spinor covariant derivatives are given by $\nabla_{AA'} = \partial/\partial x^{AA'}$.

REMARK. *We always have the freedom*

$$(\xi_A, \eta_{A'}) \mapsto (\lambda\xi_A, \lambda^{-1}\eta_{A'}) \,, \tag{2.2}$$

for any non-zero scalar function λ.

Suppose now that φ is horizontally conformal, so that (2.1) holds. Then φ is harmonic and so is a harmonic morphism, if and only if

$$\nabla^{AA'}\xi_A\eta_{A'} = 0. \tag{2.3}$$

Conversely, given a pair of spinor fields $\xi_A, \eta_{A'}$ on M, we would like conditions which ensure they determine a harmonic morphism. Now the product $\xi^A\eta^{A'}$ determines a null vector field v^a. We require $\nabla_{[a}v_{b]}$ to be zero. This is equivalent to the pair of spinor equations:

$$\begin{cases} \nabla^{B'}_{(B}\xi_{A)}\eta_{B'} = 0 \\ \nabla_{A(B'}\eta_{A')}\xi^A = 0 \,. \end{cases} \tag{2.4}$$

Combining (2.3) and (2.4) we obtain:

THEOREM 2.1. *Let $M \subset \mathbb{R}^4$ be a simply connected open subset. There is a correspondence between harmonic morphisms $\varphi : M \to \mathbb{C}$ and pairs of spinor fields $(\xi^A, \eta^{A'})$ on M satisfying the spinor equations*

$$\begin{cases} \nabla_{AA'}\xi^A\eta^{B'} = 0 \\ \nabla_{AA'}\xi^B\eta^{A'} = 0 \end{cases} \tag{2.5}$$

Proof. It is clear that (2.5) implies (2.3) and (2.4). Conversely, suppose we consider the first of equations (2.5) with $A' = B' = 0$. Then

$$\nabla_{00'}\xi^0\eta^{0'} + \nabla_{10'}\xi^1\eta^{0'} = \tfrac{1}{2}(\nabla_{00'}\xi^0\eta^{0'} + \nabla_{11'}\xi^1\eta^{1'} + \nabla_{10'}\xi^1\eta^{0'} + \nabla_{01'}\xi^0\eta^{1'}),$$

by (2.4). But this equals zero by (2.3). The other equations are obtained similarly.

REMARKS 2.2.

(i) The correspondence is one-to-one if we identify two harmonic morphisms which differ by a constant and subject the spinor fields to the equivalence (2.2) .

(ii) We may choose λ in (2.2) in such a way that the spinor fields $\xi_A, \eta_{A'}$ extend continuously to critical points. For choose any continuous scalar function λ such that $|\xi_A| = |\eta_{A'}|$ (they are now defined up to equivalence $|\lambda| = 1$). Then as in Remarks 1.2 (ii), $\|\nabla_{AA'}\varphi\|^2 = |\xi_A|^2|\eta_{A'}|^2 = |\xi_A|^4$ and both $\xi_A, \eta_{A'}$ extend continuously to 0 at a critical point.

(iii) In terms of the geometric description of Penrose & Rindler (1984), at each point $x \in M$ where $\nabla_{AA'}\varphi \neq 0$, the projectivized spinor $[\xi_A(x)]$ determines an α-plane $\alpha(x)$ (a complex line in this case) on the quadric $Q_2 \subset \mathbb{C}P^3$, and $[\eta_{A'}(x)]$ determines a β-plane $\beta(x)$. Then $\alpha(x), \beta(x)$ intersect in a point of Q_2. This point corresponds (under the identification of Q_2 with the Grassmannian of oriented 2-planes in $\mathbb{R}^4$) to a real 2-plane through the origin in $\mathbb{R}^4$. This plane is the vertical space at x (the tangent to the fibre of φ through x), translated to the origin.

(iv) A harmonic morphism on $\mathbb{R}^3$ may be viewed as a harmonic morphism on $\mathbb{R}^4$ which is invariant under some translation. Thus equation (1.5) is a special case of equations (2.5).

3. Examples. Particular examples of harmonic morphisms $\varphi : \mathbb{R}^4 \to \mathbb{C}$ are given by maps which are $\pm$-holomorphic with respect to one of the Kähler structures on $\mathbb{R}^4$. Each Kähler structure arises from the standard one obtained by identifying $\mathbb{R}^4$ with $\mathbb{C} \times \mathbb{C}$, by composing with an isometry of $\mathbb{R}^4$.

Use coordinates (z, w) for $\mathbb{C} \times \mathbb{C}$, so that $z = x^0 + ix^1, w = x^2 + ix^3$. Then $\varphi : M \to \mathbb{C}$, M open in $\mathbb{R}^4$, is holomorphic if and only if

$$\frac{\partial\varphi}{\partial\bar{z}} = \frac{\partial\varphi}{\partial\bar{w}} = 0,$$

which holds if and only if

$$\nabla_{00'}\varphi = \nabla_{10'}\varphi = 0.$$

Then clearly $\det \nabla_{AA'}\varphi = 0$ and

$$\nabla_{AA'}\varphi = \begin{pmatrix} 0 & * \\ 0 & * \end{pmatrix} = \begin{pmatrix} * \\ * \end{pmatrix} \quad (0 \ \ *)$$

so that $\eta_{B'} = (0, \lambda)$, for some scalar λ. Similarly φ is anti-holomorphic if and only if $\xi_A = (\mu, 0)$ for some scalar μ. We now consider the effect of an isometry on the spinor decomposition of $\nabla_{AA'}\varphi$.

There is a well-known double cover $SU(2) \times SU(2) \to SO(4)$. Suppose that $\theta \in SO(4)$ and define $\tilde{\varphi} = \varphi \circ \theta$. Then $\nabla\tilde{\varphi}(x) = \nabla\varphi(\theta(x)) \circ \theta$. If $(A, B) \in SU(2) \times SU(2)$ covers θ, then the induced action on spinors is given by

$$\xi_A \eta_{A'} \mapsto A\xi_A \eta_{A'} B^*,$$

where $B^* = \overline{B}^T$, so that $(\xi_A, \eta_{A'}) \mapsto (A\xi_A, \eta_{A'}B^*)$, that is $\tilde{\xi}_A(\theta(x)) = A(\xi_A(x))$ and $\tilde{\eta}_{A'}(\theta(x)) = \eta_{A'}(x)B^*$. Note that under the equivalence (2.2), this is independent of the choice of (A, B) covering θ. In particular we see that $\varphi : M \to \mathbb{C}$ is $\pm$-holomorphic with respect to a Kähler structure obtained from the standard one by an orientation preserving isometry if and only if $[\eta_{A'}] \in \mathbb{C}P^1$ is constant. Similarly φ is $\pm$-holomorphic with respect to a Kähler structure obtained by an orientation reversing isometry if and only if $[\xi_A] \in \mathbb{C}P^1$ is constant. Summarizing we have

THEOREM 3.1. *If $\varphi : N \to \mathbb{C}$ (M open in $\mathbb{R}^4$) is a harmonic morphism, then φ is $\pm$-holomorphic with respect to one of the Kähler structures on $\mathbb{R}^4$ if and only if either $[\eta_{A'}]$ or $[\xi_A]$ is constant.*

EXAMPLE 3.2. Let $\varphi : \mathbb{R}^4 \to \mathbb{C}$ be given by $\varphi(z, w) = zw$ (after identifying $\mathbb{R}^4$ with $\mathbb{C}^2$). Then we may compute the spinor fields $\eta_{A'}, \xi_A$ to be

$$\eta_{A'} = (0, 1), \quad \xi_A = \sqrt{2}\begin{pmatrix} z \\ -\mathrm{i}w \end{pmatrix}.$$

In this case φ is holomorphic with respect to the standard complex structure on $\mathbb{R}^4$ represented by $[\eta_{A'}]$ (see Section 4). There is a single critical point at the origin and both spinor fields extend smoothly over this point.

Another class of examples are those which have totally geodesic fibres. These are classified in Baird & Wood (1988) (see also II.1.13). If $\varphi : M \to \mathbb{C}$, M open in $\mathbb{R}^4$, is a harmonic morphism with totally geodesic fibres, let N denote the leaf space of the fibres. Locally, and in favourable circumstances globally, N can be given the structure of a smooth Riemann surface and φ is given implicitly by an equation

$$\alpha_a(\varphi(x))x^a = 1,$$

where $\alpha = \frac{1}{2h}\left((1 + fg),\ \mathrm{i}(1 - fg),\ f - g,\ -\mathrm{i}(f + g)\right)$ and $f, g, h : N \to \mathbb{C} \cup \infty$ are meromorphic functions. In this case it is easily checked that $\xi_A, \eta_{A'}$ are given by

$$\xi_A = \frac{1}{\sqrt{2^{1/2}(\alpha' \cdot x)h}}\begin{pmatrix} 1 \\ \mathrm{i}g \end{pmatrix} \quad \eta_{A'} = \frac{1}{\sqrt{2^{1/2}(\alpha' \cdot x)h}}(\mathrm{i}, f),$$

(here, f, g and h are evaluated at $\pi(x)$, where π is the natural projection onto N).

Note that, in general, neither of these is projectively constant and so the harmonic morphisms are not $\pm$-holomorphic. More generally, the second author described all harmonic morphisms from open sets of $\mathbb{R}^4$ locally in terms of holomorphic functions (Wood (1992)). When we translate that classification into spinor fields, we can express all solutions to equations (2.5) as follows:

Locally a harmonic morphism $\mu : M \to \mathbb{C}$ (M open in $\mathbb{R}^4$) is given implicitly by an equation

$$\psi(z - \mu\overline{w}, w + \mu\overline{z}, \mu) = 0, \tag{3.1}$$

or an equation

$$\psi(z - \mu w,\ w + \mu z, \mu) = 0, \tag{3.1$'$}$$

where $\psi = \psi(u_1, u_2, \mu)$ is a holomorphic function of three complex variables. By the Implicit Function Theorem, equation (3.1) has smooth solutions μ if

$$P \equiv -\overline{w}\,\frac{\partial\psi}{\partial u_1} + \overline{z}\,\frac{\partial\psi}{\partial u_2} + \frac{\partial\psi}{\partial\mu} \neq 0. \tag{3.2}$$

Corresponding spinor fields are given by

$$\xi_A = \frac{1}{\sqrt{P}}\begin{pmatrix} -\mathrm{i}\partial\psi/\partial u_2 \\ -\partial\psi/\partial u_1 \end{pmatrix}, \quad \eta_{A'} = \frac{1}{\sqrt{P}}(-\mu, -\mathrm{i})$$

(or any pair equivalent in the sense of (2.2)). We have a similar description in the case (3.1$'$) with the rôles of ξ and η reversed.

REMARKS 3.3.

(i) *The case with totally geodesic fibres may be obtained from this more general description by setting $\mu = -f$ and $\psi(u_1, u_2, \mu) = u_1 - gu_2 - 2h$.*

(ii) *It follows from Wood (1992) that (2.5) does not have any solutions globally defined on $\mathbb{R}^4$ apart from those with $[\xi_A]$ or $[\eta_{A'}]$ constant.*

4. Interpretation in terms of twistor bundles. Here we relate our spinor description to the description given by the second author Wood (1986) in terms of twistor bundles, thus interpreting the equations (2.5) in terms of holomorphicity properties of Gauss sections. We briefly summarize the results of Wood (1986).

Let V be a 2-dimensional distribution in an oriented 4-dimensional Riemannian manifold M, and let H be the corresponding orthogonal 2-dimensional distribution. We may locally choose orientations for each $V_x, H_x, x \in M$, so that the combined orientation of $V_x \oplus H_x = T_xM$ is that of M. We then define almost complex structures J^V, J^H on each V_x, H_x to be rotation through $+\pi/2$. Note that changing the orientation of V_x changes that of H_x, and replaces (J^V, J^H) by $(-J^V, -J^H)$. All the results below will be independent of this change, so that there is no loss of generality in assuming that J^V and J^H are globally chosen.

The Gauss section of $V, \gamma : M \to \tilde{G}_2(TM)$, defined by $\gamma(x) = V_x$, maps into the Grassmannian bundle of oriented 2-planes in TM. The almost complex structures J^V and J^H combine to give almost-complex structures $J^1 = J^1_x = (J^V, J^H)$ and $J^2 = J^2_x = (-J^V, J^H)$ on each tangent space $T_x M$. We have thus defined two almost complex structures on the manifold M. Note that J^1 is compatible with the orientation, i.e., there exists an oriented basis of the form $e_1, J^1 e_1, e_2, J^1 e_2$, whereas J^2 is incompatible. Let Z^+ (respectively Z^-) be the fibre bundle over M whose fibre at x is all metric almost complex structures on $T_x M$ which are compatible (respectively incompatible) with the orientation; these are the well-known twistor bundles of M (Eells & Salamon 1985). The distribution V defines sections $\gamma^1 : M \to Z^+$ and $\gamma^2 : M \to Z^-$ by $\gamma^1(x) = J^1_x, \gamma^2(x) = J^2_x$. Note that, if M is an open subset of Euclidean space $\mathbb{R}^4$, the twistor bundles are trivial: $Z^\pm = M \times S^2$, and there is a well-known holomorphic bijection $\tilde{G}_2(\mathbb{R}^4) \approx S^2 \times S^2$.

Given a submersive harmonic morphism $M \to N$ from an oriented Riemannian four-manifold to a Riemann surface, the tangent spaces to its fibres gives an integrable, minimal and conformal distribution (see Wood 1986). Conversely, any such distribution arises locally in this way. We recall the following characterization of such a distribution:

THEOREM 4.1. *(Wood 1986) Let V be a 2-dimensional distribution on an oriented 4-dimensional Riemannian manifold M. Then V is integrable, minimal and conformal if and only if the section $\gamma^1 : M \to Z^+$ is holomorphic with respect to the almost complex structure J^2 on M and the section $\gamma^2 : M \to Z^-$ is holomorphic with respect to the almost complex structure J^1 on M.*

REMARK. *In Wood (1986), the opposite complex structure was used on the twistor space Z^- which resulted in γ^2 being antiholomorphic with respect to J^1.*

Now let $\varphi : M \to \mathbb{C}$ be a submersive harmonic morphism from an open subset M of $\mathbb{R}^4$. Then the orthogonal complements to the tangent planes to the fibres determine an oriented 2-dimensional distribution H on M. At each point x, H_x is given by

$$[\partial_1\phi, \partial_2\phi, \partial_3\phi, \partial_4\phi] \in Q_2 \subset \mathbb{C}P^3,$$

where $Q_2 \approx \tilde{G}_2(\mathbb{R}^4)$ is the standard identification of the Grassmannian with the complex quadric. Then, if $\nabla_{AA'}\phi = \xi_A \eta_{A'}$ is the spinor decomposition, a direct computation verifies that in suitable charts on $Z^\pm$, $\gamma^1 = [\eta_{A'}], \gamma^2 = [\xi_A]$.

Write $W = \nabla\varphi$, then

$$W^a \leftrightarrow \frac{1}{\sqrt{2}} \begin{pmatrix} iW^0 + W^1 & W^2 + iW^3 \\ W^2 - iW^3 & iW^0 - W^1 \end{pmatrix} = \begin{pmatrix} \xi^0 \\ \xi^1 \end{pmatrix} (\eta^{0'}, \eta^{1'}),$$

and at each point $x \in M$,

$$[W] = [-i(\xi^0\eta^{0'} + \xi^1\eta^{1'}),\ \xi^0\eta^{0'} - \xi^1\eta^{1'}, \xi^0\eta^{1'} + \xi^1\eta^{0'},\ i(\xi^1\eta^{0'} - \xi^0\eta^{1'})] \in Q_2.$$

But the right hand side is the image of $\left([\xi^0,\xi^1],[\eta^{0'},\eta^{1'}]\right)$ under the standard identification of $\mathbb{C}P^1 \times \mathbb{C}P^1$ with Q_2. Thus identifying each $\mathbb{C}P^1$ with $\mathbb{C}\cup\infty$ by the stereographic projections $[\xi^0,\xi^1]\mapsto \xi^0/\xi^1, [\eta^{0'},\eta^{1'}]\mapsto \eta^{0'}/\eta^{1'}$, we find that

$$\gamma^1 = \frac{W^2 - \mathrm{i}W^3}{\mathrm{i}W^0 - W^1} = \frac{\mathrm{i}W^0 + W^1}{W^2 + \mathrm{i}W^3}, \quad \gamma^2 = \frac{W^2 + \mathrm{i}W^3}{\mathrm{i}W^0 - W^1} = \frac{\mathrm{i}W^0 + W^1}{W^2 - \mathrm{i}W^3}.$$

In order to show that equations (2.5) imply the holomorphicity results of Theorem 4.1, we consider a point x and suppose, without loss of generality, that ∂_2, ∂_3 span V_x. Then $W^2 + \mathrm{i}W^3 = W^2 - \mathrm{i}W^3 = 0$ at x. By horizontal conformality in a neighbourhood of x we have $\sum (W^a)^2 = 0$ so that

$$(W^0 + \mathrm{i}W^1)(W^0 - \mathrm{i}W^1) = -(W^2 + \mathrm{i}W^3)(W^2 - \mathrm{i}W^3), \tag{4.1}$$

so, at x, either $W^0 + \mathrm{i}W^1 = 0$ or $W^0 - \mathrm{i}W^1 = 0$. In fact, by choice of orientation we have that $W^0 + \mathrm{i}W^1 = 0$, so that $W^0 - \mathrm{i}W^1 \neq 0$. Differentiating (4.1) we have

$$(W^0 - \mathrm{i}W^1)\partial_a (W^0 + \mathrm{i}W^1) + (W^0 + \mathrm{i}W^1)\partial_a (W^0 - \mathrm{i}W^1) = 0,$$

for any $a = 0,1,2,3$, so that

$$\partial_a (\mathrm{i}W^0 - W^1) = 0$$

at x. Now the first of equations (2.5) (with $A' = 0, B' = 1$) implies

$$(-\mathrm{i}\partial_0 + \partial_1)(W^2 + \mathrm{i}W^3) = 0;$$

together with the formula $\dfrac{1}{\gamma_1} = \dfrac{W^2 + W^3}{\mathrm{i}W^0 + W^1}$ this implies that γ^1 is horizontally holomorphic.

A similar computation (with $A' = 1, B' = 0$) shows that $(\partial_2 - \mathrm{i}\partial_3)(W^2 + \mathrm{i}W^3) = 0$ so that γ^1 is vertically antiholomorphic. Similarly, the second of equations (2.5) shows that γ^2 is horizontally holomorphic and vertically holomorphic. These conditions combine to give the holomorphicity assertions of Theorem 4.1. We have therefore shown directly that the spinor equations (2.5) imply the holomorphicity assertions of that theorem.

Conversely, a conformal distribution determines a null vector field, which can be described by spinor fields $\xi_A, \eta_{A'}$. If the corresponding Gauss maps satisfy the holomorphicity conditions of Theorem 4.1, then by that theorem the distribution is integrable and minimal, and the spinor fields $\xi_A, \eta_{A'}$ satisfy equations (2.5).

Theorem 4.1 gives an interpretation of the spinor equations in Theorem 2.1. The advantage of Theorem 2.1 over Theorem 4.1 is that it is valid for arbitrary harmonic morphisms (i.e., those with critical points). Indeed, for the harmonic morphism φ of Example 3.2, which has an isolated critical point ($\nabla\varphi = 0$) at the origin, the distributions V, H do not extend over this critical point. Regarding the projectivized spinors, $\gamma^1 = [\eta_{A'}]$ extends across the origin, whereas $\gamma^2 = [\xi_A]$ does not. However, as in Remark 2.2 (iv), with suitable normalization, the unprojectivized spinor fields $\xi_A, \eta_{A'}$ always extend at least continuously across critical points.

5. Minkowski space. We consider a map $\varphi : U \to \mathbb{C}$, where U is an open subset in Minkowski space M^4, satisfying the equations

$$(\partial_0\varphi)^2 - (\partial_1\varphi)^2 - (\partial_2\varphi)^2 - (\partial_3\varphi)^2 = 0 \tag{5.1}$$

$$\partial_0^2\varphi - \partial_1^2\varphi - \partial_2^2\varphi - \partial_3^2\varphi = 0\,. \tag{5.2}$$

Such a map is a harmonic morphism in the sense that it pulls back harmonic functions to harmonic functions (see Parmar 1991, Lemma 2.1.7). Equation (5.1) may still be interpreted as saying that φ is conformal on the horizontal space (the orthogonal space to the fibre, now with respect to the standard Lorentz metric), while (5.2) is simply the wave equation—the harmonic equation in Lorentz metric.

The spinor correspondence is given by

$$x^a \leftrightarrow x^{AA'} = \frac{1}{\sqrt{2}}\begin{pmatrix} x^0 + x^1 & x^2 + ix^3 \\ x^2 - ix^3 & x^0 - x^1 \end{pmatrix}.$$

Proceeding exactly as we did for the $\mathbb{R}^4$ case we obtain

THEOREM 5.1. *There is a correspondence between*

(i) mappings $\varphi : U \to \mathbb{C}, U \subset M^4$ open and simply connected, satisfying equations (5.1) and (5.2), and

(ii) pairs of spinor fields $(\xi^A, \eta^{A'})$ on U satisfying the spinor equations:

$$\begin{cases} \nabla_{AA'}\xi^A\eta^{B'} = 0 \\ \nabla_{AA'}\xi^B\eta^{A'} = 0\,. \end{cases} \tag{5.3}$$

REMARK. *Given a static monopole there corresponds a harmonic morphism on $\mathbb{R}^3$. This can be thought of as a time (x^0)-independent solution of (5.3). Thus we have a different view of monopoles, either as solutions to (5.3) independent of x^0, or as solutions to (2.5) independent of translation in a particular direction. This contrasts with describing them as time-independent Yang-Mills fields.*

Comments (1994). Harmonic morphisms were studied in the late 1970's by Fuglede (1978) and Ishihara (1979), as mappings between Riemannian manifolds which preserve harmonic functions. Their investigation was taken up almost ten years later by Baird and Wood, who, noting the holomorphicity properties associated to a harmonic morphism with values in a surface (Wood 1986,

Baird 1987), classified the harmonic morphisms from 3-dimensional manifolds (Baird & Wood 1988, 1991, 1992).

Wood (1992) extended this study to 4-manifolds, classifying the harmonic morphisms on anti-self-dual Einstein 4-manifolds in terms of holomorphic sections of twistor bundles. The direct correspondence between harmonic morphisms and shear free ray congruences of relativity theory was observed by Baird (1992).

With the natural introduction of spinor fields, the article in this volume explores the links between static monopoles and harmonic morphisms defined on 3-dimensional Euclidean space. The geometrical elegance and simplicity of harmonic morphisms may lead to a greater understanding of monopoles.

References

Baird, P. (1987) Harmonic morphisms onto Riemann surfaces and generalized analytic functions, *Ann. Inst. Fourier, Grenoble* **37:1**, p. 135–173.

Baird, P. (1990) Harmonic morphisms and circle actions on 3- and 4-manifolds, *Ann. Inst. Fourier, Grenoble* **40:1**, p. 177–212.

Baird, P. (1992) Riemannian twistors and Hermitian structures on low-dimensional space forms, *J. Math. Phys.* **33**(10), p. 3340–3355.

Baird, P. & Eells, J. (1981) A conservation law for harmonic maps, *Geometry Symp. Utrecht, Proceedings, Lect. Notes in Math.* **894**, (Springer-Verlag), p. 1–25.

Baird, P. & Wood, J.C. (1988) Bernstein theorems for harmonic morphisms from $\mathbb{R}^3$ and S^3, *Math. Ann* **280**, p. 579–603.

Baird, P. & Wood, J.C. (1991) Harmonic morphisms and conformal foliations by geodesics of three-dimensional space forms, *J. Australian Math. Soc.* **(A) 51**, p. 118–153.

Baird, P. & Wood, J.C. (1992) Harmonic morphisms, Seifert fibre spaces and conformal foliations, *Proc. London Math. Soc.* **(3) 64**, p. 170–196.

Bernard, P., Campbell, E.A. & Davie, A.M. (1979) Brownian motion and generalized analytic and inner functions, *Ann. Inst. Fourier, Grenoble.* **29:1**, p. 207–228.

Eells, J. & Salamon, S. (1985) Twistorial constructions of harmonic maps of surfaces into four-manifolds, *Ann. Scuola Norm. Sup. Pisa* **(4) 12**, p. 589–640.

Fuglede, B. (1978) Harmonic morphisms between Riemannian manifolds, *Ann. Inst. Fourier, Grenoble* **28:2**, p.107–144.

Gudmundsson, S. & Wood, J.C. (1993) Multivalued harmonic morphisms, *Math. Scand.* **72** (to appear).

Hitchin, N.J. (1982) Monopoles and geodesics, *Comm. Math. Phys.* **83**, p. 579–602.

Hitchin, N.J. (1983) On the construction of monopoles, *Comm. Math. Phys.* **89**, p. 145–190.

Ishihara, T. (1979) A mapping of Riemannian manifolds which preserves harmonic functions, *J. Math. Kyoto Univ.* **19**, p. 215–229.

Parmar, V. (1991) Harmonic morphisms between semi-Riemannian manifolds, *Thesis, University of Leeds.*

Penrose, R. & Rindler, W. (1986) *Spinors and Space-Time, vol. II: Spinor and Twistor Methods in Space-Time Geometry* (Cambridge University Press, Cambridge).

Sommers, P. (1980) Space spinors, *J. Maths. Phys.* **(21)** (10), p. 2567–2571.

Wood, J.C. (1986) Harmonic morphisms, foliations and Gauss maps, *Contemporary Mathematics, American Math. Soc.* **49**, p. 145–183.

Wood, J.C. (1992) Harmonic morphisms and Hermitian structures on Einstein 4-manifolds, *International J. Math.***3**, p. 415–439.

§II.1.15 **Twistor Theory and Harmonic Maps from Riemann Surfaces**

by M.G.Eastwood (TN 15, January 1983)

Suppose M and N are oriented Riemannian manifolds with M made of rubber and N of stone. If M is constrained to lie on N by means of a smooth mapping $\phi : M \to N$ and then released it will attempt to attain an equilibrium configuration. This may be impossible, i.e. M snaps. In many cases equilibrium is always possible. If ϕ is in equilibrium it is called a *harmonic* map. Geodesics and parametrized minimal surfaces are examples. See Eells & Lemaire (1978) for a comprehensive review article. To be more precise, define the *energy* $E(\phi)$ of ϕ by

$$E(\phi) = \int_M \frac{1}{2}|d\phi|^2 d\text{vol}$$

where $|d\phi|$ is the Hilbert-Schmidt norm of $d\phi : T_xM \to T_{\phi(x)}N$, i.e. in local coordinates x^i on M, y^α on N, $|d\phi|^2 = \partial y^\alpha/\partial x^i \partial y^\beta/\partial x^j g^{ij} h_{\alpha\beta}$ where g is the metric on M and h is the metric on N. Then ϕ is said to be harmonic iff it is a critical point of E (if M is not compact then E must be computed on compact subdomains of M with only compactly supported variations allowed). In other words, ϕ is harmonic iff it satisfies the corresponding Euler-Lagrange equations

$$\text{trace}\, \nabla(d\phi) = 0$$

where ∇ is the connection on $\Omega^1(\phi^*TN)$ induced from the Levi-Civita connections on M and N. Bearing in mind the elastic nature of M, trace $\nabla(d\phi)$ is called the *tension* of ϕ. In local coordinates it becomes

$$g^{ij}\left\{\frac{\partial^2 y^\alpha}{\partial x^i \partial x^j} - \Gamma^k_{ij}\frac{\partial y^\alpha}{\partial x^k} + \Omega^\alpha_{\beta\gamma}\frac{\partial y^\beta}{\partial x^i}\frac{\partial y^\gamma}{\partial x^j}\right\}$$

where Γ_{ij}^k and $\Omega_{\beta\gamma}^\alpha$ are Christoffel symbols on M and N respectively. Without the trace, $\nabla(d\phi)$ is the second fundamental form of the mapping ϕ and behaves well with respect to composition. The tension, however, does not in general compose for the composition of mappings.

Perhaps a neater way to write the energy is as

$$E(\phi) = \int_M \frac{1}{2} \text{ trace } d\phi \wedge *d\phi$$

where $* : \Omega^1(\phi^*TN) \to \Omega^{m-1}(\phi^*TN)$ is the Hodge $*$-operator on M and the trace is with respect to h. The Euler-Lagrange equations are then

$$\nabla(*d\phi) = 0$$

where $\nabla : \Omega^{m-1}(\phi^*TN) \to \Omega^m(\phi^*TN)$ is the pull-back of the Levi-Civita connection on N. $\nabla(*d\phi)$ is the Hodge dual of the more usual tension. In local coordinates (or abstract index notation) on N, maintaining a more abstract notation on M

$$\nabla(*d\phi) = d * d\phi^\alpha + \Omega_{\beta\gamma}^\alpha d\phi^\beta \wedge *d\phi^\gamma.$$

At this point an obvious analogy with the Yang- Mills action and equations springs to mind (an analogy familiar to many mathematicians and physicists, the latter being interested in harmonic maps under the name of '$\mathbb{CP}_n$ model', 'σ-model', or 'current algebra'). For harmonic maps $d\phi$ is the analogue of the gauge field F and $\nabla(*d\phi) = 0$ replaces $\nabla(*F) = 0$. Recall that in gauge theories there is the Bianchi identity $\nabla F = 0$. The corresponding identity holds in the harmonic case because $\nabla(d\phi) = \nabla(\phi^*\delta) = \phi^*\nabla\delta$ where $\delta \in \Omega^1(TN)$ on N is the tautological section (Kronecker delta) and $\nabla\delta$ is exactly the torsion of ∇. Now recall that the Yang-Mills equations are special in dimension 4 (being conformally invariant) and that twistor theory gives an illuminating way of looking at them. It is natural to ask whether there is an analogy of this for harmonic maps.

The special dimension for harmonic maps is $\dim M = 2$ for then the $*$-operator is conformally invariant when acting on 1- forms. Thus, the harmonic field equations $\nabla(*d\phi) = 0$ are manifestly conformally invariant. For Euclidean Yang-Mills in dimension 4 there are special solutions namely $\pm$ self-dual, $*F = \pm F$, which satisfy the equations as a consequence of the Bianchi identity. In dimension 2, however, $*^2 = -1$ so the analogue of $\pm$ self-dual can only be $*d\phi = \pm i d\phi$. For this to make sense TN must be a complex bundle so N should be an almost complex manifold and more specifically it is easiest to take N complex. From now on N is made from complex stone (more rigid than the strongest granite). But now, even if N has a nicely compatible metric, i.e. an Hermitian one, ∇ is not necessarily complex linear. Indeed ∇ being complex linear is equivalent to N being Kähler, so ... Kähler stone from now on. Now, if $d\phi$ is $\pm$ self-dual,

$$\nabla(*d\phi) = \nabla(\pm i d\phi) = \pm i\nabla(d\phi) = 0$$

as required. By the Korn-Licktenstein theorem the $*$-operator on M is equivalent to a complex structure where $*$ becomes multiplication by i. Thus, $*d\phi = id\phi$ is precisely the condition that $d\phi$ be complex linear, i.e. $d\phi$ is self-dual means ϕ is holomorphic (and anti-self-dual means antiholomorphic).

Now to the question: what has this to do with twistor theory? Twistor theory can be viewed as a way of doing conformal geometry in dimension 4 (for a conformally right-flat space). This can be clearly seen after *complexification.* More precisely, if the original 4-fold is $\mathbb{R}$-analytic then the conformal structure takes on a geometric significance, namely through α-planes, after 'thickening' into the complex. The Ward correspondence, for example, becomes a simple geometric construction in the complexification. The same procedure works equally well in the 2- dimensional $\mathbb{R}$-analytic case (where there is no integrability condition analogous to 'right-flat'). In particular, the existence of the complex structure induced by a conformal structure in the $\mathbb{R}$-analytic category (due to Gauss), becomes a rather simple geometric observation. The Cauchy-Riemann equations then take on a geometric interpretation which leads to holomorphicity in exactly the same way that Maxwell's equations lead to the twisted photon — Maxwell's equations should be regarded as a 4-dimensional version of the Cauchy-Riemann equations. Without going into details it is clear that one should hope that complexification will throw geometric light on harmonic mappings from Riemann surfaces. The purpose of this article is to explain further upon this hope.

A harmonic mapping between $\mathbb{R}$-analytic Riemannian manifolds is automatically $\mathbb{R}$-analytic (by ellipticity). Thus, in our case of M a Riemann surface and N Kähler, one can complexify to obtain $\phi : \mathbb{C}M \to \mathbb{C}N$ on some neighbourhood of M. Then ϕ will satisfy a complexified version of the harmonic field equations. To identify these equations it is best to rewrite $\nabla(*d\phi) = 0$ as follows. The derivative $d\phi$ may be split into its self-dual and anti-self-dual parts $d\phi = \partial\phi + \bar{\partial}\phi$ usually called the $(1,0)$ and $(0,1)$ parts. Bearing in mind that $\nabla(d\phi) = 0$ it follows that ϕ is harmonic $\Leftrightarrow$

$$\nabla(\partial\phi) = 0 \text{ or equivalently } \nabla(\bar{\partial}\phi) = 0.$$

In this form it is straightforward to complexify. Since M and N are already complex, their complexifications are easy to identify: $\mathbb{C}M = M \times \overline{M}$, for example, where $\overline{M}$ denotes the smooth manifold M but with the conjugate holomorphic structure and $M \hookrightarrow \mathbb{C}M$ is the diagonal $(m \mapsto (m, \bar{m}))$. If $\phi : M \to N$ is holomorphic then its complexification $\mathbb{C}\phi : \mathbb{C}M \to \mathbb{C}N$ is given by $\mathbb{C}\phi(p, \bar{q}) = (\phi(p), \bar{\phi}(q))$. In general a $\mathbb{R}$-analytic $\phi : M \to N$ will complexify to only a small neighbourhood of M in $\mathbb{C}M$ so it is best to maintain $\mathbb{C}M$ as a general notation for any neighbourhood of M in $M \times \overline{M}$. A holomorphic $\psi : \mathbb{C}M \to \mathbb{C}N$ is the complexification of $\phi : M \to N$ iff ψ satisfies the 'reality' condition $\psi(p, \bar{q}) = \overline{\psi(q, \bar{p})}$. The splitting $d\phi = \partial\phi + \bar{\partial}\phi$ is now manifested geometrically: after complexification, ∂ refers to differentiation along M whilst holding the $\overline{M}$ variable fixed and $\bar{\partial}$ refers to differentiation along $\overline{M}$ with M occurring parametrically. In other words $(1,0)$-forms complexify to the cotangent bundle to $\overline{M}$. The connection ∇ on N complexifies (assuming it is $\mathbb{R}$-analytic) to a holomorphic torsion free connection on $\mathbb{C}N$. More generally, if L is any holomorphic manifold with holomorphic torsion free connection ∇ then we define $\phi : \mathbb{C}M \to L$ (holomorphic) to be *harmonic* iff $\nabla(\partial\phi) = 0$. To interpret this geometrically let $\mu : \mathbb{C}M \to M$ be projection onto the first factor. Then $\nabla : \Omega^{1,0}(\phi^*TL) \to \Omega^2(\phi^*TL)$ can be interpreted as a relative connection ∇_μ on $\mu^*\Omega^1(\phi^*TL)$ i.e. $\nabla_\mu : \mu^*\Omega^1(\phi^*TL) \to \Omega^1_\mu(\mu^*\Omega^1(\phi^*TL))$ satisfying $\nabla_\mu(fs) = f\nabla_\mu s + \bar{\partial} f \otimes s$. Thus $\mu^*\Omega^1(\phi^*TL)$ and hence ϕ^*TL maybe canonically regarded as the pull-back of a bundle on M (since, by dimension reasons (plus some simple topological restrictions on the fibres of μ) ∇_μ is relatively flat). Therefore ϕ is harmonic iff $\partial\phi \in \mu^*\Omega^1(\phi^*TL)$ is covariant constant on the fibres of μ or, in other words, pushes down to a section of the bundle on M. In particular, this proves

PROPOSITION. *The zero set of $\partial\phi$ consists of the fibres of $\mu : \mathbb{C}M \to M$ over a discrete set of points and, in particular, by rescaling, $\partial\phi$ defines a complex direction at $\phi(x)$ even if $\partial\phi(x) = 0$ (unless $\partial\phi \equiv 0$).* □

This proposition is one of the key steps in the recent classification of harmonic isotropic maps from Riemann surfaces into complex projective spaces (Din & Zakrzewski (1980) General classical solutions in the $\mathbb{CP}^{n-1}$ model, Nucl. Phys. **B174**, 397–406. Din & Zakrzewski (1980) Properties of the general classical $\mathbb{CP}^{n-1}$ model, Phys. Lett. **95B**, 419–422. D. Burns, unpublished. Eells & Wood, (1983) Harmonic maps from surfaces to complex projective spaces, Advances in Math. **49**, 217-263). Actually, the whole classification theorem complexifies rather well. Because it would make this article rather too lengthy to go into detail I will just describe some of the key points. Firstly the complexification of complex projective space. An Hermitian form on V gives rise to the Fubini-Study metric on $\mathbb{P}(V)$. The Hermitian form on V is the same as an isomorphism $\overline{V} \cong V^*$ and so $\mathbb{P}(V)$ can be complexified to $\mathbb{P}(V) \times \mathbb{P}(V^*)$. The natural pairing $\langle\ ,\ \rangle : V \otimes V^* \to \mathbb{C}$ gives rise to a $\mathbb{C}^2$-valued metric on $\mathbb{P}(V) \times \mathbb{P}(V^*)$ induced by $\langle(a,b),(c,d)\rangle = (\langle a,c\rangle, \langle b,d\rangle)$ on $V \otimes V^*$. This is the complexification of the Fubini-Study metric and there is a corresponding connection. In general, a mapping $\phi : \mathbb{C}M \to L$ is said to be isotropic iff $\partial\phi$ and higher derivatives along M are orthogonal to $\bar{\partial}\phi$ and higher derivatives along $\overline{M}$. This is the case for example if $\bar{\partial}\phi = 0$. In the case of $L = \mathbb{P}(V) \times \mathbb{P}(V^*)$ it is essentially just algebra (albeit cunning algebra) to show that if $\phi : \mathbb{C}M \to \mathbb{P}(V) \times \mathbb{P}(V^*)$ is harmonic and isotropic then so is $D\phi$ defined by $D(R,S) = (\eth R - \langle \eth R, S\rangle R/\langle R,S\rangle,\ \bar{\eth} S - \langle R, \bar{\eth} S\rangle/\langle R,S\rangle)$ where $(R,S) : \mathbb{C}M \to V \times V^*$ is a lift of ϕ (make local choices and patch). The proposition above is used to show that D is well-defined provided ϕ avoids the quadric $\{(R,S) \in \mathbb{P}(V) \times \mathbb{P}(V^*) \text{ s.t. } \langle R,S\rangle = 0\}$ (D is essentially ∂). Hence the classification.

In the best of worlds there would be a harmonic analogue of the Isenberg-Yasskin-Green & Witten ambitwistor description of Yang-Mills. $\mathbb{C}M$ is the ambitwistor space for M.

References

Burns, D. unpublished.

Din & Zakrzewski (1980) General classical solutions in the $\mathbb{CP}^{n-1}$ model, Nucl. Phys. **B174**, 397–406.

Din & Zakrzewski (1980) Properties of the general classical $\mathbb{CP}^{n-1}$ model, Phys. Lett. **95B**, 419–422.

Eells & Lemaire (1978) A report on harmonic maps, Bull. L.M.S. **10**, 1–68.

Eells & Wood, (1983) Harmonic maps from surfaces to complex projective spaces, Advances in Math. **49**, 217-263).

§II.1.16 **Contact birational correspondences between twistor spaces of Wolf spaces** *by P.Z.Kobak*

Introduction. The task of classifying minimal 2-spheres in various classes of Riemannian manifolds can be often accomplished by translating the problem to the language of complex geometry and specifying holomorphic data which completely characterise such maps. One of several approaches is to use twistor theory. Indeed, twistor methods have been successful to the extent that some authors even *define* twistor spaces as fibrations of almost complex manifolds which can be used to produce harmonic maps (cf. Burstall and Rawnsley 1990). For example one can construct minimal 2-spheres in S^4 from twistor projections of holomorphic horizontal curves in $\mathbb{CP}^3$ (a curve is horizontal iff it is tangent to the horizontal distribution on the twistor space, i.e. is orthogonal to the twistor fibres). Such curves can be constructed as follows: if f, g are meromorphic functions on a Riemann surface and $g \neq$ const then the curve $[1 : f - g\frac{f'}{2g'} : g : \frac{f'}{2g'}]$ is horizontal in $\mathbb{CP}^3$ (Bryant 1982). The horizontal distribution on $\mathbb{CP}^3$ arises from a complex contact structure and Lawson (1983) reinterpreted Bryant's result by defining a contact birational correspondence between $\mathbb{CP}^3$ and $F_{12}(\mathbb{C}^3)$ (the manifold of complete flags in $\mathbb{C}^3$). We shall call this the Bryant-Lawson correspondence, it is defined by the following formulae (Gauduchon 1984):

$$b : F_{12}(\mathbb{C}^3) \ni (x_0 : x_1 : x_2 ; \xi_0 : \xi_1 : \xi_2) \mapsto [x_2\xi_2 - x_0\xi_0 : 2x_0\xi_2 : x_0\xi_1 : -2x_1\xi_2] \in \mathbb{CP}^3 \qquad (1)$$

where $\sum x_i\xi_i = 0$. The flag manifold $F_{12}(\mathbb{C}^3)$ is the twistor space of $\mathbb{G}\mathrm{r}_2(\mathbb{C}^3) \simeq \mathbb{CP}^{2^*}$ and can be identified with $\mathbb{P}T^*\mathbb{CP}^2$. Moreover, all non-vertical contact curves in $\mathbb{P}T^*\mathbb{CP}^2$ are Gauss lifts of curves in $\mathbb{CP}^2$. The birational map b can be used to transfer these curves to the twistor space of S^4. This map is a biholomorphism away from the surfaces S_1 and S_2 given by the equations $x_0 = 0$ and $\xi_2 = 0$ and is well defined away from the lines $l_1 = p_1^{-1}([0:0:1])$ and $l_2 = p_2^{-1}([1:0:0])$ (here p_1 and p_2 denote the canonical projections of $F_{12}(\mathbb{C}^3)$ to $\mathbb{CP}^2$ and $\mathbb{CP}^{2^*}$ respectively). This imposes conditions on the curves in $\mathbb{CP}^2$ which are used to construct horizontal holomorphic curves in $F_{12}(\mathbb{C}^3)$: Gauss lifts give rise to minimal immersions in S^4 if they stay away from $L_1 \cup L_2$ and on the divisor $S_1 \cup S_2$ are transversal to the kernel of the differential of b (cf. Lawson 1983).

Burstall (1990) found a very interesting generalisation of the Bryant-Lawson correspondence to twistor spaces of compact symmetric quaternion-Kähler spaces. These symmetric spaces, known as Wolf spaces, have many properties in common with S^4 and $\mathbb{CP}^2$: being quaternion-Kähler they have twistor spaces which are fibrations of complex contact manifolds with fibre $\mathbb{CP}^1$. In general, Riemannian symmetric spaces of compact type are endowed with a family of twistor spaces recruited from among generalised flag manifolds (cf. Burstall and Rawnsley 1990); we reserve the term ***quaternionic twistor space*** for the twistor spaces which arise from quaternion-Kähler structures (see Salamon 1982 for more details on quaternion-Kähler geometry).

THEOREM 1 (BURSTALL 1990). *Quaternionic twistor spaces of Wolf spaces of the same dimension are birationally equivalent as complex contact manifolds.*

The proof of Theorem 1 is based on the fact that quaternionic twistor spaces of Wolf spaces are flag manifolds with isomorphic nilradicals. (With conventions as in Baston and Eastwood 1989 these flag manifolds correspond to extended Dynkin diagrams with crosses through the nodes linked to the added node; their nilradicals are generalised Heisenberg Lie algebras.) Our aim is to present a geometric description of such correspondences. We shall use the fact that the twistor spaces considered here can be represented as projectivised nilpotent orbits. This makes it possible to construct contact birational correspondences between quaternionic twistor spaces of Wolf spaces from projectivised linear maps from a highest root nilpotent orbit to its tangent space. Moreover, the singular sets of these correspondences can be described in terms of the Bruhat decomposition. For example the divisors S_1, S_2 and lines l_1, l_2 in $F_{12}(\mathbb{C}^3)$, defined above, are the Schubert varieties in a Bruhat decomposition of $F_{12}(\mathbb{C}^3)$; the corresponding Hasse diagram is shown below. We also derive explicit formulae for contact birational maps between $\mathbb{CP}^{2n+1}$ and the flag manifold $\mathbb{P}T^*\mathbb{CP}^{n+1}$ of which (1) is a special case. Another interpretation of Burstall correspondences can be derived from the coordinatisations of nilpotent Lie groups described in Warner (1972), §1.1.4. For more details and examples of applications of the correspondences to constructions of harmonic maps see Kobak (1993), Chapter 3.

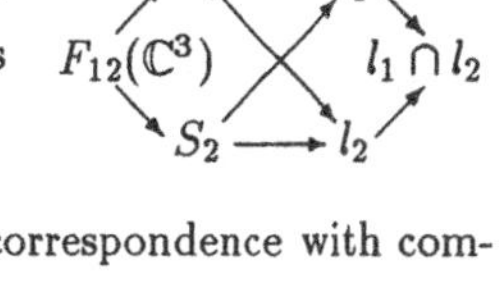

Twistor fibrations over Wolf spaces. Wolf spaces are in a bijective correspondence with compact simple Lie algebras. For example the Wolf spaces arising from $\mathfrak{su}(n)$ and $\mathfrak{sp}(n)$ are the Grassmann manifolds $\mathbb{G}r_2(\mathbb{C}^n)$ and $\mathbb{HP}^{n-1}$ with twistor spaces given by the flag manifolds $F_{1,n-1}(\mathbb{C}^n)$ (lines in hyperplanes in $\mathbb{C}^n$) and $\mathbb{CP}^{2n-1}$ respectively. All flag manifolds have natural projective realisations (cf. Baston and Eastwood 1989). In particular twistor spaces of Wolf spaces are projectivised nilpotent orbits. More precisely, let Z denote the quaternionic twistor space of a Wolf space arising from a complex simple Lie algebra $\mathfrak{g}^{\mathbb{C}} = \mathrm{Lie}(G^{\mathbb{C}})$. Let $\mathfrak{j} \subset \mathfrak{g}^{\mathbb{C}}$ be a Cartan subalgebra with root system Δ and root spaces $\mathfrak{g}^{\mathbb{C}}_{\alpha}$, $\alpha \in \Delta$. We choose a decomposition $\Delta = \Delta_+ \cup \Delta_-$ into positive and negative roots. If $\rho \in \Delta_+$ is the highest root then there is an identification $Z = \mathbb{P}\mathcal{N} \subset \mathbb{P}\mathfrak{g}^{\mathbb{C}}$, where $\mathcal{N} = G^{\mathbb{C}}o$ and $0 \neq o \in \mathfrak{g}^{\mathbb{C}}_{\rho}$. $\mathcal{N}$ is called the highest root nilpotent orbit; this is in fact the smallest nontrivial nilpotent orbit in $\mathfrak{g}^{\mathbb{C}}$. Adjoint orbits carry a natural symplectic form (the Kostant-Kirillov-Soriau 2-form), denoted here by ω. For a nilpotent orbit $\mathcal{N}$ the contraction of ω with the radial vector field gives a $G^{\mathbb{C}}$-equivariant complex contact structure θ on $\mathbb{P}\mathcal{N}$. We have

$$\omega([e,A],[e,B]) = \langle e,[A,B]\rangle\,, \qquad \theta([e,X]) = \langle e,X\rangle\,, \tag{2}$$

where $T_e\mathcal{N}$ is identified with $[e, \mathfrak{g}^{\mathbb{C}}]$ and $\langle\cdot,\cdot\rangle$ is the Killing form on $\mathfrak{g}^{\mathbb{C}}$ (for more details and further references see Kobak 1994).

We shall use the following conventions. We shall write $\mathbb{P}\mathcal{N} = G^{\mathbb{C}}\mathfrak{g}^{\mathbb{C}}_{\rho}$, viewing $\mathfrak{g}^{\mathbb{C}}_{\rho}$ as a point of $\mathbb{P}\mathfrak{g}^{\mathbb{C}}$, where $\mathfrak{g}^{\mathbb{C}} = \mathrm{Lie}(G^{\mathbb{C}})$ and $G^{\mathbb{C}}$ is connected and simply connected. If w lies in the Weyl group W of $\mathfrak{g}^{\mathbb{C}}$ then $\ell(w)$ denotes the least number of factors in the decomposition of w into a product of simple root reflexions. Finally, $\mathfrak{m}_-$ denotes the nilpotent algebra $\bigoplus_{\alpha\in\Delta_-}\mathfrak{g}^{\mathbb{C}}_{\alpha}$. The following lemma lists the basic properties of the Bruhat decomposition of flag manifolds. (We consider here projectivised highest root nilpotent orbits but the general case is similar, see Baston and Eastwood 1989.)

LEMMA 2. *(1) The projectivised highest root nilpotent orbit $\mathbb{P}\mathcal{N} \subset \mathbb{P}\mathfrak{g}^{\mathbb{C}}$ has a Bruhat decomposition*

$$\mathbb{P}\mathcal{N} = \bigsqcup_{\alpha\in L} \exp \mathfrak{m}_- \mathfrak{g}^{\mathbb{C}}_{\alpha} = \bigsqcup_{\alpha\in L} N^{\alpha}\mathfrak{g}^{\mathbb{C}}_{\alpha} \quad \textit{(disjoint union)},$$

where $L \subset \Delta$ denotes the set of long roots and $N^{\alpha} = \exp \mathfrak{n}^{\alpha}$ is the nilpotent Lie group with Lie algebra $\mathfrak{n}^{\alpha} \subset \mathfrak{m}_-$ defined by the formula

$$\mathfrak{n}^{\alpha} = \bigoplus_{\gamma\in\Delta_-,\ \langle\alpha,\gamma\rangle<0} \mathfrak{g}^{\mathbb{C}}_{\gamma}.$$

(2) If $\alpha \in L$ then the map $\mathfrak{n}^{\alpha} \ni x \mapsto e^{x}\mathfrak{g}^{\mathbb{C}}_{\alpha} \in \mathbb{P}\mathcal{N}$ is a complex diffeomorphism between $\mathbb{C}^k$ and the Bruhat cell $N^{\alpha}\mathfrak{g}^{\mathbb{C}}_{\alpha}$ where $k = \#\{\gamma \in \Delta_- : \langle\alpha,\gamma\rangle < 0\} = \dim \mathbb{P}\mathcal{N} - \ell(w)$, and $w \in W$ is the element of the shortest length such that $w\alpha = \rho$.

Let us consider the following special case of a grading of $\mathfrak{g}^{\mathbb{C}}$, defined by Burstall and Rawnsley (1990). Let $\Delta^k = \{\alpha \in \Delta \cup \{0\} : \langle\alpha, \rho^{\vee}\rangle = k\}$ and put

$$\mathfrak{g}^{(k)} = \bigoplus_{\alpha\in\Delta^k} \mathfrak{g}^{\mathbb{C}}_{\alpha}\,; \qquad \text{then} \qquad \mathfrak{g}^{\mathbb{C}} = \bigoplus_{k=-2}^{k=2} \mathfrak{g}^{(k)} \qquad \text{and} \qquad [\mathfrak{g}^{(i)}, \mathfrak{g}^{(j)}] \subset \mathfrak{g}^{(i+j)}. \tag{3}$$

Here $\mathfrak{g}_0^{\mathbb{C}}$ is by definition the Cartan algebra $\mathfrak{j}$. We have the following equalities:

$$\mathfrak{g}^{(\pm 2)} = \mathfrak{g}_{\pm\rho}^{\mathbb{C}}, \qquad \mathfrak{g}^{(0)} = \mathfrak{l}, \quad \text{and} \quad \mathfrak{g}^{(\pm 1)} \oplus \mathfrak{g}^{(\pm 2)} = \mathfrak{n}_{\pm}$$

where $\mathfrak{n}_- \stackrel{\text{df}}{=} \mathfrak{n}^{\rho}$; $n_+ \stackrel{\text{df}}{=} \overline{n}_-$ is the nilradical and $\mathfrak{l}$ the reductive part of $\mathfrak{p}$.

A geometric interpretation. According to Theorem 1 any projectivised minimal nilpotent orbit of dimension $2n+1$ is birationally equivalent as a complex contact manifold to $\mathbb{CP}^{2n+1}$ (the complex contact structure on $\mathbb{CP}^{2n+1}$ comes from the contraction of the constant complex symplectic structure on $\mathbb{C}^{2n}$ with the radial vector field). One might hope for a more geometric and a more explicit description of such equivalences. Indeed, for $z \in \mathcal{N}$ we have a contact $\mathbb{CP}^{2n+1}$ at hand, namely $\mathbb{P}T_z\mathcal{N}$ with the contact structure induced by the symplectic structure on $T_z\mathcal{N}$. In fact contact birational maps from $\mathbb{P}\mathcal{N}$ to $\mathbb{P}T_z\mathcal{N}$ can be constructed from linear maps $\mathfrak{g}^{\mathbb{C}} \to T_z\mathcal{N}$ which are defined as follows. Let $R \subset \Delta_+$ denote the set of simple roots and choose a Weyl basis $\{H_\alpha \in \mathfrak{j},\ X_\beta \in \mathfrak{g}_\beta^{\mathbb{C}}\}_{\alpha\in R,\ \beta\in\Delta}$, for $\mathfrak{g}^{\mathbb{C}}$. This means that $\langle H, H_\alpha\rangle = \alpha(H)$ for all $H \in \mathfrak{j}$ and $\langle X_\alpha, X_{-\alpha}\rangle = 1$ (so $[X_\alpha, X_{-\alpha}] = H_\alpha$). The grading (3) shows that the tangent space $T_z\mathcal{N} = [z, \mathfrak{g}^{\mathbb{C}}]$ can be written as

$$T_z\mathcal{N} = \mathfrak{n} \oplus \mathbb{C}H_\rho = \mathfrak{g}^{(2)} \oplus \mathfrak{g}^{(1)} \oplus \mathbb{C}H_\rho \tag{4}$$

(here $z = X_\rho \in \mathfrak{g}_\rho^{\mathbb{C}} = \mathfrak{g}^{(2)}$). This gives a decomposition $\mathfrak{g}^{\mathbb{C}} = T_z\mathcal{N} \oplus V_z$ where

$$V_z = (\mathfrak{g}^{(0)} \ominus \mathbb{C}H_\rho) \oplus \mathfrak{g}^{(-1)} \oplus \mathfrak{g}^{(-2)} = (\mathfrak{j} \cap H_\rho^{\perp}) \oplus \bigoplus_{\alpha\in\Delta,\langle\alpha,\rho\rangle\leqslant 0} \mathfrak{g}_\alpha^{\mathbb{C}}.$$

Let $p : \mathfrak{g}^{\mathbb{C}} = T_z\mathcal{N} \oplus V_z \to T_z\mathcal{N}$ denote the linear projection to the first summand. The map

$$\pi : \mathfrak{g}^{\mathbb{C}} \ni x \mapsto p(x) + \langle x, X_{-\rho}\rangle z \in T_z\mathcal{N} \tag{5}$$

is simply a composition of p with a linear endomorphism of $T_z\mathcal{N}$ which stretches $\mathfrak{g}^{(2)}$ by the factor of 2 and is constant on the remaining components in the direct sum (4). There is a simple relation between the canonical contact structures on $\mathbb{P}\mathcal{N}$ and $\mathbb{P}T_z\mathcal{N}$:

LEMMA 3. *Let $C = \exp \mathfrak{n}_-\mathfrak{g}_\rho^{\mathbb{C}} \subset \mathbb{P}\mathcal{N}$ be the big cell in the Bruhat decomposition of $\mathbb{P}\mathcal{N}$. We assume that $z = X_\rho \in \mathfrak{g}_\rho^{\mathbb{C}}$ and let $\pi : \mathfrak{g}^{\mathbb{C}} \to T_z\mathcal{N}$ be defined by Formula (5). If $x \in \mathfrak{n}_-$ then*

$$\pi(e^x z) = 2z + [x, z]. \tag{6}$$

Moreover, π induces a $1:1$ contact map from C to $\mathbb{P}T_z\mathcal{N}$.

Proof. We begin with the proof of Formula (6). If x is an element of $\mathfrak{n}_-$ then

$$e^x z = \mathrm{Ad}_{e^x} z = e^{\mathrm{ad}_x}(z) = \sum_{n=0}^{4} \frac{1}{n!}(\mathrm{ad}_x)^n(z) = z + [x, z] + \frac{1}{2}[x, [x, z]] + \dots \tag{7}$$

where the omitted terms belong to $\mathfrak{n}_- = \mathfrak{g}^{(-1)} \oplus \mathfrak{g}^{(-2)}$. We can write $x = x_\rho X_{-\rho} + \sum_{\alpha\in\Delta^1} x_\alpha X_{-\alpha}$ where $x_\rho, x_\alpha \in \mathbb{C}$. Since $[X_{-\rho}, [x, z]]$ and $[x, [X_{-\rho}, z]]$ lie in $\mathfrak{n}_- \subset V_z$ we can ignore the summand $x_\rho X_{-\rho}$ when calculating $\pi([x, [x, z]])$. The remaining summands give

$$\sum_{\alpha\in\Delta^1} N_{-\alpha,\rho} x_\alpha [x, X_{\rho-\alpha}] = \sum_{\alpha,\beta\in\Delta^1} N_{-\alpha,\rho} x_\alpha x_\beta [X_{-\beta}, X_{\rho-\alpha}].$$

The constants $N_{\alpha,\beta}$, $\alpha+\beta \neq 0$, are defined by the formula $[X_\alpha, X_\beta] = N_{\alpha,\beta} X_{\alpha+\beta}$. The summands with $\alpha+\beta \neq \rho$ are either zero or lie in $\mathfrak{g}^{\mathbb{C}}_{\rho-\alpha-\beta} \subset V_z$ and the summands with $\alpha+\beta=\rho$ give

$$-\frac{1}{2} \sum_{\substack{\alpha,\beta\in\Delta^1 \\ \alpha+\beta=\rho}} x_\alpha x_\beta (N_{-\alpha,\alpha+\beta} H_\beta + N_{-\beta,\alpha+\beta} H_\alpha) = \frac{1}{2} \sum_{\substack{\alpha,\beta\in\Delta^1 \\ \alpha+\beta=\rho}} N_{\alpha,\beta} x_\alpha x_\beta (H_\alpha - H_\beta)$$

since $N_{-\alpha,\alpha+\beta} = -N_{-\beta,\alpha+\beta} = N_{\alpha,\beta}$ (cf. Warner 1972, p. 3). But $H_\alpha - H_\beta$ is perpendicular to H_ρ since $\langle H_\alpha - H_\beta, H_{\alpha+\beta}\rangle = \langle \alpha-\beta, \alpha+\beta\rangle$ and $\|\alpha\| = \|\beta\|$. This shows that $p(e^x z) = z + [x, z]$. Since $\langle X_{-\rho}, X_\rho\rangle = 1$, Formula (7) implies that $\langle e^x z, X_{-\rho}\rangle z = z$, and (6) follows.

It remains to show that π induces a contact map from C to $\mathbb{P}T_z\mathcal{N}$. We shall represent vectors tangent to $\mathbb{P}\mathcal{N}$ and to $\mathbb{P}T_z\mathcal{N}$ as projections of vectors tangent to $\mathcal{N}$ and to $T_z\mathcal{N}$ respectively. Let $x, y \in \mathfrak{n}_-$ and let $a = e^x \in G^{\mathbb{C}}$. The vector $[y, az] \in T_{az}\mathcal{N}$ projects to a contact vector in $T_{\mathbb{C}(az)}\mathbb{P}\mathcal{N}$ if and only if $\langle y, az\rangle = 0$ (Formula (2)), i.e. $\langle a^{-1}y, z\rangle = 0$. But $a^{-1}y = \mathrm{Ad}_{e^{-x}}\, y = y - \mathrm{ad}_x\, y$ since $(\mathrm{ad}_x)^k y = 0$ if $k > 1$. As a result

$$[y, az] \in T_{az}\mathcal{N} \text{ projects to a contact vector in } T_{\mathbb{C}(az)}\mathbb{P}\mathcal{N} \Leftrightarrow \langle y - [x, y], z\rangle = 0. \tag{8}$$

We shall now calculate $\pi_*[y, az]$. We have $[y, az] = \frac{d}{dt}\big|_{t=0} e^{ty} e^x z = \frac{d}{dt}\big|_{t=0} e^{ty+x+\frac{t}{2}[y,x]} z$ by the Baker-Campbell-Hausdorff formula, so the differential of π is given by the formula

$$\pi_* : T_{az}\mathcal{N} \ni [y, az] \mapsto [y + \frac{1}{2}[y, x], z] \in T_{(2z+[x,z])}(T_z\mathcal{N}).$$

It follows from (2) that $X = [A, z] \in T_z\mathcal{N}$ projects to a contact vector in $T_{(\mathbb{C}Y)}\mathbb{P}T_z\mathcal{N}$ iff $\langle A, Y\rangle = 0$. Consequently, the vector $\pi_*[y, az]$ is contact in $\mathbb{P}T_z\mathcal{N}$ iff $\langle 2z + [x, z], y + \frac{1}{2}[y, x]\rangle = 0$. We have:

$$\begin{aligned}\langle 2z + [x, z], y + \frac{1}{2}[y, x]\rangle &= \langle [x, z], y\rangle + 2\langle z, y\rangle + \langle z, [y, x]\rangle + \frac{1}{2}\langle [x, z], [y, x]\rangle \\ &= 2\langle z, y + [y, x]\rangle + \frac{1}{2}\langle [x, z], [y, x]\rangle = 0\end{aligned}$$

(the last equality follows from (8) and the fact that $[x, z] \in \mathfrak{g}^{(1)} \oplus \mathfrak{g}^{(0)}$ and $[x, y] \in \mathfrak{g}^{(-2)}$). □

The map $\mathbb{P}\pi : \mathbb{P}\mathcal{N} \to \mathbb{P}T_z\mathcal{N}$ is a candidate for a contact birational correspondence. We recall that a map is birational if it is rational and has rational inverse; a map $r : M \to N \subset \mathbb{CP}^n$ is rational if it can be written as $z \mapsto r(z) = [1 : f_1 : \ldots : f_n]$ where f_i are global meromorphic functions on M. Birational maps are generically biholomorphic and well defined away from a subvariety of codimension 2.

It is clear that $\mathbb{P}\pi$ is well defined on the set $\mathbb{P}\mathcal{N} \setminus \mathbb{P}(\mathcal{N} \cap V_z)$. This set can be bigger than the big cell C. It turns out that $\mathbb{P}(\mathcal{N} \cap V_z)$ is a union of Bruhat cells.

Figure 1. Rank two root systems

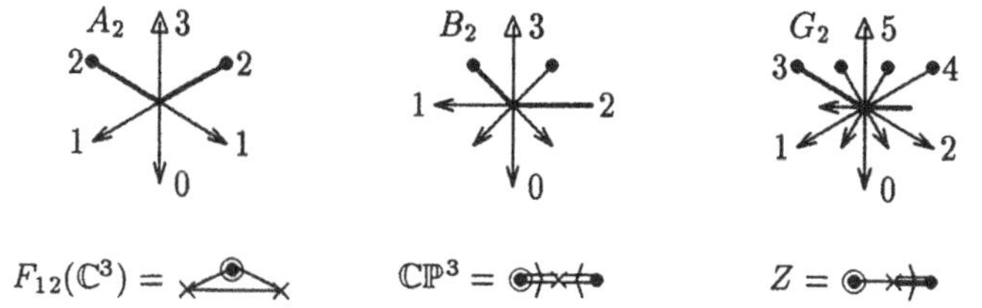

Arrows denote negative roots, the roots in Δ^1 are marked by •, the highest root ρ is marked by ▵ (and $-\rho$ in Dynkin diagrams by ◉). Thick lines represent simple roots. The numbers next to the long roots indicate the dimensions of the corresponding Bruhat cells.

LEMMA 4. *With the above notation we have:*

$$\mathbb{P}(\mathcal{N} \cap V_z) = \bigsqcup_{\gamma \in L,\ \langle \gamma, \rho \rangle \leqslant 0} N^\gamma \mathfrak{g}^{\mathbb{C}}_\gamma \quad \textit{(disjoint union).}$$

Moreover, if $\gamma \in \Delta^1 \cap L$ *then the map* π *is singular on the Bruhat cell* $N^\gamma \mathfrak{g}^{\mathbb{C}}_\gamma$.

Proof. First note that if $y \in \mathfrak{g}^{\mathbb{C}}_\tau$ and $x \in \mathfrak{m}_-$ then the component of $\mathrm{Ad}_{e^x} y$ in $\mathfrak{g}^{\mathbb{C}}_\tau$ is equal to y. Let τ be a long root. If $\tau \in \Delta^1 \cup \{\rho\}$ then $\mathfrak{g}^{\mathbb{C}}_\tau \subset \mathfrak{g}^{(2)} \oplus \mathfrak{g}^{(1)} \subset T_z\mathcal{N}$ so all elements of the Bruhat cell $N^\tau \mathfrak{g}^{\mathbb{C}}_\tau = \exp \mathfrak{m}_- \mathfrak{g}^{\mathbb{C}}_\tau$ have a nonzero component in $T_z\mathcal{N}$. If $\tau \notin \Delta^1 \cup \{\rho\}$ then $\mathfrak{g}^{\mathbb{C}}_\tau \subset V_z$ and, since V_z is an $\mathfrak{m}_-$-module, the Bruhat cell $\exp \mathfrak{m}_- \mathfrak{g}^{\mathbb{C}}_\tau$ lies entirely in $\mathbb{P}V_z$.

It remains to show that $\mathbb{P}\pi$ is singular on each Bruhat cell of codimension at least 1. Let $\tau \in \Delta^1$ be a long root. Consider the vector field $\Phi : y \mapsto [X_{-\rho}, y]$ on $N^\tau \mathfrak{g}^{\mathbb{C}}_\tau$. Elements of $N^\tau \mathfrak{g}^{\mathbb{C}}_\tau$ have no components in $\mathfrak{g}^{(2)}$ so nonzero components of $[X_{-\rho}, y]$ lie in $\mathfrak{g}^{(-1)} \oplus \mathfrak{g}^{(-2)} \subset V_z$. As a result $\pi_*(\Phi) = 0$. This ends the proof since Φ is not radial. □

Let us have a look at the variety $\mathbb{P}(\mathcal{N} \cap V_z)$ in the simplest cases, when $\mathfrak{g}^{\mathbb{C}}$ has rank 2. Diagrams in Fig. 1 show that $\mathbb{P}(\mathcal{N} \cap V_z)$ is the union of two 1-dimensional Schubert varieties (codimension 2) for the root system A_2 and is a 3-dimensional Schubert variety (codimension 3) for G_2 (Schubert varieties are by definition the closures of Bruhat cells). For C_2 the set $\mathbb{P}(\mathcal{N} \cap V_z)$ is a 2-dimensional Schubert variety (codimension 1), so $\mathbb{P}\pi$ is defined only on the big cell. In fact this situation occurs precisely for root systems C_n. To see this note that, according to Lemma 2, the cells of codimension 1 in the Bruhat decomposition of $\mathbb{P}\mathcal{N}$ are parametrised by the roots $\sigma_\alpha(\rho)$ for which $\alpha \in R$ and $\sigma_\alpha(\rho) \neq \rho$ ($\sigma_\alpha \in W$ is the reflexion corresponding to α). If α is such a root then, by Lemma 4, the corresponding Bruhat cell is disjoint from $\mathbb{P}(\mathcal{N} \cap V_z)$ precisely when $\sigma_\alpha(\rho) \notin \Delta^1$. Now $\sigma_\alpha(\rho) = \rho - \langle \rho, \alpha^\vee \rangle \alpha$ and consider the rank 2 root system $\mathrm{span}_{\mathbb{R}}(\alpha, \rho) \cap \Delta$. This root system cannot be semisimple (i.e. of type $A_1 \oplus A_1$) since $\sigma_\alpha(\rho) \neq \rho$. All simple root systems of rank 2 are shown in Fig. 1 and it is clear that $\sigma_\alpha(\rho)$ does not lie in Δ^1 precisely when the extended Dynkin diagram for $\mathfrak{g}^{\mathbb{C}}$ contains the configuration . This is possible only for the root systems C_n, i.e. when $\mathfrak{g}^{\mathbb{C}} = \mathfrak{sp}(n, \mathbb{C})$. This might seem surprising as we expect $\mathbb{P}\pi$ to be birational. Moreover, the quaternionic twistor space $\mathbb{P}\mathcal{N}$ is in such case the projective space $\mathbb{CP}^{2n+1}$ itself. We shall see, however, that in this case the map $\mathbb{P}\pi$ trivially extends to $\mathbb{P}\mathcal{N}$. To summarise, we have the following

PROPOSITION 5. *If $\mathfrak{g}^{\mathbb{C}}$ is a complex simple Lie algebra and $\mathcal{N} \subset \mathfrak{g}^{\mathbb{C}}$ is the minimal nilpotent orbit then the linear map $\pi : \mathfrak{g}^{\mathbb{C}} \to T_z\mathcal{N}$, defined in Formula (5), induces a contact birational map $\mathbb{P}\pi$ from $\mathbb{P}\mathcal{N}$ to $\mathbb{P}T_z\mathcal{N}$. Moreover, the map $\mathbb{P}\pi$ is well defined on the set*

$$\bigoplus_{\alpha \in \Delta^1 \cap L \cup \{\rho\}} N^{\alpha}\mathfrak{g}^{\mathbb{C}}_{\alpha},$$

and if $\mathfrak{g}^{\mathbb{C}}$ is not of type C_n then $\mathbb{P}\pi$ is singular beyond the big cell $N^{\rho}\mathfrak{g}^{\mathbb{C}}_{\rho} = N_{-}\mathfrak{g}^{\mathbb{C}}_{\rho}$.

REMARK 6. The cell structure of flag manifolds is encoded in Hasse diagrams (see Baston and Eastwood 1989); from these diagrams one can find which Bruhat cells lie in the singular set $\mathbb{P}(\mathcal{N} \cap V_z)$. For example one gets the following diagram for the group F_4. In the diagram weights corresponding to cells in $\mathbb{P}(\mathcal{N} \setminus V_z)$ are marked with a "$*$"; the set $\mathbb{P}(\mathcal{N} \cap V_z)$ is a 10-dimensional variety (codimension 5 in $\mathbb{P}\mathcal{N}$).

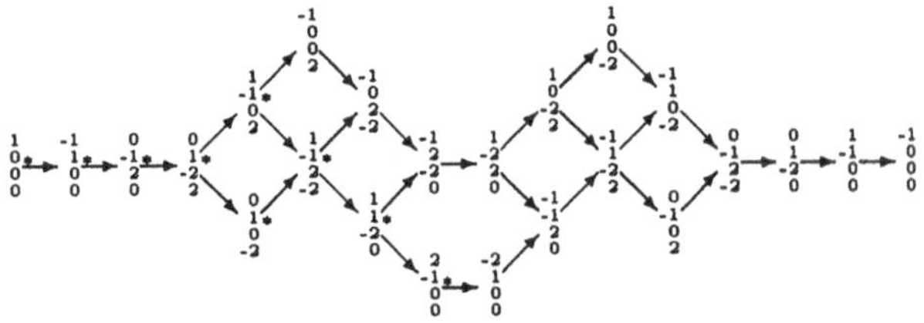

Highest root nilpotent orbit in $\mathfrak{sp}(n, \mathbb{C})$. We shall find explicit formulae for the projection $\pi : \mathfrak{sp}(n, \mathbb{C}) \to T_z\mathcal{N}$. Let ω be a nondegenerate skew-symmetric 2-form on $\mathbb{C}^{2n}$. We choose a basis $(e_i)_{i=1\ldots 2n}$ in which $\omega = \sum_{i=1}^{n} e_i \wedge e_{2n+1-i}$. Let $z = e_1 \otimes e^{2n}$ where $(e^i)_{i=1\ldots 2n}$ is the dual basis for $(\mathbb{C}^{2n})^*$. We have $\mathfrak{sp}(n, \mathbb{C}) = \{A \in Gl(2n, \mathbb{C}) : \forall_{X,Y \in \mathbb{C}^{2n}}\ \omega(AX, Y) + \omega(X, AY) = 0\}$ and we can use the Cartan algebra consisting of the diagonal matrices to define a decomposition $\mathfrak{sp}(n, \mathbb{C}) = T_z\mathcal{N} \oplus V_z$ and the associated map $\pi : \mathcal{N} \to T_z\mathcal{N}$. The map

$$\varphi : \mathbb{C}^{2n} \setminus \{0\} \ni x \to x \otimes \omega(x, \cdot) \in \mathcal{N}$$

parametrises the minimal nilpotent orbit in $\mathfrak{sp}(n, \mathbb{C})$ (this shows that $\mathcal{N} = (\mathbb{H}^n \setminus \{0\})/\mathbb{Z}_2$ and $\mathbb{P}\mathcal{N} = \mathbb{CP}^{2n-1}$). We have $z = \varphi(e_1)$ and $T_z\mathcal{N} = \varphi_*(T_{e_1}\mathbb{C}^{2n}) = \{e_1 \otimes \omega(v, \cdot) + v \otimes \omega(e_1, \cdot) : v \in \mathbb{C}^{2n}\}$. Take $x = \sum_{i=1}^{2n} x_i e_i$ and put $w = x - x_1 e_1$. We get:

$$\begin{aligned}\varphi(x) &= (x_1 e_1 + w) \otimes \omega(x_1 e_1 + w, \cdot) \\ &= x_1^2 e_1 \otimes \omega(e_1, \cdot) + x_1\big(e_1 \otimes \omega(w, \cdot) + w \otimes \omega(e_1, \cdot)\big) + w \otimes \omega(w, \cdot).\end{aligned}$$

Now $w \otimes \omega(w, \cdot) \in V_z$ and, since $z = e_1 \otimes \omega(e_1, \cdot)$, we have $\langle \varphi(x), X_{-\rho}\rangle z = \langle \varphi(x), e_{2n} \otimes e^1\rangle z = x_1^2 z$ (we can use the scalar product $\langle A, B\rangle = \operatorname{trace} AB$ since it is proportional to the Killing form).

In the basis $(\varphi_*(e_i))_{i=1\dots 2n}$ for $T_z\mathcal{N}$ the map $\pi\circ\varphi$ is given by the formula $\pi\circ\varphi(x_1,\dots,x_{2n}) = (x_1^2, x_1x_2,\dots,x_1x_{2n})$ so π vanishes if $x_1 = 0$ (this condition defines the complement of the big cell $\mathcal{C}$) but $\mathbb{P}\pi$ obviously extends to $\mathbb{CP}^{2n-1}$.

Highest root nilpotent orbit in $\mathfrak{sl}(n+1,\mathbb{C})$. In a similar way one can find formulae for a contact birational correspondence between the quaternionic twistor spaces of $\mathrm{Gr}_2(\mathbb{C}^{n+1})$ and $\mathbb{HP}^{n-1}$, i.e. between $\mathbb{CP}^{2n-1}$ and $\mathbb{P}\mathcal{N}$, where $\mathcal{N}\subset\mathfrak{sl}(n+1,\mathbb{C})$ is the highest root nilpotent orbit (we assume that $n\geqslant 2$). In this case we have a surjective map

$$\varphi : \mathbb{C}^{n+1}\times(\mathbb{C}^{n+1})^* \supset D \ni (x,\xi)\mapsto x\otimes\xi\in\mathcal{N}$$

where $D = \{(x,\xi)\in\mathbb{C}^{n+1}\times(\mathbb{C}^{n+1})^* : \xi(x) = 0,\ x\neq 0,\ \xi\neq 0\}$ (note that there is a bijection $\mathbb{P}\mathcal{N}\ni[A]\to(\operatorname{im}A\subset\ker A)\in F_{1,n}(\mathbb{C}^{n+1})$). We adopt the same notation as in the previous example (but basis elements in $\mathbb{C}^{n+1}$ are indexed now by the set $\{0,\dots,n\}$) and take $z = \varphi(e_0,e^n) = e_0\otimes e^n$. Then $T_z\mathcal{N} = \varphi_*(T_{(e_0,e^n)}D)$ consists of the tensors $e_0\otimes\eta + v\otimes e^n$ where $\eta\in(\mathbb{C}^n)^*$, $v\in\mathbb{C}^n$ and $\eta(e_0)+e^n(v) = 0$. We choose the following basis for $T_z\mathcal{N}$:

$$e_0\otimes e^n,\ e_1\otimes e^n,\dots,\ e_{n-1}\otimes e^n,\ e_0\otimes e^{n-1},\ e_0\otimes e^{n-2},\dots,\ e_0\otimes e^1,\ e_0\otimes e^0 - e_n\otimes e^n.$$

The Cartan algebra $\mathfrak{j}$ consisting of the diagonal matrices in $\mathfrak{sl}(n,\mathbb{C})$ gives a decomposition $\mathfrak{sl}(n,\mathbb{C}) = T_z\mathcal{N}\oplus V_z$ where V_z is spanned by matrices $e_i\otimes e^j$ with $i\neq j$, $i\neq 0$ and $j\neq n$, and by diagonal matrices perpendicular to $H_\rho = e_0\otimes e^0 - e_n\otimes e^n$. Now take $x = \sum x_ie_i\in\mathbb{C}^{n+1}$ and $\xi = \sum\xi_ie^i\in(\mathbb{C}^{n+1})^*$ such that $\sum x_i\xi_i = 0$. We have

$$x\otimes\xi = \sum_{i=1}^{n} x_0\xi_ie_0\otimes e^i + \sum_{i=1}^{n-1} x_i\xi_ne_i\otimes e^n + \sum_{i=0}^{n} x_i\xi_ie_i\otimes e^i + \sum_{\substack{i=1,\dots,n\\ i\neq j}}\ \sum_{j=0}^{n-1} x_i\xi_je_i\otimes e^j.$$

The last summand in the formula above belongs to V_z and the diagonal matrix $\sum_{i=0}^n x_i\xi_ie_i\otimes e^i$ projects orthogonally in $\mathfrak{j}$ to the matrix $\frac{1}{2}(x_0\xi_0 - x_n\xi_n)(e_0\otimes e^0 - e_n\otimes e^n)\in\mathbb{C}H_\rho\subset T_z\mathcal{N}$. Taking into account the summand $\langle x\otimes\xi, X_{-\rho}\rangle z = x_0\xi^n e_0\otimes e^n$ we get

PROPOSITION 7. *The formula*

$$(x_0:\dots:x_n;\xi_0:\dots:\xi_n)\mapsto[2x_0\xi_n : x_1\xi_n:\dots:x_{n-1}\xi_n : x_0\xi_{n-1}:\dots:x_0\xi_1 : \tfrac{1}{2}(x_0\xi_0 - x_n\xi_n)]$$

defines a contact birational map from $F_{1,n}(\mathbb{C}^{n+1})$ to $\mathbb{CP}^{2n-1}$. This map is well defined away from the set $\{x_0 = \xi_n = 0\}$ and biholomorphic if and only if $x_0\xi_n\neq 0$.

REMARK 8. In the case when $n = 2$ one recovers Gauduchon's Formulae (1). To see this use the basis $-\frac{1}{2}(e_0\otimes e^0 - e_2\otimes e^2)$, $e_0\otimes e^2$, $e_0\otimes e^1$, $-\frac{1}{2}e_1\otimes e^2$ for $T_{(e_0\otimes e^2)}\mathcal{N}$.

References

Baston, R.J. and Eastwood, M.G. (1989) *The Penrose transform, its interaction with representation theory*, Clarendon Press, Oxford Mathematical Monographs.

Bryant, R.L. (1982) *Conformal and minimal immersions of compact surfaces into the 4-sphere*, J. Diff. Geom. **17**, 455–473.

Burstall, F.E. (1990) *Minimal surfaces in quaternionic symmetric spaces*, Geometry of low-dimensional manifolds, Cambridge University Press, 231–235.

Burstall, F.E. and Rawnsley, J.H. (1990) *Twistor theory for Riemannian symmetric spaces*, Springer-Verlag, Lect. Notes Math. 1424.

Gauduchon, P. (1987) *La correspondance de Bryant*, Astérisque, 181–208.

Kobak, P.Z. (1993) *Quaternionic geometry and harmonic maps*, D. Phil. Thesis, University of Oxford.

Kobak, P.Z. (1994) *Twistors, nilpotent orbits and harmonic maps*, Harmonic Maps and Integrable Systems (Fordy, A.P. and Wood, J.C., eds.), Vieweg, Braunschweig/Wiesbaden.

Lawson, H.B., Jr, (1983) *Surfaces minimales et la construction de Calabi-Penrose*, Sém. Bourbaki **624**, Astérisque 121–122 (Soc. Math. France, 1985, 197–211).

Salamon, S.M. (1982) *Quaternionic Kähler manifolds*, Invent. Math. **67** 143–171.

Swann, A.F. (1991) *HyperKähler and quaternionic Kähler geometry*, Math. Ann. **289**, 421–450.

Warner, G. (1972) *Harmonic analysis on semi-simple Lie groups*, vol. I, Springer-Verlag.

Chapter 2

Applications to conformal geometry

§II.2.1 **Introduction** *by M.G.Eastwood, L.P.Hughston & L.J.Mason*

There are several significant ways in which spinors and twistors impinge on conformal differential geometry. The links are particularly strong in four dimensions. The underlying reason for this is the isomorphism of Lie algebras

$$\mathfrak{so}(4) \cong \mathfrak{su}(2) \times \mathfrak{su}(2)$$

and the fact that the irreducible representations of $\mathfrak{su}(2)$ are especially easy to describe (as the symmetric powers of the fundamental spinor representations on $\mathbb{C}^2$). As a consequence the group $SO(4)$ is not simple (it is the only such orthogonal group), and the Weyl curvature decomposes into a self-dual and an anti self-dual part (whereas in other dimensions, the Weyl curvature is irreducible). This allows one to define metrics with self-dual Weyl curvature and bundles with self-dual connections, the subjects of two important achievements of twistor theory, Penrose's non-linear graviton construction for self-dual solutions of Einstein's equations and Ward's twistor construction for self-dual Yang-Mills fields.

A similar coincidence of low-dimensional Lie algebras can be regarded as the basis of twistor theory, namely

$$\mathfrak{so}(6,\mathbb{C}) \cong \mathfrak{sl}(4),$$

and its various real forms, such as

$$\mathfrak{so}(4,2) \cong \mathfrak{su}(2,2).$$

The spin representation of $\mathfrak{so}(6,\mathbb{C})$ is precisely the representation on $\mathbb{C}^4$ induced by the above isomorphism. These isomorphisms give a powerful computational tool for differential geometry in six dimensions, just as do spinors in four dimensions (cf. §§II.2.2-5). Part of the special utility here arises from the fact that in six dimensions spinors are automatically *pure*, whereas this is not the case in any higher dimension. (A spinor in $2n$ dimensions is said to be 'pure' when the associated n-form is totally null, cf. the appendix of Penrose & Rindler 1986, and S.B.Petrack in §I.3.9.)

The second isomorphism also underpins the use of twistors in four-dimensional conformal geometry, the point being that $SO(4,2)$ is the group of conformal motions of compactified Minkowski space. This is familiar in that natural constructions on flat twistor space (such as forming various sheaf cohomology groups) give rise to conformally invariant objects (such as differential operators) on Minkowski space. In this context 'conformally invariant' may be taken to mean 'invariant under conformal motions'.

There is also a much wider sense in which twistors enter into conformal differential geometry. On compactified Minkowski space, there is an exact sequence of vector bundles, usually denoted

$$0 \longrightarrow \mathcal{O}_{A'} \longrightarrow \mathcal{O}^{\alpha} \longrightarrow \mathcal{O}^{A} \longrightarrow 0.$$

The bundles at either end are spin bundles and the bundle in the middle is the trivial bundle with twistor space as fibre. Surprisingly, this situation persists on a general four-dimensional conformal manifold (if spin). Moreover, the bundle in the middle (called the bundle of 'local twistors') comes naturally equipped with a conformally invariant connection called *local twistor transport* (Penrose & MacCallum 1972). This is the spin connection associated with Cartan's $SO(4,2)$ conformal connection.

The local twistor connection may be defined quite explicitly in terms of the Levi-Civita connection of a metric in the conformal class (see Penrose & Rindler 1986 for full details). A choice of conformal factor induces a splitting

$$\mathcal{O}^{\alpha} = \mathcal{O}^{A} \oplus \mathcal{O}_{A'}$$

of the above exact sequence. Under a conformal rescaling $\hat{g}_{ab} \longrightarrow \Omega^2 g_{ab}$ the splitting changes so that a twistor Z^{α} represented by $(\omega^A, \pi_{A'})$ with respect to g_{ab} is represented by

$$(\hat{\omega}^A, \hat{\pi}_{A'}) = (\omega^A, \pi_{A'} + \mathrm{i}\Upsilon_{AA'}\omega^A)$$

with respect to $\hat{g}_{ab}$ where $\Upsilon_a = \nabla_a \log \Omega$. The covariant derivative can be expressed as

$$\nabla_{BB'}(\omega^A, \pi_{A'}) = (\nabla_{BB'}\omega^A + \mathrm{i}\epsilon_B^A \pi_{B'}, \nabla_{BB'}\pi_{A'} + \mathrm{i}\mathrm{P}_{bAA'}\omega^A)$$

where

$$\mathrm{P}_{ab} = \Phi_{ab} - \Lambda g_{ab} = \frac{1}{12}Rg_{ab} - \frac{1}{2}R_{ab}.$$

Here R_{ab} is the Ricci tensor and R is the scalar curvature. It can be checked that this connection as defined is invariant under conformal rescalings. Local twistors at a point can be thought of as the flat space twistors associated to the flat conformal Minkowski space that best approximates the space-time to second order at that point.

Many of the articles in the chapter use this local twistor construction, for example in the construction of conformally invariant differential operators (cf. §§II.2.13-19). Several other articles are concerned with the more geometric realization of twistors as 'α-surfaces' (and dual twistors as β-surfaces). More generally, in $2n$ dimensions, an α-surface (resp. β-surface) is a totally null n-surface that is anti-self-dual (resp. self-dual).

The interest in foliations of space-times by such surfaces goes back in effect more than three decades to 1961, when Ivor Robinson published a result that was to have a significant influence on the subsequent development of general relativity. Robinson's theorem showed that associated with any shearfree congruence of null geodesic rays on a real curved space-time with hyperbolic signature there is a null (algebraically degenerate) solution of Maxwell's equations. If one complexifies the space-time, one discovers that the shearfree congruence is equivalent to a foliation by α-surfaces and a conjugate foliation consisting of β-surfaces.

This result ties in naturally with twistor theory in a variety of non-trivial ways, and as a consequence the illumination of various aspects of this result (including generalizations and related constructions, such as the Kerr theorem and the Goldberg-Sachs theorem) has from the outset been an important recurrent theme in the development of twistor theory. Several of the articles in this chapter touch on this theme and explore various aspects of the geometry of shearfree ray congruences or foliations by α-surfaces. Such foliations also arise when a space-time admits a *Killing spinor* (cf. the articles by Hughston and Jeffreyes in §§II.3.4–6).

Summary of chapter. The articles in §§II.2.2–4 by Hughston and Jeffryes apply *flat* twistor space as an aid to analysing the differential geometry of *curved six-dimensional spaces.* Here the twistors live in the tangent space, and play the same role relative to the underlying six-dimensional manifold as do the familiar two-component spinors to space-time. The resulting formalism, which is flexible as regards the signature of the associated Riemannian (or pseudo-Riemannian) metric, is of some interest in its own right, and may ultimately be of use in the study of certain global problems in the geometry of six dimensional manifolds (e.g. in connection with the existence of complex structures). See §I.4.17 (Minimal surfaces and strings in six dimensions, by Hughston & Shaw) for a related application.

In §II.2.5 the same methods (using twistors as the spinors for geometry in six dimensions) are employed to prove an analogue of the Kerr theorem in six dimensions. The result is then extended to eight dimensions (and was later extended to $2n$ dimensions in Hughston & Mason 1988).

In §II.2.6 a spinor method is used to produce an elementary (i.e. algebraic) proof of Robinson's theorem. In §II.2.7 these ideas are extended to embrace the Sommers-Bell-Szekeres theorem on zero rest mass fields (see Sommers 1976, Penrose & Rindler §7.3) which says that if a spinor field satisfies the shearfree ray condition and is a p-fold spinor of a zero rest mass field of valence $p+q$, then it is also a repeated spinor of the Weyl spinor, unless $p = 3p + q$.

In §II.2.8 null self-dual Maxwell fields are characterized in terms of integrable distributions of alpha-planes, and a natural invariant formulation is provided for Penrose's original identification of null zero rest mass fields in terms of holomorphic functions with *simple poles* (Penrose 1968, 1969).

In §II.2.9–10 Bailey studies a conformally invariant connection obtained by decomposing the local twistor connection in the presence of a pair of spinor fields. This essentially yields Penrose's conformally invariant eth and thorn operators. This is first used to study Killing spinors. Then Bailey goes on to investigate the structure of the space S of leaves of a foliation by α-planes. In flat space the space S would naturally sit inside twistor space and therefore have a formal neighbourhood. When the 'Goldberg Sachs' condition is satisfied, it is shown that this structure survives in the curved case also. In §II.2.11 relative cohomology on S is studied and shown to give rise to a multipole-like structure. Robinson's theorem in flat space is interpreted in terms of these structures.

The relationship between conformal circles and parameters on curves in conformal manifolds is explored in §II.2.12.

The articles §§II.2.13–17 are concerned with the study of conformally invariant operators, tensors and scalars. One of the goals in studying conformal invariants is to construct a complete list of polynomial invariants. A key step is contained in Eastwood & Rice (1987) in which all the conformally invariant differential operators in Minkowski space are classified using representation theory, and many of them generalized to curved space (using the local twistor connection). In §II.2.13 an operator is constructed that acts on spinors, such that, when applied to the Weyl spinor in four dimensions, it gives rise to a highly non-trivial scalar conformal invariant originally discovered by Fefferman & Graham (1985) using alternative machinery. A subtext of §II.2.13 is that conformally invariant tensors and scalars can be generated from the theory of conformally invariant operators in curved space. This is pursued in further detail in §II.2.15. In §II.2.14 it is shown that the assumption in Eastwood & Rice that the conformal weights of the bundles on which the invariant operators act are integral is not essential—no new operators are obtained by allowing non-integral weights in even dimensions (and flat space).

Differential polynomial conformal invariants can be thought of as conformal polynomial invariants on jet bundles (that is, the bundle of sections at a point together with all derivatives perhaps up to some finite order). In §II.2.16 the structures of bundles of infinite jets of functions (and more generally bundles) are studied in terms of their representation theory and compared to their counterparts in flat space. In §II.2.17 a simpler case is studied in which one considers a projective structure rather than a conformal structure. In the flat case one can generate many invariants for weighted functions on projective space by regarding them as homogeneous functions on the corresponding non-projective space, taking their derivatives, and constructing all the affine invariants of the corresponding tensors (which were classified by Weyl). In §II.2.17 it is shown that certain 'exceptional' invariants do not arise in this way.

Finally in §II.2.18-19 some applications are given. In §II.2.18 a condition that a conformal manifold admits a conformal factor with vanishing trace free Ricci tensor is given. This condition degenerates and fails when the Weyl tensor is self-dual (the result requires a nondegeneracy condition on the Weyl tensor). However, in §II.2.19 a conformally invariant tensor is presented that distinguishes between space-times with self-dual Weyl tensors that have vanishing Ricci tensors, and those that cannot.

References

Eastwood, M.G. & Rice, J. (1987) Conformally invariant differential operators on Minkowski space and their curved analogues, Comm. Math. Phys. **109**, 207–228, and Erratum, Comm. Math. Phys., **144** (1992), 213.

Fefferman, C. & Graham, C.R. (1985) Conformal invariants, in *Élie Cartan et les Mathématiques d'Aujourdui*, Astérisque, 95–116.

Hughston, L.P. & Ward, R.S. (1979) *Advances in twistor theory*, Pitman.

Hughston, L.P. & Mason, L.J. (1988) A generalised Kerr-Robinson theorem, Class. Quant. Grav. **5**, 275-285.

Hughston, L.P. & Mason, L.J. (1990) *Further advances in twistor theory, Volume I: The Penrose transform*, Pitman research notes in mathematics series, **231**, Longmans.

Penrose, R. (1968) Twistor quantisation and curved space-time, Int. J. Theor. Phys., **1**, 61-99.

Penrose, R. (1969) Solutions of the zero rest-mass equations, J. Math. Phys., **10**, 38-39.

Penrose, R. & MacCallum, M.A.H. (1972) Twistor theory: an approach to the quantization of fields and space-time, Phys. Repts., **6C**, 241-315.

Penrose, R. & Rindler, W. (1986) *Spinors and space-time*, Vol. 2, CUP.

Robinson, I. (1961) Null electromagnetic fields, J. Math. Phys. **2**, 290-291.

Sommers. P.D. (1976) Properties of shearfree congruences of null geodesics, Proc. Roy. Soc. **A349**, 309-318.

§II.2.2 **Differential Geometry in Six Dimensions** *by L.P.Hughston* ($\mathbb{TN}$ 19, January 1985)

Twistors are useful and illuminating in the analysis of manifolds of dimension six. This is on account of the fact that twistors are the spinors for the group $O(6,\mathbb{C})$. Thus twistors play a role in the geometry of six dimensional spaces similar in many respects to the role played by two-component spinors in the geometry of four-manifolds. Whether these considerations are of any physical interest remains to be seen—my purposes here are primarily geometrical, and I shall summarise a number of results in outline form.

Conventions: i, j, k etc. $= 0, 1, 2, 3, 4, 5$; α, β, γ etc. $= 0, 1, 2, 3$

Point: $X^i = X^{\alpha\beta}$ (skew, abstract index convention)

Metric: $g_{ij} = \varepsilon_{\alpha\beta\gamma\delta}$ $(\varepsilon_{\alpha\beta\gamma\delta} = \varepsilon_{\gamma\delta\alpha\beta})$

Vector field: $V^i(X) = V^{\alpha\beta}(X)$ (skew)

Null vector field: $V^{\alpha\beta} = P^{[\alpha}Q^{\beta]}$, where $P^\alpha(X)$ and $Q^\alpha(X)$ are 'spinor fields'.

$$\Omega_{ij}V^iV^j = 0 \Longleftrightarrow V^{\alpha\beta} = P^{[\alpha}Q^{\beta]}$$

Two-forms: $F^{ij} = F^{\alpha\beta\rho\sigma} = -F^{\rho\sigma\alpha\beta} \approx E^\alpha_\beta$ with $E^\alpha_\alpha = 0$

Three-forms: $F^{ijk} = F^{[\alpha\beta][\rho\sigma][\xi\eta]} \approx \phi^{\alpha\beta} \oplus \psi_{\alpha\beta}$, $\phi^{\alpha\beta} = \phi^{(\alpha\beta)}$, $\psi_{\alpha\beta} = \psi_{(\alpha\beta)}$

Self-dual 3-forms: $F^{ijk} \sim \phi^{\alpha\beta}$

Anti-self-dual 3-forms: $F^{ijk} \sim \psi_{\alpha\beta}$

Curvature tensor: $R_{ijkl} \sim \Psi^{\alpha\beta}_{\rho\sigma} \oplus \Phi_{\mu\nu\xi\eta} \oplus \Lambda$ (105 components)

Conformal spinor: $\Psi^{\alpha\beta}_{\rho\sigma} = \Psi^{(\alpha\beta)}_{(\rho\sigma)}$ $(\Psi^{\alpha\beta}_{\alpha\sigma} = 0)$ (84 components)

Ricci spinor: $R_{ij} - \frac{1}{6}g_{ij}R \cong \Phi_{\mu\nu\xi\eta}$ ($\boxplus$ symmetry, 20 components)

Vacuum Bianchi identities: $\nabla_{\xi\alpha}\Psi^{\alpha\beta}_{\rho\sigma} = 0$, $\nabla^{\eta\rho}\Psi^{\alpha\beta}_{\rho\sigma} = 0$.

Ricci identities: $\nabla_{[i}\nabla_{j]}\xi_k = R_{ijk}{}^l\xi_l$.

Define $\Box^\alpha_\beta \sim \nabla_{[i}\nabla_{j]}$ $(\Box^\alpha_\alpha = 0)$. Then we find the following relations:

$$\Box^\alpha_\beta\xi_\gamma = \Psi^{\alpha\delta}_{\beta\gamma}\xi_\delta + \Phi^{\alpha\delta}_{\beta\gamma}\xi_\delta + \delta^\alpha_\gamma\Lambda\xi_\beta$$
$$\Box^\alpha_\beta\eta^\delta = -\Psi^{\alpha\delta}_{\beta\gamma}\eta^\gamma - \Phi^{\alpha\delta}_{\beta\gamma}\eta^\gamma - \delta^\delta_\beta\Lambda\eta^\alpha$$

where

$$\Phi^{\alpha\delta}_{\beta\gamma} = \Phi^{[\alpha\delta]}_{[\beta\gamma]} = \Phi_{[\beta\gamma][\rho\sigma]}\varepsilon^{\alpha\delta\rho\sigma}.$$

Note that these formulae are actually simpler in form than their four-dimensional analogues!

'Maxwellian' equations: $F^{ijk} = F^{[ijk]} \sim \phi^{\alpha\beta} \oplus \psi_{\alpha\beta}$. Set $\nabla^{[i}F^{jkl]} = 0$ and $\nabla_i F^{ijk} = 0$. These are equivalent to: $\nabla_{\alpha\beta}\phi^{\beta\gamma} = 0$ and $\nabla^{\alpha\beta}\psi_{\beta\gamma} = 0$. In flat 6-space these imply $\Box\phi^{\beta\gamma} = 0$ and $\Box\psi_{\beta\gamma} = 0$ where $\Box = \nabla_i\nabla^i$.

General solution of $\nabla_{\alpha\beta}\phi^{\beta\gamma} = 0$ in flat space:

$$\phi^{\alpha\beta}(X^{\rho\sigma}) = \oint Z^\alpha Z^\beta f(X_{\rho\sigma}Z^\sigma, Z^\sigma)\,\mathcal{D}^3 Z,$$

where $\mathcal{D}^3 Z = \varepsilon_{\alpha\beta\gamma\delta}Z^\alpha\, dZ^\beta \wedge dZ^\gamma \wedge dZ^\delta$ and $F(W_\rho, Z^\sigma)$ is homogeneous of degree -6, defined on a suitable region of the space $W_\alpha Z^\alpha = 0$. Note that the pair $\{W_\alpha, Z^\alpha\}$ is a spinor for the group $O(8, \mathbb{C})$, i.e. is in effect a 'twistor' for the flat six space.

Comment (1994). For further discussion of this formula see my article (Hughston 1986) in the I. Robinson Festschrift. (Cf. also Penrose & Rindler 1986, pp. 462-464.)

Algebraic classification of symmetric spinor fields in six dimensions is a more intricate matter than in four dimensions. Reality conditions aside, a field $\phi^{\alpha\beta}$ can at each point be one of four essentially distinct types. The most degenerate of these (which I shall call 'null') is when $\phi^{\alpha\beta}$ is of the form $P^\alpha P^\beta$ for some spinor field $P^\alpha(X)$.

1. LEMMA. *If* $\nabla_{\alpha\rho}\phi^{\alpha\beta} = 0$ *and* $\phi^{\alpha\beta} = P^\alpha P^\beta$ *then* $P^\alpha(X)$ *satisfies*

$$(P^\alpha \nabla_{\alpha\beta} P^{[\gamma})P^{\delta]} = 0. \qquad (*)$$

2. REMARK. This condition is analogous to the geodesic shear free condition

$$(o^A \nabla_{A'A} o^{[B})o^{C]} = 0$$

for a spinor field in four dimensions.

3. PROBLEM. Suppose $P^\alpha(X)$ satisfies $(*)$ as above. Does there necessarily exist a field $\phi^{\alpha\beta} = \phi P^\alpha P^\beta$ such that $\nabla_{\alpha\beta}\phi^{\beta\gamma} = 0$ for a suitable choice of the scalar $\phi(X)$?

Solutions of $(*)$ can be generated by consideration of analytic varieties of appropriate codimension:

4. THEOREM. *Suppose* $F_r(W_\alpha, Z^\alpha)$, $r = 1, 2, 3$ *is a triple of holomorphic functions, homogeneous of some degree, defined on regions of the quadric* $W_\alpha Z^\alpha = 0$. *Then the variety* $F_r = 0$ *determines a spinor field* $P^\alpha(X)$ *according to the scheme*

$$F_r(X_{\alpha\beta}P^\beta(X), P^\alpha(X)) = 0,$$

and $P^\alpha(X)$ *satisfies* $(P^\alpha \nabla_{\alpha\beta} P^{[\gamma})P^{\delta]} = 0$.

5. PROBLEM. Does every analytic $P^\alpha(X)$ satisfying $(*)$ arise in this way?

6. LEMMA. *Suppose a curved six-dimensional space satisfies* $R_{ij} = 0$ *and has a degenerate Weyl spinor to the extent that*

$$\Psi^{\alpha\beta}_{\gamma\delta} = P^\alpha P^\beta Q_{\gamma\delta} \qquad (**)$$

for some P^α, $Q_{\alpha\beta}$. *Then* $(P^\alpha \nabla_{\alpha\beta} P^{[\gamma})P^{\delta]} = 0$.

7. PROBLEM. Does the converse hold, in the sense that if P^α satisfies $(*)$ and $R_{ij} = 0$ is the Weyl spinor necessarily of the form $(**)$?

8. DEFINITION. (Robinson & Trautman) In a manifold $\mathscr{M}$ of n dimensions (signature unimportant) let $k^a(x)$ $(a = 1 \ldots n)$ be a vector field which is *conformally geodesic*, i.e. if $\hat{g}_{ab}$ is the metric of $\mathscr{M}$ then for suitable $\Omega(x)$ we have $g_{ab} = \Omega^2 \hat{g}_{ab}$ such that $k^a \nabla_a k_b = 0$, where ∇_a is the

connection associated with g_{ab} and indices are raised and lowered with g_{ab}. Then k^a is *shear-free* if $\mathscr{L}_{k^a} k_{[a}g_{b][c}k_{d]} = \phi\, k_{[a}g_{b][c}k_{d]}$ for some ϕ, or equivalently $\nabla_{(a}k_{b)} = \psi\, g_{ab} + \xi_{(a}k_{b)}$ for some ψ, ξ_a.

9. REMARK. If $\mathscr{M}$ is space-time then this definition reduces to the 'standard' ones if k^a is null or timelike.

10. LEMMA. *Suppose the spinor fields A^α and B^α each satisfy $(*)$. Then the vector field $K^{\alpha\beta} = A^{[\alpha}B^{\beta]}$ is geodesic and shearfree in the sense noted above.*

11. REMARK. To show $K^{\alpha\beta}$ is *geodesic* is straightforward enough: Since A^α and B^α satisfy $(*)$ we have $A^\alpha\nabla_{\alpha\beta}A^\rho = \lambda_\beta A^\rho$ and $B^\alpha\nabla_{\alpha\beta}B^\rho = \mu_\beta B^\rho$ for some λ_β, μ_β. Thus $A^\alpha B^\beta\nabla_{\alpha\beta}A^\rho = \lambda A^\rho$ and $A^\alpha B^\beta\nabla_{\alpha\beta}B^\rho = \mu B^\rho$ for suitable λ, μ. Whence

$$A^\alpha B^\beta\nabla_{\alpha\beta}A^{[\rho}B^{\sigma]} = (\lambda+\mu)A^{[\rho}B^{\sigma]}. \qquad \square$$

To show $A^{[\alpha}B^{\beta]}$ is shearfree is more intricate.

12. PROBLEM. Show that the converse to Lemma 10 does not hold.

13. REMARK. Reality conditions: for signature $++----$ we impose the 'usual' twistor conjugation rules, i.e. $\overline{Z^\alpha} = \overline{Z}_\alpha$ with a Hermitian correlation of signature $++--$. For signature $+-----$ we impose standard conjugation, i.e. $\overline{Z^\alpha} = \overline{Z}^\alpha$ component by component.

14. PROBLEM. In a real six-dimensional curved space-time of signature $++----$ determine all 'null' solutions of the vacuum equations, i.e. for which $\Psi^{\alpha\beta}_{\gamma\delta} = Z^\alpha Z^\beta \overline{Z}_\gamma \overline{Z}_\delta$, with $Z^\alpha\overline{Z}_\alpha = 0$.

References

Hughston, L.P. (1986) Applications of $SO(8)$ spinors, in *Gravitation and Geometry* (I. Robinson Festschrift volume), eds. W. Rindler and A. Trautman (Bibliopolis, Naples) pp. 253–287.

Penrose, R. & Rindler, W. (1986) *Spinors and Space-Time*, Vol. 2, Cambridge University Press.

§II.2.3 **A Theorem on Null Fields in Six Dimensions** *by L.P.Hughston* (TN 20, September 1985)

In what follows I shall outline a rather striking result, holding in six dimensions, which can be regarded as a generalization of Robinson's theorem on null electromagnetic fields in four dimensions.

By a 'massless field' in six dimensions I mean a symmetric spinor field $\phi^{\alpha\beta\ldots\gamma}$ which satisfies $\nabla_{\delta\alpha}\phi^{\alpha\beta\ldots\gamma} = 0$; to be a 'totally null' field it must satisfy $\phi^{\alpha\beta\ldots[\gamma}P^{\delta]} = 0$ for some spinor P^α.

LEMMA. *Suppose $\phi^{\alpha\beta\ldots\gamma}$ satisfies these conditions; then P^α must satisfy*

$$(P^\alpha\nabla_{\alpha\beta}P^{[\gamma})P^{\delta]} = 0. \tag{1}$$

Proof. It will be easily seen that $\phi^{\alpha\beta\ldots\gamma}$ is totally null iff there exists a scalar ψ such that

$$\phi^{\alpha\beta\ldots\gamma} = e^\psi P^\alpha P^\beta \ldots P^\gamma. \tag{2}$$

The zero rest mass condition then implies, and indeed is equivalent to:

$$P^\beta P^\gamma \ldots P^\delta \nabla_{\alpha\beta}\psi + P^{\gamma\cdots}P^\delta\nabla_{\alpha\beta}P^\beta + (n-1)P^\beta(\nabla_{\alpha\beta}P^{(\gamma})\ldots P^{\delta)} = 0, \tag{3}$$

where n is the valence of $\phi^{\alpha\beta\ldots\gamma}$. If one multiplies (3) by P^ε and skews over γ and ε the condition (1) follows at once ($n \geqslant 2$). □

Now I shall establish, locally, a result that is essentially a converse to this lemma.

THEOREM. *Let P^α be a holomorphic spinor field satisfying* (1) *on a region of a six-dimensional complex manifold endowed with a non-degenerate holomorphic metric tensor and a Riemann-compatible holomorphic connection. Then locally there exists a totally null massless field of valence n, with principal spinor P^α, providing that*

$$(n-2)P^\alpha P^\beta \Psi^{\rho[\sigma}_{\alpha\beta}P^{\tau]} = 0, \tag{4}$$

where $\Psi^{\rho\sigma}_{\alpha\beta}$ is the Weyl spinor (conformal curvature spinor).

Proof. (1) is equivalent to the existence of a spinor Λ_α such that

$$P^\alpha\nabla_{\alpha\beta}P^\gamma = \Lambda_\beta P^\gamma, \tag{5}$$

whence

$$P^\beta\nabla_{\alpha\beta}\psi + \nabla_{\alpha\beta}P^\beta - (n-1)\Lambda_\alpha = 0 \tag{6}$$

as follows from (3). Now consider an equation of the form

$$P^{\beta}\nabla_{\alpha\beta}\psi + A_{\alpha} = 0 \tag{7}$$

with ψ unknown, A_{α} specified, and P^{β} satisfying (1). Such an equation admits solutions, locally, by the Frobenius theorem, iff A_{α} satisfies

$$P^{\alpha}\nabla_{\alpha[\beta}A_{\gamma]} = \Lambda_{[\beta}A_{\gamma]}. \tag{8}$$

To see the *necessity* of (8) operate on (7) with $P^{\rho}\nabla_{\sigma\rho}$, and skew over α and σ; (8) then follows by use of (5).

We wish to see whether there exists a scalar ψ such that (6) holds; thus we examine the expression

$$P^{\alpha}\nabla_{\alpha[\beta}A_{\gamma]} - \Lambda_{[\beta}A_{\gamma]} =: I_{\beta\gamma} \tag{9}$$

with

$$A_{\alpha} = \nabla_{\alpha\beta}P^{\beta} - (n-1)\Lambda_{\alpha}\,. \tag{10}$$

A straightforward calculation gives

$$I_{\beta\gamma} = -(n-2)P^{\alpha}\nabla_{\alpha[\beta}\Lambda_{\gamma]}\,. \tag{11}$$

To arrive at (11) use is made of the Ricci identity

$$\underbrace{\nabla_{\alpha[\beta}\nabla_{\gamma]\delta}}P^{\delta} = R\varepsilon_{\alpha\beta\gamma\delta}P^{\delta}\,, \tag{12}$$

where R is the scalar curvature; furthermore we require the simple identity

$$(\nabla_{\delta[\beta}P^{\alpha})(\nabla_{\gamma]\alpha}P^{\delta}) = 0. \tag{13}$$

Now we wish to examine the expression appearing in (11). Suppose we operate on (5) with $P^{\xi}\nabla_{\xi\eta}$, skewing over η and β. A short calculation gives

$$P^{\xi}P^{\alpha}\underbrace{\nabla_{\xi\eta}\nabla_{\alpha\beta}}P^{\gamma} = (P^{\xi}\nabla_{\xi[\eta}\Lambda_{\beta]})P^{\gamma}; \tag{14}$$

but the vanishing of the left side of this equation is, by another Ricci identity, equivalent to

$$P^{\alpha}P^{\beta}\Psi_{\alpha\beta}^{\rho[\sigma}P^{\tau]} = 0. \tag{15}$$

Therefore the vanishing of $I_{\beta\alpha}$, the desired integrability condition, is equivalent to (4). □

Note that for $n = 2$, the case corresponding to the classical Robinson theorem in dimension four, no restrictions are imposed on the curvature beyond those already implied by (1); these conditions, incidentally, are $P^{\alpha}P^{[\eta}\Psi^{\nu][\sigma}_{\alpha\beta}P^{\tau]}P^{\beta} = 0$, as follows from (14) directly by skew-symmetrization with P^{ϵ} over γ and ε. In flat space, given a solution of (1), null fields of any valence can be constructed: these may be generated via a contour integral formula with a holomorphic function showing an appropriately simple pole structure.

Gratitude is expressed to Lionel Mason and Ben Jeffryes, both of whom in discussion and correspondence made contributions to these results.

Comment (1994). For a generalization of these results to higher dimensions see Hughston, L.P. & Mason, L.J. (1988) A generalized Kerr-Robinson theorem, *Class. Quant. Grav.* **5**, 275–285.

§II.2.4 **A Six Dimensional 'Penrose diagram'** *by B.P.Jeffryes* (TN 21, February 1986)

One of the areas in which the use of spinors greatly simplifies four-dimensional general relativity is in the algebraic classification of the Weyl tensor (see Penrose & Rindler 1986 for details of various approaches). Rather than looking for eigenspinors ϕ_C^A and eigenvalues λ of the Weyl spinor Ψ_{ABCD} such that

$$\phi_C^A\Psi_{AB}^{CD} = \lambda\phi_B^D \tag{1}$$

(classification then being with regard to the multiplicity of the eigenvalues and the dimension of the space spanned by the eigenspinors), a classification by the multiplicity of the principal null directions (pnd's) is used. o^A is a pnd of Ψ_{ABCD} if $o^A o^B o^C o^D \Psi_{ABCD} = 0$, or written alternatively

$$o^A o^B o^{[C}\Psi_{AB}^{D][E}o^{F]} = 0; \tag{2}$$

o^A is a double pnd if

$$o^A o^B o^{[C}\Psi_{AB}^{D]E} = 0 \tag{3}$$

and so on.

The reason for the curious position of the indices is for easier comparison with the six-dimensional case; here of course anti-symmetrisation with an upstairs index is equivalent to contraction with a downstairs index.

We might hope for a similar simplification in studying curved six-dimensional spaces (see L.P. Hughston's article II.2.2 for the spinor notation), within which the Weyl tensor is represented by the totally trace-free Weyl spinor $\Psi^{(\alpha\beta)}_{(\gamma\delta)}$ which has 84 components! Everything is much more complicated to classify than in four dimensions; as an example consider the use of equation (1) to classify Ψ_{ABCD} by the multiplicity of its eigenvalues. This gives ({partitions of 2}+1 = 3) different algebraic classes. The use of the analogous scheme in six dimensions

$$\phi^{\alpha}_{\gamma}\Psi^{\gamma\delta}_{\alpha\beta} = \lambda\phi^{\delta}_{\beta} \tag{4}$$

leads to ({partitions of 14} + 1 = 136) different classes. Luckily a concept similar to principal null directions can still be used.

By way of analogy with equations (2) and (3) we wish to consider an algebraic operation $\mathcal{O}(\Psi^{\alpha\beta}_{\gamma\delta}, P^{\alpha})$ on the Weyl spinor $\Psi^{\alpha\beta}_{\gamma\delta}$ and a number of copies of a spinor P^{α} such that $\mathcal{O}(\Psi^{\alpha\beta}_{\gamma\delta}, P^{\alpha}) = 0$ does not imply that either $\Psi^{\alpha\beta}_{\gamma\delta} = 0$ or $P^{\alpha} = 0$. It is clear that all that is possible is some combination of anti-symmetrising P^{α}s with the upstairs indices of $\Psi^{\alpha\beta}_{\gamma\delta}$ or contracting P^{α}s with the downstairs indices of $\Psi^{\alpha\beta}_{\gamma\delta}$. Should $\mathcal{O}(\Psi^{\alpha\beta}_{\gamma\delta}, P^{\alpha}) = 0$ then there will be some algebraic relationship between $\Psi^{\alpha\beta}_{\gamma\delta}$ and P^{α}. Unlike in four dimensions the operations of anti-symmetrisation and contraction are not equivalent.

Now for the diagram. Given $\Psi^{\alpha\beta}_{\gamma\delta}$ and P^{α} we abbreviate

$$P^{[\epsilon}\Psi^{\alpha]\beta}_{\gamma\delta}P^{\delta} = 0 \quad \text{as} \quad \Psi^{]\bullet}_{\bullet\star}$$

and

$$\Psi^{\alpha\beta}_{\gamma\delta} = 0 \quad \text{as} \quad \Psi^{\bullet\bullet}_{\bullet\bullet}$$

and so on. Then

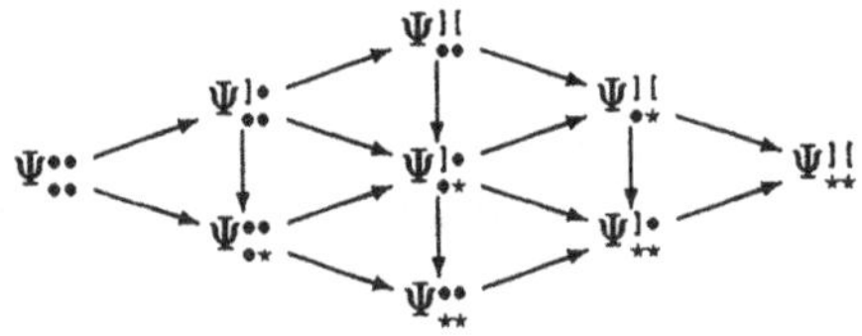

(5)

The diagonal relationships are obvious; the vertical ones arise because $\Psi^{\alpha\beta}_{\gamma\delta}$ is totally trace-free. This classification scheme just involves an upstairs spinor P^{α}; clearly the whole procedure can be carried out inverted with a downstairs spinor instead. The corresponding number of freely specifiable components for each class is

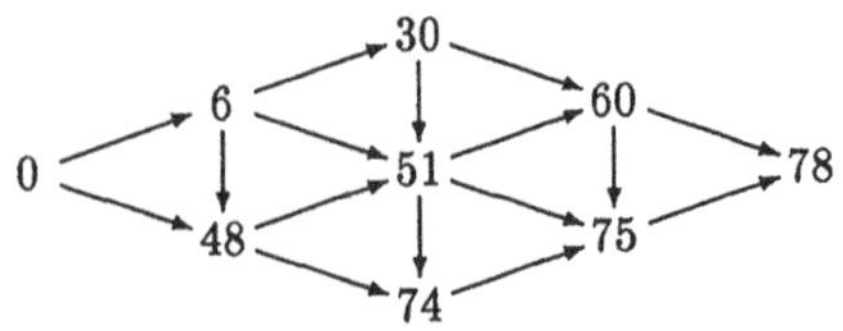

(6)

Reality conditions may restrict the algebraic types possible, for instance if the space is Riemannian then $\Psi^{]\bullet}_{\bullet\bullet}$ is impossible. Unlike in four dimensions a generic Weyl spinor will not have pnd's.

For a given $\Psi^{\alpha\beta}_{\gamma\delta}$ in a chosen basis the condition $\Psi^{!!}_{**}$ is six homogeneous quartic equations on the four components of P^α. The question arises "What use is all this!". This classification scheme turns out to be connected to the existence of special spinor fields. If P^α satisfies

$$(P^\alpha \nabla_{\alpha\beta} P^{[\gamma}) P^{\delta]} = 0 \tag{7}$$

then the Weyl spinor is of type $\Psi^{!!}_{**}$ with respect to P^α (see II.2.2 and II.2.3 for the significance of this equation). If the space is a Kähler manifold (not necessarily vacuum) it implies the existence of a P^α such that

$$P^{[\gamma} \nabla_{\alpha\beta} P^{\delta]} = 0 \tag{8}$$

forcing the Weyl spinor to be of type $\Psi^{!\bullet}_{**}$ with respect to P^α. If the space *is* vacuum it is of type $\Psi^{!\bullet}_{\bullet *}$. If in addition there is a constant holomorphic 3-form (as there would be in the case of currently fashionable Calabi-Yau spaces), this P^α is constant and the Weyl spinor is of type $\Psi^{\bullet\bullet}_{\bullet *}$.

References

Hughston, L.P., §II.2.2 and §II.2.3.

Penrose, R. & Rindler, W. (1986) *Spinors and Space-Time, vol. II: Spinor and Twistor Methods in Space-Time Geometry*, Cambridge University Press.

§II.2.5 **Null Surfaces in Six and Eight Dimensions** *by L.P.Hughston* (TN 22, September 1986)

1. This note is concerned with the construction of null surfaces of dimension n in complex flat space of dimension $2n$. For the construction of null 2-surfaces in four dimensions we have the well-known 'Kerr theorem' (see Penrose 1967, §8; Penrose & Rindler 1986, §7.4) which shows how to specify the general analytic spinor field ξ^A which satisfies $\xi^A \xi^B \nabla_{A'A} \xi_B = 0$. The solution is given in terms of an essentially arbitrarily specified analytic surface in projective twistor space. Remarkably, analogues of this construction exist in six and eight dimensions as well.

In dimension six the general analytic spinor field $\xi^\alpha(X^i)$ ($\alpha = 1 \ldots 4$, $i = 1 \ldots 6$) satisfying $\xi^\alpha \nabla_{\alpha\beta} \xi^\gamma = \Lambda_\beta \xi^\gamma$ for some Λ_β is given in terms of an analytic variety of dimension three in the associated 'twistor space' (which in this case is the space of 'pure' spinors for $SO(8)$—a six-dimensional quadric).

In dimension eight the general analytic spinor field $\xi^\alpha(X^i)$ (here $\alpha = 1\ldots8, i = 1\ldots8$) satisfying $\xi^\alpha\Gamma^i_{\alpha\alpha'}\nabla_i\xi^\beta = \Lambda_{\alpha'}\xi^\beta$ and $\xi^\alpha\xi^\beta\Omega_{\alpha\beta} = 0$ is given by an arbitrary analytic variety of dimension four in the associated twistor space—in this case the twistor space in question is the space of pure spinors (of a given helicity) for the group $SO(10)$.

The pattern of proof is similar in each case, and involves the 'purity' conditions on the relevant spinors and twistors in an essential way.

2. Dimension Six. We wish to solve the equation

$$(\xi^\alpha\nabla_{\alpha\beta}\xi^{[\gamma})\xi^{\delta]} = 0, \tag{1}$$

where $\alpha = 1\ldots4$. Here $\nabla_{\alpha\beta} = \partial/\partial X^{\alpha\beta}$, where $X^{\alpha\beta} = -X^{\beta\alpha}$ coordinatises $\mathbb{C}^6$. The significance of equation (1) is that it is the integrability condition for the family of null vector fields of the form $\xi^{[\alpha}\eta^{\beta]}$, η^β arbitrary, to be closed under commutation, hence providing for a family of null 3-surfaces.

A twistor for $\mathbb{C}^6$ can be represented by a pair Z^α, W_α satisfying $Z^\alpha W_\alpha = 0$: the projective twistor space for $\mathbb{C}^6$ is thus a projective quadric Q^6_+ in $\mathbb{CP}^7$. It will be shown that each solution of (1) corresponds (locally) to an analytic variety of dimension 3 in Q^6_+.

THEOREM 1. *Let $F^r(Z^\alpha, W_\alpha) = 0$ ($r = 1,2,3$) be an analytic variety of dimension 3 in the projective quadric $Z^\alpha W_\alpha = 0$, and suppose a spinor field $\xi^\alpha(X^i)$ defined on an open region U of $\mathbb{C}^6$ satisfies*

$$F^r(\xi^\alpha, X_{\alpha\beta}\xi^\beta) = 0 \tag{2}$$

for each value of $X^i \subset U$. Then $\xi^\alpha(X^i)$ satisfies (1).

Proof. Since $F^r(Z^\alpha, W_\alpha)$ is by hypothesis homogeneous of degree (say) n_r in Z^α and W_α jointly we have $(Z^\alpha\hat{Z}_\alpha + W_\alpha\hat{W}^\alpha)F^r = n_rF^r = 0$ on the variety (where $\hat{Z}_\alpha = \partial/\partial Z^\alpha, \hat{W}^\alpha = \partial/\partial W_\alpha$); whence $\xi^\alpha(\hat{Z}_\alpha - X_{\alpha\beta}\hat{W}^\alpha)F^r = 0$ by the substitution $(Z^\alpha, W_\alpha) = (\xi^\alpha, X_{\alpha\beta}\xi^\beta)$. Writing

$$R^r_\alpha := (\hat{Z}_\alpha - X_{\alpha\beta}\hat{W}^\alpha)F^r \tag{3}$$

we have

$$\xi^\alpha R^r_\alpha = 0, \qquad (r = 1,2,3). \tag{4}$$

Furthermore, by differentiation of (2) we get $\nabla_{\rho\sigma}F^r(\xi^\alpha, X_{\alpha\beta}\xi^\beta) = 0$,

whence $$(\nabla_{\rho\sigma}\xi^\alpha)\hat{Z}_\alpha F^r + (\nabla_{\rho\sigma}X_{\alpha\beta}\xi^\beta)\hat{W}^\alpha F^r = 0;$$

so $$(\nabla_{\rho\sigma}\xi^\alpha)\hat{Z}_\alpha F^r + \varepsilon_{\rho\sigma\alpha\beta}\xi^\beta\hat{W}^\alpha F^r + (\nabla_{\rho\sigma}\xi^\beta)X_{\alpha\beta}\hat{W}^\alpha F^r = 0,$$

whence $$(\nabla_{\rho\sigma}\xi^\alpha)R^r_\alpha + \varepsilon_{\rho\sigma\alpha\beta}\xi^\beta\hat{W}^\alpha F^r = 0;$$

and by transvection of this relation with ξ^ρ we get:

$$(\xi^\rho \nabla_{\rho\sigma}\xi^\alpha)R^r_\alpha = 0, \qquad (r = 1,2,3). \tag{5}$$

Now since R^r_α $(r = 1,2,3)$ are linearly independent vectors at generic points of the variety (i.e. $R^r_\alpha R^s_\beta R^t_\gamma \varepsilon_{rst} \neq 0$), it follows from (4) and (5) that $\xi^\rho \nabla_{\rho\sigma}\xi^\alpha = \lambda_\sigma \xi^\alpha$ for some λ_σ. □

3. Dimension Eight. Here we use the notation set out in my article for the I. Robinson Festschrift. In this case to generate a family of null four-surfaces in $\mathbb{C}^8$ we require a spinor field $\xi^\alpha(X^i)$ $(\alpha = 1\ldots 8,\ i = 1\ldots 8)$ satisfying $\xi^\alpha\xi^\beta\Omega_{\alpha\beta} = 0$ (where $\Omega_{\alpha\beta}$ is the natural 'metric' induced on the spin space in that dimension—thus ξ^α is taken to be a 'pure' spinor field), and

$$\xi^\alpha \nabla_{\alpha\alpha'}\xi^\beta = \Lambda_{\alpha'}\xi^\beta, \tag{6}$$

where $\nabla_{\alpha\alpha'} = \Gamma^i_{\alpha\alpha'}\nabla_i$ $(\nabla_i = \partial/\partial X^i)$. One can verify that (6) together with $\xi^\alpha\xi_\alpha = 0$ are necessary and sufficient conditions for all vector fields of the form $V^i = \Gamma^i_{\alpha\alpha'}\xi^\alpha\eta^{\alpha'}$ ($\eta^{\alpha'}$ arbitrary) to be null and closed under commutation with one another.

A 'twistor' for dimension eight is a pair of pure spinors $(Z^\alpha,\ W^{\alpha'})$ with $Z^\alpha Z^\beta\Omega_{\alpha\beta} = 0$ and $W^{\alpha'}W^{\beta'}\Omega_{\alpha'\beta'} = 0$ satisfying the incidence relation $\Gamma^i_{\alpha\alpha'}Z^\alpha W^{\alpha'} = 0$. These conditions are the 'purity' relations for the $SO(10)$ spinor defined by Z^α, $W^{\alpha'}$. (Cf. Cartan 1937, Petrack 1982, Hughston §I.2.8 and Petrack §I.3.9). The space of such projective pure twistors is a complex manifold of dimension ten, which we shall denote $\mathscr{S}^{10}$.

In what follows we consider an analytic variety V^4 of dimension four in $\mathscr{S}^{10}$, given locally by a set of equations $F^r(Z^\alpha, W^{\alpha'}) = 0$, with $r = 1\ldots 6$. For each value of r we require that F^r be homogeneous of some degree jointly in Z^α and $W^{\alpha'}$. Thus $F^r(\lambda Z^\alpha, \lambda W^{\alpha'}) = \lambda^{n_r}F^r(Z^\alpha, W^{\alpha'})$ for a suitable set of integers n_r.

THEOREM 2. *Let $F^r(Z^\alpha, W^{\alpha'}) = 0$ $(r = 1\ldots 6)$ define an analytic variety of dimension four in the space $\mathscr{S}^{10}$ (given by $Z^\alpha Z_\alpha = 0$, $W^\alpha W_\alpha = 0$, $Z^\alpha W^{\alpha'}\Gamma^i_{\alpha\alpha'} = 0$), and suppose a spinor field $\xi^\alpha(X^i)$ $(\alpha = 1\ldots 8, i = 1\ldots 8)$ defined on a region U of $\mathbb{C}^8$ satisfies $F^r(\xi^\alpha, \xi^\alpha X^i\Gamma^{\alpha'}_{\alpha i}) = 0$ for each value of $X^i \subset U$. Then ξ^α satisfies (1), i.e. $(\xi^\alpha\nabla_{\alpha\alpha'}\xi^{[\beta})\xi^{\gamma]} = 0$.*

Proof. By homogeneity we have $(Z^\alpha\hat{Z}_\alpha + W^{\alpha'}\hat{W}_{\alpha'})F^r = n_r F^r = 0$ on V^4, where $\hat{Z}_\alpha = \partial/\partial Z^\alpha$, $\hat{W}_{\alpha'} = \partial/\partial W^{\alpha'}$, $Z_\alpha = \Omega_{\alpha\beta}Z^\beta$, $W_{\alpha'} = \Omega_{\alpha'\beta'}W^{\beta'}$; whence

$$\xi^\alpha R^r_\alpha = 0 \tag{7}$$

where $R^r_\alpha = (\hat{Z}_\alpha + X_{\alpha\alpha'}\hat{W}^{\alpha'})F^r$, $(r = 1\ldots 6)$ with $X_{\alpha\alpha'} = X^i\Gamma_{i\alpha\alpha'}$. Furthermore by differentiation of $F^r(\xi^\alpha, \xi^\alpha X^{\alpha'}_\alpha) = 0$ we get:

$$\nabla_{\alpha\alpha'}F^r(\xi^\beta, X^{\beta'\beta}\xi_\beta) = 0,$$
$$(\nabla_{\alpha\alpha'}\xi^\beta)\hat{Z}_\beta F^r + \nabla_{\alpha\alpha'}(X^{\beta'\beta}\xi_\beta)\hat{W}_{\beta'}F^r = 0,$$
$$(\nabla_{\alpha\alpha'}\xi^\beta)\hat{Z}_\beta F^r + \Gamma_{\alpha\alpha' i}\Gamma^i_{\beta\beta'}\xi^\beta\hat{W}^{\beta'}F^r + (\nabla_{\alpha\alpha'}\xi_\beta)X^{\beta'\beta}\hat{W}_{\beta'}F^r = 0.$$

Thus $(\nabla_{\alpha\alpha'}\xi^{\beta})R^r_{\beta} + \xi^{\beta}\Gamma_{\alpha\alpha' i}\Gamma^i_{\beta\beta'}\hat{W}^{\beta'}F^r = 0$, and therefore

$$(\xi^{\alpha}\nabla_{\alpha\alpha'}\xi^{\beta})R^r_{\beta} + (\xi^{\alpha}\xi^{\beta}\Gamma_{\alpha\alpha' i}\Gamma^i_{\beta\beta'})\hat{W}^{\beta'}F^r = 0.$$

But in eight dimensions there is the remarkable identity $\Gamma^i_{\alpha'(\alpha}\Gamma_{\beta)\beta' i} = \Omega_{\alpha\beta}\Omega_{\alpha'\beta'}$; thus

$$\xi^{\alpha}\xi^{\beta}\Gamma_{\alpha\alpha' i}\Gamma^i_{\beta\beta'} = (\xi^{\alpha}\xi^{\beta}\Omega_{\alpha\beta})\Omega_{\alpha'\beta'} = 0.$$

Therefore

$$(\xi^{\alpha}\nabla_{\alpha\alpha'}\xi^{\beta})R^r_{\beta} = 0. \tag{8}$$

Thus ξ^{α} and $\xi^{\alpha}\nabla_{\alpha\alpha'}\xi^{\beta}$ are each orthogonal by equations (7) and (8) to six linearly independent vectors, i.e. the vectors R^r_{β} $(r = 1\ldots 6)$. But they are also each orthogonal to the vector $\Omega_{\beta\gamma}\xi^{\gamma}$; and are therefore each orthogonal to seven independent vectors; and are therefore proportional. □

The result outlined here for six dimensions is mentioned briefly in §II.2.2 without proof. For a discussion of the relation of equation (1) to 'null' fields in six dimensions see §II.2.3.

Comment (1994). See Hughston & Mason (1988) for the higher-dimensional analogues of this work. See also Hughston (1979) page 146.

References

Cartan, E. (1937) *The Theory of Spinors* (reprinted by Dover 1966).

Hughston, L.P. (1979) *Twistors and Particles*, Springer Lecture Notes on Physics **97**.

Hughston, L.P. A Remarkable Connection between the Wave Equation and Spinors in Higher Dimensions, §I.2.8.

Hughston, L.P. (1986) Applications of $SO(8)$ spinors, in *Gravitation and Geometry* (I. Robinson Festschrift volume), eds. W. Rindler and A. Trautman (Bibliopolis, Naples) pp. 253–287.

Hughston, L.P. & Mason, L.J. (1988) A generalized Kerr-Robinson theorem, Class. Quant. Grav. **5**, 275–285.

Penrose, R. (1967) Twistor Algebra, *J. Math. Phys.* **8**, pp. 345–366.

Penrose, R. & Rindler, W. (1986) *Spinors and Space-Time, vol. II: Spinor and Twistor Methods in Space-Time Geometry* (Cambridge University Press).

Petrack, S.B. An Inductive Approach to Higher Dimensional Spinors, §I.3.9.

Petrack, S.B. (1982) Spinors and Complex Geometry in Arbitrary Dimensions, Qualifying Thesis (Oxford University).

§II.2.6 **A Proof of Robinson's Theorem** *by L.P.Hughston* (TN 20, September 1985)

In 1976 Paul Sommers published an elegant simplified proof of Robinson's theorem (1961) on shear-free congruences. In what follows I shall outline a new proof that tightens up and improves on some aspects of his argument and of the development of the same material as outlined in Penrose & Rindler (1986), §7.3. The methods are of some interest in their own right.

THEOREM (Robinson). *Suppose $\mathscr{M}$ is a complex manifold of dimension four with a holomorphic metric g_{ab}. Let κ^A be a spinor field defined on an open set $U \subset \mathscr{M}$ satisfying*

$$\kappa^A \kappa^B \nabla_{A'A} \kappa_B = 0. \tag{1}$$

Then for each point $p \in U$ there exists a neighbourhood $V \subset U$ such that there exists a scalar ψ on V with $\nabla_{A'A} \phi^{AB} = 0$, where

$$\phi^{AB} = e^{\psi} \kappa^A \kappa^B. \tag{2}$$

Proof. Note that $\nabla_{A'A}\phi^{AB} = 0$ is equivalent, by (2), to

$$\kappa^B \kappa^A \nabla_{A'A} \psi + \kappa^B \nabla_{A'A} \kappa^A + \kappa^A \nabla_{AA'} \kappa^B = 0. \tag{3}$$

Since (1) is equivalent to the existence of a spinor $\lambda_{A'}$ such that

$$\kappa^A \nabla_{AA'} \kappa^B = \lambda_{A'} \kappa^B \tag{4}$$

it follows by insertion of (4) in (3) that we seek a scalar ψ such that

$$\kappa^A \nabla_{AA'} \psi + \nabla_{AA'} \kappa^A + \lambda_{A'} = 0. \tag{5}$$

Now consider an equation of the form

$$\kappa^A \nabla_{AA'} \psi + \alpha_{A'} = 0, \tag{6}$$

where κ^A satisfies (4). As a lemma we require the fact that if $\alpha_{A'}$ is specified then there exist solutions of (6) locally if and only if $\alpha_{A'}$ satisfies

$$\kappa^A \nabla_{AA'} \alpha^{A'} = \lambda_{A'} \alpha^{A'}. \tag{7}$$

The proof of this lemma follows as an application of the Frobenius theorem. (To see how (7) arises as a necessary condition we transvect (6) with $\kappa^B \nabla_B^{A'}$ to obtain

$$\kappa^B \nabla_B^{A'} (\kappa^A \nabla_{AA'} \psi) + \kappa^B \nabla_B^{A'} \alpha_{A'} = 0;$$

whence,

$$(\kappa^B \nabla_B^{A'} \kappa^A) \nabla_{AA'} \psi + \kappa^B \kappa^A \nabla_B^{A'} \nabla_{AA'} \psi = \kappa^B \nabla_{BA'} \alpha^{A'},$$

which by (4) gives $\lambda^{A'} \kappa^A \nabla_{A'A} \psi = \kappa^A \nabla_{AA'} \alpha^{A'}$, which by use of (6) gives (7). Frobenius' theorem shows that (7) is also sufficient.)

We wish to determine whether there exists a scalar ψ such that (5) is satisfied. Thus we must examine the expression $I := \kappa^A \nabla_{AA'} \alpha^{A'} - \lambda_{A'} \alpha^{A'}$ with $\alpha^{A'} = \nabla_A^{A'} \kappa^A + \lambda^{A'}$. We have:

$$\begin{aligned} I &= \kappa^A \nabla_{AA'} (\nabla_B^{A'} \kappa^B + \lambda^{A'}) - \lambda_{A'} (\nabla_A^{A'} \kappa^A + \lambda^{A'}) \\ &= \kappa^A \nabla_{AA'} \nabla_B^{A'} \kappa^B + (\kappa^A \nabla_{AA'} \lambda^{A'} + \lambda^{A'} \nabla_{A'A} \kappa^A) \\ &= -\kappa^A \nabla_{BA'} \nabla_A^{A'} \kappa^B + 2\kappa^A \nabla_{A'(A} \nabla_{B)}^{A'} \kappa^B + (\nabla_{AA'} \kappa^A \lambda^{A'}) \\ &= -[\nabla_{BA'} (\kappa^A \nabla_A^{A'} \kappa^B) - (\nabla_{BA'} \kappa^A)(\nabla_A^{A'} \kappa^B)] + 2\kappa^A \square_{AB} \kappa^B + \nabla_{AA'} \kappa^A \lambda^{A'}. \end{aligned}$$

But $\kappa^A \square_{AB} \kappa^B$ vanishes for any κ^A since $\square_{AB} \kappa^B = -3\Lambda \kappa_A$. Furthermore we have

$$(\nabla_{A'B} \kappa^A)(\nabla_A^{A'} \kappa^B) = 0$$

for any κ^A. Thus:

$$\begin{aligned} I &= -\nabla_{BA'} (\kappa^A \nabla_A^{A'} \kappa^B) + \nabla_{AA'} \kappa^A \lambda^{A'} \\ &= -\nabla_{BA'} (\lambda^{A'} \kappa^B) + \nabla_{AA'} \kappa^A \lambda^{A'} \\ &= 0. \end{aligned}$$

Since I vanishes the integrability condition for ψ is satisfied, and the theorem is proved. □

THEOREM (Sommers-Bell-Szekeres generalization of Robinson's result). *Suppose, in the venue as above, κ^A satisfies* (1). *Furthermore let κ^A be a p-fold principal spinor ($p \geqslant 1$) of a massless field of valence $p + q$. Then:*

$$(p - 2 - 3q) \Psi_{ABCD} \kappa^A \kappa^B \kappa^C = 0,$$

where Ψ_{ABCD} is the Weyl spinor.

The proof follows essentially the same line of reasoning as in the first theorem (see II.2.7). The Goldberg-Sachs theorem follows if we note that a spinor satisfying (1) is automatically a 1-fold principal spinor of the Weyl spinor, and that in a vacuum the Weyl spinor satisfies the zero-rest-mass equations.

Comment (1994). The idea that it should be possible to formulate a purely 'covariant' spinorial approach to Robinson's theorem and the Goldberg-Sachs theorem was suggested to Paul Sommers and me in conversation in the early 1970's by Martin Walker, who in turn attributed the idea to Robert Geroch.

References

Robinson, I. (1961) Null electromagnetic fields, *J. Math. Phys.* **2**, p. 290–291.

Sommers, P.D. (1976) Properties of shear-free congruences of null geodesics, *Proc. Roy. Soc. Lond.*, **A349**, p. 309–318.

§II.2.7 **A Simplified Proof of a Theorem of Sommers** *by L.P.Hughston* (TN 22, September 1986)

1. Introduction. In 1976 an interesting theorem on zero-rest-mass fields in curved space-time was published by P. D. Sommers. His main result is as follows:

THEOREM 1. *If a spinor field* ξ^A *which satisfies* $\xi^A\xi^B\nabla_{A'A}\xi_B = 0$ *is a p-fold principal spinor* ($p \geqslant 1$) *of a zero-rest-mass field of valence* $p+q$, *then it is also a repeated princpal spinor of the Weyl spinor, unless* $p = 3q+2$.

The purpose of this note is to present a condensed proof of this theorem, improving in some respects on the original argument devised by Sommers. In what follows I employ as far as possible the notation and conventions of Penrose & Rindler (1986) (cf. in particular §7.3).

2. The SFR Condition. The shear-free ray condition on ξ^A, given by

$$\xi^A\xi^B\nabla_{A'A}\xi_B = 0 \tag{2.1}$$

can be expressed alternatively in either of the forms

$$\nabla_{A'(A}\xi_{B)} = \Lambda_{A'(A}\xi_{B)} \tag{2.2}$$

or

$$\xi^A\nabla_{AA'}\xi^B = \eta_{A'}\xi^B. \tag{2.3}$$

It is helpful to be able to use both of these expressions in computations, and as a consequence we shall find it useful to have at our disposal a relationship between $\eta_{A'}$ and $\Lambda_{A'A}$:

LEMMA 1. $\nabla_{A'A}\xi^A + \Lambda_{A'A}\xi^A = 2\eta_{A'}$ (where $\xi^A \neq 0$).

Proof.

$$\begin{aligned}\nabla_{A'A}\xi_B &= \nabla_{A'(A}\xi_{B)} + \nabla_{A'[A}\xi_{B]}\\ &= \Lambda_{A'(A}\xi_{B)} + \frac{1}{2}\varepsilon_{AB}\nabla_{A'C}\xi^C\\ &= \Lambda_{A'A}\xi_B - \Lambda_{A'[A}\xi_{B]} + \frac{1}{2}\varepsilon_{AB}\nabla_{A'C}\xi^C\\ &= \Lambda_{A'A}\xi_B + \frac{1}{2}\varepsilon_{AB}(\nabla_{A'C}\xi^C - \Lambda_{A'C}\xi^C).\end{aligned}$$

Contraction of each side with ξ^A, followed by use of (2.3), then yields the desired result. □

3. Principality. It will be useful to have another lemma at our disposal which relates ξ^A to the Weyl spinor:

LEMMA 2. *A spinor field ξ^A which satisfies the SFR condition is necessarily a principal spinor of the Weyl spinor. A necessary and sufficient condition for ξ^A to be a repeated principal spinor of the Weyl spinor is $\xi^A\nabla_{AA'}\eta^{A'} = 0$.*

Proof. We have $\xi^C\nabla_C^{B'}\xi_A = \eta^{B'}\xi_A$; whence

$$\begin{aligned}
\xi^B\nabla_{BB'}(\xi^C\nabla_C^{B'}\xi_A) &= \xi^B\nabla_{BB'}(\eta^{B'}\xi_A),\\
(\xi^B\nabla_{BB'}\xi^C)\nabla_C^{B'}\xi_A + \xi^C\xi^B\nabla_{BB'}\nabla_C^{B'}\xi_A &= (\xi^B\nabla_{BB'}\eta^{B'})\xi_A + \eta^{B'}(\xi^B\nabla_{BB'}\xi_A),\\
(\eta_{B'}\xi^C)\nabla_C^{B'}\xi_A + \xi^B\xi^C\,\square_{BC}\,\xi_A &= (\xi^B\nabla_{BB'}\eta^{B'})\xi_A + \eta^{B'}(\eta_{B'}\xi_A);
\end{aligned}$$

whence $\Psi_{ABCD}\xi^B\xi^C\xi^D = (\xi^B\nabla_{BB'}\eta^{B'})\xi_A$, from which it follows at once that $\Psi_{ABCD}\xi^A\xi^B\xi^C\xi^D = 0$, and that $\Psi_{ABCD}\xi^B\xi^C\xi^D = 0$ iff $\xi^B\nabla_{BB'}\eta^{B'} = 0$. □

4. Proof of Theorem 1. Let ξ_A be a p-fold principal spinor of a valence $p+q$ zero-rest-mass field $\phi_{A...E}$. If we transvect $\phi_{A...E}$ with q ξs we get

$$\overbrace{\xi^A\ldots\xi^B}^{q}\phi_{A...BCD...E} = e^{\psi}\,\overbrace{\xi_C\xi_D\cdots\xi_E}^{p}$$

for some scalar ψ. Therefore, given $\nabla^{C'C}\phi_{A...BCD...E} = 0$, we have:

$$\begin{aligned}
&\xi^A\ldots\xi^B\nabla^{C'C}\phi_{A...BCD...E} = 0,\\
&\nabla^{C'C}(\xi^A\ldots\xi^B\phi_{A...BCD...E}) - \phi_{A...BCD...E}\nabla^{C'C}(\xi^A\ldots\xi^B) = 0,\\
&\nabla^{C'C}(e^{\psi}\xi_C\xi_D\cdots\xi_E) - q\phi_{A...BCD...E}\Lambda^{C'C}\xi^A\ldots\xi^B = 0,\\
&e^{\psi}(\xi_C\nabla^{C'C'}\psi)\xi_D\cdots\xi_E + e^{\psi}(\nabla^{C'C}\xi_C)\xi_D\cdots\xi_E\\
&\qquad\qquad + e^{\psi}\xi_C\nabla^{C'C}(\xi_D\cdots\xi_E) - qe^{\psi}\Lambda^{C'C}\xi_C\xi_D\cdots\xi_E = 0;
\end{aligned}$$

which gives

$$e^{\psi}\xi_D\cdots\xi_E\,[\xi_C\nabla^{CC'}\psi + \nabla^{C'C}\xi_C - (p-1)\eta^{C'} - q\Lambda^{C'C}\xi_C] = 0,$$

whence

$$\xi^C\nabla_{CC'}\psi + \nabla_{C'C}\xi^C - q\Lambda_{C'C}\xi^C + (p-1)\eta_{C'} = 0.$$

By lemma 1 we have $-\Lambda_{C'C}\xi^C = -2\eta_{C'} + \nabla_{C'C}\xi^C$, so the term involving $q\Lambda_{C'C}\xi^C$ can be eliminated from the equation above to give us

$$\xi^C \nabla_{CC'}\psi + (q+1)\nabla_{C'C}\xi^C + (p-2q-1)\eta_{C'} = 0.$$

Now this is an equation of the form $\xi^C\nabla_{CC'}\psi + \alpha_{C'} = 0$ from which it follows at once, by differentiation with respect to $\xi^B\nabla^{C'}_B$ and use of the SFR condition, that $\alpha_{C'}$ must satisfy $\xi^A\nabla_{AA'}\alpha^{A'} = \eta_{A'}\alpha^{A'}$. Therefore we investigate the consequences of this relation with $\alpha_{A'} = r\nabla_{A'A}\xi^A + s\eta_{A'}$, where $r = q+1$, $s = p-2q-1$. We have:

$$\xi^A\nabla_{AA'}\alpha^{A'} - \eta_{A'}\alpha^{A'} = 0,$$

$$r\xi^A\nabla_{AA'}\nabla^{A'}_B\xi^B + s\xi^A\nabla_{AA'}\eta^{A'} - r\eta_{A'}\nabla^{A'}_A\xi^A = 0,$$

$$2r\xi^A\nabla_{A'(A}\nabla^{A'}_{B)}\xi^B - r\xi^A\nabla_{A'B}\nabla^{A'}_A\xi^B + s\xi^A\nabla_{AA'}\eta^{A'} + r\eta^{A'}\nabla_{A'A}\xi^A = 0,$$

$$2r\xi^A\Box_{AB}\xi^B - r[\nabla_{A'B}(\xi^A\nabla^{A'}_A\xi^B) - (\nabla_{BA'}\xi^A)(\nabla^{A'}_A\xi^B)]$$
$$+ (s-r)\xi^A\nabla_{AA'}\eta^{A'} + r[\xi^A\nabla_{AA'}\eta^{A'} + \eta^{A'}\nabla_{A'A}\xi^A] = 0,$$

from which it follows that $(s-r)\xi^A\nabla_{AA'}\eta^{A'} = 0$, by use the Ricci relation $\Box_{AB}\xi^B = -3\Lambda\xi_A$, the algebraic identity $(\nabla_{A'A}\xi_B)(\nabla^{A'B}\xi^A) = 0 \quad \forall\xi^A$, and the SFR condition $\xi^A\nabla^{A'}_A\xi^B = \eta^{A'}\xi^B$. And thus we see that if $s \neq r$ (i.e. $p \neq 3q+2$) we have $\xi^A\nabla_{AA'}\eta^{A'} = 0$, from which it follows according to lemma 2 that ξ^A is a repeated principal spinor of the Weyl spinor. □

As was pointed out by Sommers, the Goldberg-Sachs theorem (and its generalization due to Robinson & Schild and to Kundt & Thompson) follows immediately as a consequence of Theorem 1 (cf. Penrose & Rindler 1986, §7.3).

Comment (1994). These results have in the meanwhile appeared in Hughston (1987). It would be an interesting challenge to come up with yet a shorter proof. Note in particular that whereas Robinson's theorem can easily be geometrized (cf. §II.2.8) the Goldberg-Sachs result is more refractory and peculiar to four dimensions.

References

Goldberg, J.N. & Sachs, R.K. (1962) A theorem on Petrov types, *Acta. Phys. Polonica*, Suppl. **22**, p. 13–23.

Hughston, L.P. (1987) Remarks on Sommers theorem, *Class. Quant. Grav.*, **4**, 1809-1811.

Kundt, W. & Thompson, A.H. (1962) Le tenseur de Weyl et une congruence associée de géodésiques isotropes sans distorsion, *C.R. Acad. Sci. Paris* **254**, p. 4257–4259.

Penrose, R. & Rindler, W. (1986) *Spinors and Space-Time, vol. II: Spinor and Twistor Methods in Space-Time Geometry* (Cambridge University Press).

Robinson, I. & Schild, A. (1963) Generalisation of a theorem by Goldberg and Sachs, *J. Math. Phys.* **4**, p. 484–489.

Sommers, P.D. (1976) Properties of shear-free congruences of null geodesics, *Proc. Roy. Soc. Lond.*, **A349**, p. 309–318.

§II.2.8 **A Twistor Description of Null Self-dual Maxwell Fields** *by M.G.Eastwood* (TN 20, September 1985)

Robinson's theorem: After complexification, Robinson's theorem (Robinson, 1961, an important motivation in the birth of twistors) states that if M is a complex conformal manifold and $E \subset TM$ is an integrable distribution of α-planes, then E may be locally defined by a closed self-dual 2-form, i.e. a self-dual Maxwell field. This can be proved with the aid of spinors (e.g. Sommers 1976): E may be defined by a spinor field $\alpha^{A'}$ and integrability is the condition $\alpha^{A'}\alpha^{B'}\nabla_{AA'}\alpha_{B'} = 0$ or, equivalently, $\alpha^{A'}\nabla_{AA'}\alpha^{B'} = \lambda_A\alpha^{B'}$. A curvature computation then shows that $\alpha^{A'}\nabla_{AA'}\psi + \nabla_{AA'}\alpha^{A'} + \lambda_A = 0$ may be locally solved for ψ whence $\nabla_{AA'}(e^{\psi}\alpha^{A'}\alpha^{B'}) = 0$, as required. Without spinors, however, the theorem is immediate from Frobenius' integrability criterion: a simple 2-form defining a distribution of α-planes is, by definition, necessarily self-dual and the integrability of E is equivalent to being able to choose this form to be closed. In fact, more precisely, suppose E is integrable and let S denote the space of leaves (locally). Then there is a 1–1 correspondence between 2-forms on S and closed 2-forms on M defining E (given by pull-back under $\pi : M \to S$). Finally, since a null ($\equiv$ simple) 2-form defines a plane distribution it follows that a null self-dual Maxwell field is exactly specified by E, a congruence of α-surfaces, and a 2-form ω on S, the 2-dimensional parameter space for the congruence.

A twistor description: Suppose now that M is conformally right flat so that it has a 3-dimensional twistor space T parameterizing the α-surfaces. Then a self-dual null Maxwell field gives S as above, a submanifold of T:

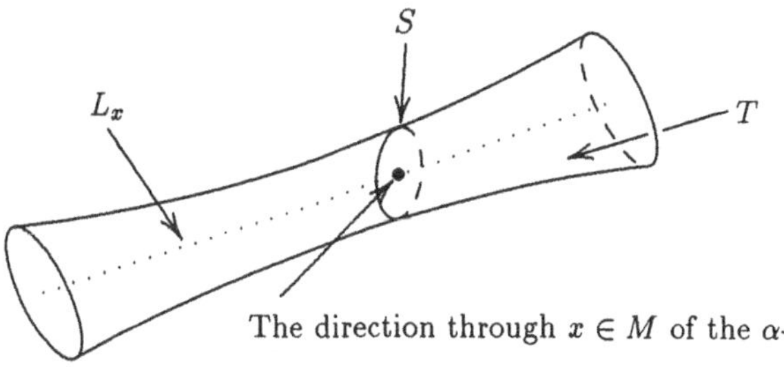

As noted above, the Maxwell field specifies a 2-form ω on S. The usual Penrose transform, however, identifies the space of all self-dual Maxwell fields with the cohomology $H^1(T,\kappa)$ where κ is the canonical bundle Ω^3. Hence there should be a natural homomorphism

$$\Gamma(S,\Omega^2) \longrightarrow H^1(T,\kappa). \qquad (*)$$

It is somewhat easier to see what is going on in 'real' twistor theory:

'Real' twistors: The usual twistor correspondence between $Gr_2(\mathbb{C}^4)$ and $\mathbb{CP}_3$ has a real form, namely the correspondence between $\mathbb{RM} \equiv Gr_2(\mathbb{R}^4)$ and $\mathbb{RP} \equiv \mathbb{RP}_3$. $\mathbb{RM}$ has a conformal metric of signature $(+,+,-,-)$ and $\mathbb{RP}$ is the space of one family of totally null 2-planes in $\mathbb{RM}$, the "α -planes". Victor Guillemin has recently been investigating this 'black-and-white' version of twistor theory and it seems that most twistor constructions have an (often simpler) black-and-white analogue. The Penrose transform is replaced by the Gelfand-Radon transform e.g.

$$\Gamma(\mathbb{RP},\kappa) \xrightarrow{\simeq} \{\omega \in \Gamma(\mathbb{RM},\Omega^2_+) \text{ s.t. } d\omega = 0\}$$

which is simpler in that functions have replaced cohomology. Although the transform can be defined locally (i.e. in a neighbourhood of a line in $\mathbb{RP}$), it is then never an isomorphism. V. Guillemin has shown, however, that globally it is an isomorphism (for all helicities and so on) and suggests that the crucial property of $\mathbb{RM}$ is that it is a Zoll manifold (all its null geodesics are closed). Suppose now that ω is a null self-dual Maxwell field on $\mathbb{RM}$ (it will have singularities but ignore such technicalities). Then, as for the holomorphic case, one obtains a surface $S \hookrightarrow \mathbb{RP}$ and a 2-form on it, which may also be denoted by ω. The analogue of $(*)$ is thus

$$\Gamma(S,\Omega^2) \longrightarrow \Gamma(\mathbb{RP},\kappa). \qquad \mathbb{R}(*)$$

A function f on $\mathbb{RP}$ may be integrated over S against ω. The linear functional

$$f \longrightarrow \int_S f\omega$$

therefore defines a distribution-valued 3-form supported on S. This is Guillemin's identification of $\mathbb{R}(*)$. S determines the singularities of ω.

Ordinary twistors: As a divisor, S gives rise to a line bundle ξ over T with a section $s \in \Gamma(T, \xi)$ defining S, and therefore an exact sequence:

$$0 \longrightarrow \xi^* \xrightarrow{\times s} \mathcal{O}_T \longrightarrow \mathcal{O}_S \longrightarrow 0. \qquad (**)$$

Moreover, $\xi|_S = N$ the normal bundle of S in T and the sequence may therefore be rewritten:

$$0 \longrightarrow \mathcal{O}_T \longrightarrow \xi \longrightarrow N \longrightarrow 0.$$

Finally, noting that $N \otimes \kappa|_S = \Omega^2$ on S, it may be rewritten

$$0 \longrightarrow \kappa \longrightarrow \kappa \otimes \xi \longrightarrow \Omega^2_S \longrightarrow 0$$

and $(*)$ is given by the connecting homomorphism of the corresponding long exact sequence:

$$\Gamma(S, \Omega^2) \longrightarrow H^1(T, \kappa) \xrightarrow{\times s} H^1(T, \kappa \otimes \xi).$$

Note also that the null fields obtained in this way are exactly those annihilated by $\times s$: an invariant way of saying that the cohomology class has a 'simple pole' along S [Penrose's (1969) original identification of null fields].

Googly photon? : The usual improved way to regard a Maxwell field is as a connection on a line-bundle. A null self-dual Maxwell field is then flat along an integrable distribution of α-planes and the covariant constant sections along the leaves give rise to a line-bundle on S. The residual information of the original connection (normal to the leaves) equips this line bundle on S with a connection of its own. This is surely the googly photon for a null self dual field. This may give clues as to what to do about the general googly photon and also suggests trying to combine this construction with the usual Ward twisted photon construction in an attempt to describe 'half algebraically special' Maxwell fields. A similar construction for the non-linear version (i.e. Yang-Mills bundles) gives vector bundles with connection flat along a congruence of α-surfaces. These are very special indeed (stronger than null and self-dual).

Higher dimensions: The Frobenius approach to the Robinson theorem evidently extends to higher (even) dimensions. L.P. Hughston has also shown (II.2.3 and lecture, Oxford 30 April 1985) how a spinor method works in dimension 6 and presumably a general spinor proof is available. In any case, a simple closed self-dual n-form on a conformal $2n$-fold gives rise to a congruence of α-n-folds (by definition orthogonal to self-dual n-forms) parameterized by an n-dimensional space S, and an n-form ω on S. Conversely, every such arises in this way. Higher dimensional twistors only exist in an obvious way for conformally flat space Q_{2n}, the complex quadric of dimenion $2n$. This has a twistor space Z_n consisting of one system of $\mathbb{P}_n$'s lying therein (the α-$\mathbb{P}_n$'s). Z_n has dimension $n(n+1)/2$. Letting M denote an open subset of Q_n and T the corresponding subset Z_n, then (subject to mild topological restrictions on M) M.F. Atiyah has shown (lecture, Oxford, 7 November 1983) that

$$H^{n(n-1)/2}(T,\kappa) \xrightarrow{\simeq} \{\omega \in \Gamma(M,\Omega^n_+) \text{ s.t. } d\omega = 0\}.$$

The homomorphism

$$\Gamma(S,\Omega^n) \longrightarrow H^{n(n-1)/2}(T,\kappa)$$

is given by composing a series of connecting homomorphisms or by a spectral sequence construction induced by the appropriate Koszul complex instead of (**). There is a corresponding black-and-white version for real split quadrics. Note that $n(n-1)/2$ is the codimension of S in T as one would expect.

Many thanks to Victor Guillemin for much interesting conversation.

References

Penrose, R. (1969) Solutions of the zero-rest-mass equations, J. Math. Phys. **10**, p. 38–39.

Robinson, I. (1961) Null electromagnetic fields, J. Math. Phys. **2**, p. 290–291.

Sommers, P.D. (1976) Properties of shear-free congruences of null geodesics, Proc. Roy. Soc. Lond., **A349**, p. 309–318.

§II.2.9 **A conformally invariant connection and the space of leaves of a shear free congruence** *by T.N.Bailey* (TN 26, March 1988)

Introduction. This is a report on work in progress, studying the structure of the complex surface which is the space of leaves of a (complexified) shear free congruence. I will show below that in conformal vacuum space-times, the surface has the first formal neighbourhood of an embedding in a complex three manifold (which in the flat space would be dual projective twistor space).

In order to describe this structure, I will first show that a conformal complex space-time with two spinor fields has a natural conformally invariant connection, which is essentially given by Penrose's 'conformally invariant edth and thorn operators'. This construction seems to have some geometric interest in its own right.

It is hoped that these these structures will help to explain the separation of various equations in the Kerr metric, and there may be other applications.

The conformally invariant connection. Let $\mathcal{M}$ be a complex conformal space-time, with two independent spinor fields o^A and ι^A, defined up to scale. Equivalently we have a splitting

$$\mathcal{O}^A = O \oplus I \tag{1}$$

of the spin bundle. Assume also that we are given an identification of the primed and unprimed conformal weights

$$\mathcal{O}[-1] \stackrel{\text{def}}{=} \mathcal{O}_{[AB]} \cong \mathcal{O}_{[A'B']}.$$

This is equivalent to allowing conformal transformations only of the form

$$\epsilon_{AB} \mapsto \Omega\,\epsilon_{AB} \qquad\qquad \epsilon_{A'B'} \mapsto \Omega\,\epsilon_{A'B'}$$

which is a natural condition if $\mathcal{M}$ is the complexification of a real space-time. Given a metric in the conformal class, the splitting in equation (1) allows us to define a one-form

$$Q_a := -2\,o^{(B}\iota^{C)}\partial_{A'B}\,(o_{(A}\iota_{C)}) = \rho' l_a + \rho n_a - \tau' m_a - \tau\overline{m}_a$$

where ∂_a is the metric connection, and we adopt the convention that $o_A\iota^A = 1$ whenever a particular metric has been chosen. Under conformal transformation

$$Q_a \mapsto Q_a - \Upsilon_a \qquad \text{where} \qquad \Upsilon_a = \Omega^{-1}\partial_a\Omega. \tag{2}$$

The significance of Q_a is that it enables us to split the *local twistor* bundle as a direct sum.

Recall the local twistor exact sequence

$$\begin{array}{ccccccc} 0 \longrightarrow & \mathcal{O}_{A'} & \longrightarrow & \mathcal{O}^\alpha & \longrightarrow & \mathcal{O}^A & \longrightarrow 0 \\ & \pi_{A'} & \longmapsto & (0, \pi_{A'}) & & & \\ & & & (\omega^A, \pi_{A'}) & \longmapsto & \omega^A & \end{array}$$

and the conformal transformation rule

$$\omega^A \mapsto \omega^A \qquad\qquad \pi_{A'} \mapsto \pi_{A'} + \mathrm{i}\Upsilon_a\omega^A.$$

If we set

$$\alpha_{A'} = \pi_{A'} + \mathrm{i}Q_a\omega^A$$

then from equation (2) there is a conformally invariant splitting

$$\begin{array}{ccc} \mathcal{O}^\alpha & \xrightarrow{\simeq} & \mathcal{O}^A \oplus \mathcal{O}_{A'} \\ (\omega^A, \pi_{A'}) & \longmapsto & \omega^A \oplus \alpha_{A'} \end{array} \tag{3}$$

of $\mathcal{O}^\alpha$ and I will use the 'split co-ordinates' $(\omega^A, \alpha_{A'})$ henceforth.

The *local twistor connection* splits to give connections, which I will denote by ∇_a, on the various spin bundles. A brief calculation shows these to be

$$\begin{array}{llllll} \mathcal{O}^A & : \nabla_b \mu^A & = & \partial_b \mu^A & + & \epsilon_B{}^A Q_{CB'} \mu^C \\ \mathcal{O}^{A'} & : \nabla_b \mu^{A'} & = & \partial_b \mu^{A'} & + & \epsilon_{B'}{}^{A'} Q_{BC'} \mu^{C'} \\ \mathcal{O}_A & : \nabla_b \mu_A & = & \partial_b \mu_A & - & Q_{AB'} \mu_B \\ \mathcal{O}_{A'} & : \nabla_b \mu_{A'} & = & \partial_b \mu_{A'} & - & Q_{BA'} \mu_{B'} \\ \mathcal{O}_{[AC]} & : \nabla_b \nu_{AC} & = & \partial_b \nu_{AC} & - & Q_b \nu_{AC} \end{array}$$

If $Z^\alpha = (\omega^A, \alpha_{A'})$ is a local twistor, we can write the local twistor connection as

$$\nabla_b Z^\alpha = (\nabla_b \omega^A + \mathrm{i}\epsilon_B{}^A \alpha_{B'}, \nabla_b \alpha_{A'} + \mathrm{i} D_{ab} \omega^A) \tag{4}$$

where D_{ab} is a conformally invariant modification of P_{ab} defined by

$$D_{ab} = P_{ab} - \partial_b Q_a + Q_{AB'} Q_{BA'}.$$

(For a definition and discussion of the modified curvature spinor $P_{AA'BB'}$, see Penrose & Rindler (1986) §6.8 or §II.2.1.)

The splitting in equation (1) allows us to define the bundles

$$\langle -r', -r \rangle := O^{r'} \otimes I^r$$

(note that $\langle 1, 1 \rangle = [1]$). The connection ∇_a can be projected on to these. For example, if λ^A is a section of $\langle -1, 0 \rangle$, so that $\lambda^A o_A = 0$,

$$\lambda^A \longmapsto -o^A \iota_C \nabla_b \lambda^C$$

is a connection, and its components are given by 'conformally invariant edth and thorn', in just the same way as the same expression with the metric connection ∂_b replacing ∇_b has components that can be computed with ordinary edth and thorn.

Since ∇_a agrees with ∂_a if you form any of the well known conformally invariant parts of the metric connection, there is scope here for producing a complete 'conformally invariant GHP formalism'. The expressions which arise as curvatures when one commutes conformal edths and thorns are components of D_{ab}. The geometrical significance of these connections will be discussed in a later section.

Shear free congruences in Minkowski space. Before starting on the general case, I will review the situation in flat space-time. In real Minkowski space, a shear free congruence of null geodesics (hereafter SFR) is given by a spinor field satisfying

$$o^A o^B \partial_a o_B = 0 \tag{5}$$

If o_A is analytic, it can be complexified, and it then determines a distribution of β-planes. This distribution is integrable, and so gives a foliation of Minkowski space by complex surfaces, precisely when o_A is shear free. The space of leaves S of this foliation is the hypersurface in dual projective twistor space $\mathbf{P}^*$, which describes the congruence, according to Kerr's Theorem.

The surface S inherits some structure from its embedding, in particular there is the tangent bundle of $\mathbf{P}^*$, the *normal bundle sequence*, and the restrictions of the line bundles $\mathcal{O}(n)$. The analysis in the accompanying article, §II.2.11, shows how massless fields of various orders along the congruence are isomorphic to sections of sheaves on S. I will now describe how this generalises to curved space.

SFRs in curved space-times. In a general space-time, an SFR is still given by a solution of equation (5), and gives a foliation in the complexification. The space of leaves still defines a complex surface S, but there is in general no twistor space in which it is embedded.

The SFR defines a Maxwell field, which in Minkowski space is the Ward transform of the line bundle defined by S considered as a divisor. This follows from the fact that equation (5) is equivalent to the existence of a one-form Φ_a with

$$\partial_{A'(A} o_{B)} = \Phi_{A'(A} o_{B)},$$

and it is easy to see that Φ_a has precisely the freedom to be the potential for a Maxwell field. The left-handed part $\phi_{AB} = \partial_{A'(A} \Phi^{A'}_{B)}$ satisfies $\Psi_{ABCD} o^D = -\phi_{(AB} o_{C)}$, and so vanishes as expected in a conformally flat space-time.[1]

The structures I shall describe on S only exist under certain conditions. In particular, I will say that the SFR o_A in the space-time $\mathcal{M}$ satisfies the *Goldberg–Sachs condition* (hereafter GS) if

$$o^A o^B o^C \Psi_{ABCD} = 0.$$

We assume the GS condition holds henceforth, since no significant part of the structure on S seems to exist otherwise. The Goldberg-Sachs Theorem implies that the GS condition is equivalent to $o^A o^B o^C \partial^D_{D'} \Psi_{ABCD} = 0$ and it is therefore satisfied by all conformally vacuum space-times. To construct bundles on S, we make use of ∇_a, the conformally invariant connection. First choose a

1 An SFR is thus a charged twistor coupled to its own canonically defined Maxwell field.

spinor direction ι_A to complement the SFR o_A, and deduce from the SFR and GS conditions that on all the bundles $\langle r', r\rangle$ and $\mathcal{O}^{A'}$, the part $o^A\nabla_a$ of the connection that differentiates up the leaves of the foliation is both independent of the choice of ι_A and flat.[2] We can thus define line bundles $\langle r', r\rangle_S$ and a rank two vector bundle $\mathcal{O}(S)^{A'}$ over S, whose sections are *by definition* sections of the corresponding bundle on $\mathcal{M}$ with vanishing conformal derivitive up the foliation.

The dual local twistor bundle also defines a vector bundle on S. We have an injection of the spinors proportional to o_A into $\mathcal{O}_\alpha$

$$0 \longrightarrow \langle 0,1\rangle \longrightarrow \mathcal{O}_\alpha \longrightarrow E \longrightarrow 0$$

defining the quotient E. The part $o^A\nabla_a$ of the local twistor connection preserves $\langle 0,1\rangle$ and hence is well defined on E. Furthermore, it is flat on the leaves and so defines a rank three vector bundle $\mathcal{E}$ on S.

Sections of $\mathcal{E}$ can be realised as spinor fields $\xi^{A'}$ satisfying a tangential twistor equation[3]

$$o^A\nabla_A{}^{(A'}\xi^{B')} = 0$$

and given that sections of $\mathcal{O}(S)^{A'}$ are spinor fields satisfying

$$o^A\nabla_{AA'}\xi^{B'} = 0$$

we get an injection $\mathcal{O}(S)^{A'} \to \mathcal{E}$ which extends to give a short exact sequence

$$0 \longrightarrow \mathcal{O}(S)^{A'} \longrightarrow \mathcal{E} \longrightarrow \langle 1,0\rangle \longrightarrow 0$$

given, in terms of equations, by

$$\begin{array}{ccccc} o^A\nabla_{AA'}\xi^{B'} = 0 & \longmapsto & o^A\nabla_A{}^{(A'}\xi^{B')} = 0 & \longmapsto & \begin{pmatrix} o^Ao^B\nabla_{BB'}\eta_A = 0 \\ \iota^A\eta_A = 0 \end{pmatrix} \\ & & \xi^{A'} & \longmapsto & \iota_A o^B\nabla_{BB'}\xi^{B'} \end{array}$$

If $\mu^{A'}$ is a section of $\mathcal{O}^{A'}\langle 0,-1\rangle$, a calculation reveals that the condition $o^A\nabla_a\mu^{B'} = 0$ is what is required[4] to make $\iota^A\mu^{A'}$ a connecting vector to a nearby leaf of the foliation. Thus, $\mathcal{O}(S)^{A'}\langle 0,-1\rangle_S$ can be identified with the tangent bundle $T(S)$ of S. The exact sequence above can be tensored through by $\langle 0,-1\rangle_S$ to give what in flat space would be the *normal bundle sequence* of S

$$0 \longrightarrow T(S) \longrightarrow \mathcal{E}\langle 0,-1\rangle_S \longrightarrow \langle 1,-1\rangle_S \longrightarrow 0.$$

---

[2] It is helpful to note that $o^A Q_a$ is independent of ι_A if o^A is SFR.

[3] To see this, note that GS and SFR imply $o^Ao^BD_{ab} = 0$ and use the conjugate version of equation (4). When writing down the splitting and connection on the dual local twistors, simply write down the conjugate *pretending that D_{ab} and Q_a are real.*

[4] The connection here is the tensor product of the conformally invariant ones on the factors.

If one is given a hypersurface in a complex manifold, then knowing the normal bundle sequence is equivalent to knowing the *first formal neighbourhood* of the embedding. I will now briefly describe how one can realise the first formal neighbourhood of an embedding of S directly.

The spinor field o_A defines a natural embedding of the space-time $\mathcal{M}$ in the *projective spin bundle* $\mathbf{P}\mathcal{O}_A$. Now realise S by choosing a two-surface $\tilde{S}$ transverse to the foliation, and note that $\tilde{S}$ has a natural embedding in the restriction of $\mathbf{P}\mathcal{O}_A$. The *first formal neighbourhood* of this embedding is independent of the choice of $\tilde{S}$, and so defines a first formal neighbourhood sheaf $\mathcal{O}^{(1)}$ on S.

In slightly more detail; recall that $\mathbf{P}\mathcal{O}_A$ has a naturally defined differential operator $\pi^A\partial_a$ which defines a two-plane distribution, the integral surfaces of which (if it has any) are lifts of β-surfaces.

When o_A is an SFR, there is a two complex parameter family of β-surfaces parametrised by S, and functions f on $\mathbf{P}\mathcal{O}_A$ which obey $\pi^A\partial_a f = 0$ *on the lift of* $\mathcal{M}$ are precisely functions on S.

A calculation shows that, given the GS condition, there are two functions of two complex variables worth of functions g on $\mathbf{P}\mathcal{O}_A$ that obey $\pi^A\partial_a g = 0$ *to first order in a neighbourhood of the lift of* $\mathcal{M}$. These form the formal neighbourhood sheaf $\mathcal{O}^{(1)}$ on S.

In terms of the conformally invariant connections, a function on the first formal neighbourhood of the lift of $\mathcal{M}$ can be written

$$g(x, \pi_A) = f(x) + \iota^A \chi_A{}^B \pi_B \qquad \text{where} \qquad o^A \chi_A{}^B = 0 = \chi_A{}^B o_B$$

If the spinor field $\chi_A{}^B$ satisfies

$$\nabla_{BA'}\chi_A{}^B = \nabla_{AA'} f$$

then it defines a section of $\mathcal{O}^{(1)}$.

Massless fields. One result of this analysis is a minor generalisation of Robinson's Theorem, which states that if o_A is an SFR, then, for each helicity, there is precisely one holomorphic function of two complex variables worth of left handed massless fields null along it. If the field has n indicees, then remembering that it has conformal weight -1, it is easy to check that these fields are in one to one correspondence with sections over S of $\langle 1, n+1\rangle_S$.

In my accompanying article §II.2.11 I show how *in flat space* fields of various orders along o_A correspond to sections of sheaves over S. Provided, as usual, that the GS condition holds, it turns out that sections of the formal neighbourhood sheaf $\mathcal{O}^{(1)} \otimes \langle 1, 1\rangle_S$ on S do give left handed Maxwell fields which have a principal null direction along the congruence. Thus there are two holomorphic functions of two complex variables worth of such things, just as in the flat case.

Apart from that case, however, more severe curvature restrictions appear. To get three functions worth of order three Maxwell fields one requires $o^A o^B \Psi_{ABCD} = 0$ in which case it seems that S has a second formal neighbourhood sheaf.

Killing spinors. Suppose $\mathcal{M}$ admits a *Killing spinor*, and choose o_A and ι_A to be along its principal null directions. The Killing spinor equation

$$\partial_{A'}{}^{(A}\omega^{BC)} = 0$$

then implies that both o_A and ι_A are SFRs. The remaining parts of the equation reduce to solving $\nabla_a \omega = 0$ where ω is a section of $\langle 1, 1\rangle$. This is only possible if the conformally invariant connection on $\langle 1, 1\rangle$ is flat, which implies

$$\partial_{[a}Q_{b]} = 0$$

This has a number of consequences. Firstly, it provides an isomorphism $\langle 1, 1\rangle \cong \langle 0, 0\rangle$ which carries over to S, thereby giving a natural trivialisation of $\langle 1, 1\rangle_S$. Secondly, the fact that Q_a is closed means that locally it is exact, and equation (2) shows that it can thus be made to vanish by a conformal transformation. In the special metric thus constructed, *all* the curvature information is contained in the single line bundle $\langle 1, 0\rangle$ and its (conformally invariant) connection; cf. Jeffryes (1984). Further work is in progress on all this, since it seems likely that, combined with the ideas in the next section, it will be possible to explain the separation of various differential equations in the Kerr solution.

Geometrical significance. To finish, I will mention some ideas due to R.Penrose and K.P.Tod which I have just started to follow up in collaboration with M.A.Singer. The conformally invariant connection constructed above is an example of a unique connection determined by a geometrical structure, and the structure one has (in the complex space-time) seems to be that which would be obtained on the complexification of a real four-manifold X with an almost complex structure $J_a{}^b$ and a compatible conformal Hermitian metric. The eigenspaces of $J_a{}^b$ are the two-plane distributions defined by o_A and ι_A so that

$$J_a{}^b = \mathrm{i}(o_A \iota^B + \iota_A o^B)\epsilon_{A'}{}^{B'}$$

The almost complex structure will be integrable when both o_A and ι_A are SFRs. Further, the suggestion is that the existence of a Killing spinor is equivalent to the Kähler condition on the Hermitian metric. This seems very likely since something very similar has been given by Flaherty (1976), whose view-point is somewhat different. I would like to thank M.A.Singer, R.Penrose, and K.P.Tod for discussions and suggestions.

Comment (1994). This work eventually became Bailey (1991a) and Bailey (1991b).

References

Bailey, T.N. (1991) Complexified conformal almost Hermitian structures and the conformally invariant edth and thorn operators, Class. Quantum Grav. **8**, 1–4.

Bailey, T.N. (1991) The space of leaves of a shear-free congruence, multipole expansions and Robinson's Theorem, J. Math. Phys. **32**, 1465–1469.

Flaherty, E.J., Jr. (1976) *Hermitian and Kählerian geometry in relativity*, Lecture Notes in Physics **46** (Springer Verlag, Berlin).

Jeffryes, B.P. (1984) Space-times with two-index Killing spinors, Proc. Roy. Soc. London **A392**, p. 323–341.

Penrose, R. & Rindler, W. (1986) *Spinors and Space-Time, vol. II: Spinor and Twistor Methods in Space-Time Geometry* (Cambridge University Press).

§II.2.10 **A conformally invariant connection** *by T.N.Bailey* (TN 27, December 1988)

This note is a postscript to the last section of my article §II.2.9 on the conformally invariant connection associated to a direct sum decomposition of one of the spin bundles. The general result lying behind the observations in that article, stated for convenience in the holomorphic category, is as follows:

THEOREM. *Let $\mathcal{M}$ be a complex conformal manifold with conformal metric g_{ab} and a given tensor field $J_a{}^c$ with $J_{(ab)} = 0$ and $J_a{}^c J_c{}^b = -\delta_a{}^b$. Then there exists a unique torsion-free connection ∇_a satisfying $\nabla_a J_b{}^a = 0$ and $\nabla_a g_{ab} = X_a g_{bc}$ for some X_a. The second condition is simply that the conformal metric is preserved.*

If one is given a direct sum decomposition of $\mathcal{O}^A$ in a complex space-time then

$$J_a{}^b = \mathrm{i}(o_A \iota^B + \iota_A o^B)\epsilon_{A'}{}^{B'},$$

where o_A, ι_A constitute a spin-frame defining the decomposition, satisfies the above conditions and the resulting connection is given in components by R. Penrose's 'conformally invariant edth and thorn' operators. The significance of the rather strange condition on the derivitive of $J_a{}^b$ which defines the connection is unclear, and work continues on the use of this connection in type D conformal space-times and related areas. Thanks to M.G.Eastwood and M.A.Singer.

§II.2.11 **Relative cohomology power series, Robinson's Theorem and multipole expansions** *by T.N.Bailey* (TN 26, March 1988)

Introduction. In my original articles on the twistor description of fields with sources on a world-line (see §I.6.6 and §I.6.8 and Bailey 1985) I gave some expressions for 'multipoles' based on a world-line. In this note, I will show how a first cohomology class, relative to a hypersurface, can be expanded in a sort of 'power series', which seems to be the twistor version of the multipole expansion. The power series also gives a precise version of the twistor description of algebraically special fields.

The relative cohomology power series. Let S be a hypersurface in a complex manifold X, and let $\mathcal{F}$ be a locally free sheaf of $\mathcal{O}_X$ modules on X. The *relative cohomology group* $H^1_S(X,\mathcal{F})$ can be described by a relative Čech cocycle, but a good intuitive picture is as follows: Choose an open cover U_i of a neighbourhood of S in X; then a representative is given by a set f_i of sections of $\mathcal{F}$ over U_i that 'blow up' on S, with the restriction that $f_i - f_j$ is holomorphic on *all* of $U_i \cap U_j$. The freedom in each f_i is the addition of a holomorphic section of $\mathcal{F}$.

Now let g_i be *defining functions* for S, then one might try and expand the relative class defined by the f_i as a power series

$$f_i = \frac{f_i^{(1)}}{g_i} + \frac{f_i^{(2)}}{g_i^2} + \cdots \frac{f_i^{(n)}}{g_i^n} + \cdots \tag{1}$$

To understand this we need the *divisor bundle* L of S, which is defined to be the line bundle with transition functions g_i/g_j on $U_i \cap U_j$. The functions g_i then give a *distinguished section* s of L which has a simple zero on S. The section s gives us a map

$$s^k : \mathcal{F} \longrightarrow \mathcal{F} \otimes L^k$$

which induces a map on the relative cohomology.

DEFINITION 1. *The k-th order relative cohomology $H^1_S(X,\mathcal{F};k)$ is defined by the exactness of*

$$0 \longrightarrow H^1_S(X,\mathcal{F};k) \longrightarrow H^1_S(X,\mathcal{F}) \xrightarrow{\times s^k} H^1_S(X,\mathcal{F}\otimes L^k)$$

The k-th order cohomology is thus the part which has a pole of order k or less on S, and it therefore corresponds to the first k terms in equation (1) above.

If $\mathcal{E}$ is a sheaf on X, and $\mathcal{I}^{(p)}\mathcal{E}$ is the ideal of sections of $\mathcal{E}$ which vanish to p-th order on S we can define the *p-th formal neighbourhood sheaf* $(\mathcal{E})^{(p)}$ by the short exact sequence

$$0 \longrightarrow \mathcal{I}^{(p+1)}\mathcal{E} \longrightarrow \mathcal{E} \longrightarrow (\mathcal{E})^{(p)} \longrightarrow 0 \tag{2}$$

so that $(\mathcal{E})^{(0)}$ is just $\mathcal{E}$ restricted to S.

LEMMA 1. *There is a natural isomorphism*

$$H^1_S(X, \mathcal{F}; k) \cong \Gamma(S, (\mathcal{F} \otimes L^k)^{(k-1)}).$$

The proof is simply to observe that in equation (1) above, the $f_i^{(k)}$ must give a section of $\mathcal{F} \otimes L^k$ with the freedom as given by equation (2).

Thus we have strictly a *filtration* of the relative cohomology (rather than an infinite direct sum), with the quotient at each stage given by the exact sequence

$$0 \longrightarrow \Gamma(S, (\mathcal{F} \otimes L^{k-1})^{(k-2)}) \xrightarrow{\times s} \Gamma(S, (\mathcal{F} \otimes L^k)^{(k-1)} \longrightarrow \Gamma(S, \mathcal{F} \otimes L^k) \longrightarrow 0.$$

Algebraically special fields. The above analysis can be applied when S is a hypersurface in a region X in projective twistor space, corresponding to a shear free congruence. We can define cohomology of order k on S just as for the relative case, and we will say that a right handed massless field is of order k on the congruence if its twistor function is in $H^1(X, \mathcal{O}(-n-2); k)$. Thus, order 1 means null, order n means the field has a principal null direction. along the congruence, and higher orders correspond to certain differential relations between the field and the congruence.

If one writes down the commutative diagram whose rows are the relative cohomology sequences, and whose columns are induced by $s^k : \mathcal{F} \to \mathcal{F} \otimes L^k$, it is easy to see that if $H^1(X, \mathcal{F}) = 0$ then there is an exact sequence

$$0 \longrightarrow \frac{\Gamma(X, \mathcal{F} \otimes L^k)}{\Gamma(X, \mathcal{F})} \longrightarrow H^1_S(X, \mathcal{F}; k) \longrightarrow H^1(X, \mathcal{F}; k) \longrightarrow 0$$

Since L has the section s which has a simple zero on S, which intersects every line in X exactly once, we can write $L = M(1)$ where M is a line bundle trivial on every line in X (M is the Ward bundle of the 'Maxwell field of the congruence'—see my accompanying article §II.2.9). Thus if $k < n+2$

$$\Gamma(X, L^k(-n-2)) = \Gamma(X, M^k(k-n-2)) = 0$$

and so

$$H^1_S(X, \mathcal{O}(-n-2); k) \cong H^1(X, \mathcal{O}(-n-2); k); \quad k < n+2.$$

The result of all this is a statement of the (generalised) flat space Robinson Theorem: The space of helicity $n/2$ right handed massless fields of order k ($k < n+2$) along the congruence is isomorphic to $\Gamma(S, (L^k(-n-2))^{(k-1)})$. This is precisely the 'k holomorphic functions of two complex variables' described by Penrose & Rindler (1986), p. 206.

The particular case where $k = 1$ and $n = 2$ was examined by M.G.Eastwood in §II.2.8. We get that these fields are given by sections of $L(-4)$ over S, but $\mathcal{O}(-4) = \Omega^3$ and L restricted to S is just the normal bundle. Thus $L(-4) = \Omega^2_S$, we get an isomorphism of the null right handed Maxwell fields with holomorphic 2-forms on S.

The null Maxwell fields inject into the order 2 fields, and give a quotient sheaf

$$0 \longrightarrow \Gamma(S, L(-4)) \longrightarrow \Gamma(S, (L^2(-4))^{(1)}) \longrightarrow \Gamma(S, L^2(-4)) \longrightarrow 0$$

The quotient corresponds in space-time to neutrino fields of order 1, coupled to the Maxwell field of the congruence. The map onto this group is 'helicity lowering', where the congruence is regarded as a charged twistor.

Multipole expansions. If S is the *ruled surface* corresponding to a world-line in Minkowski space, the first relative cohomology describes massess fields with sources on the world-line (and the Maxwell field of the congruence is the left handed part of the field of a unit charge on the world-line). We can use the analysis given above to get a filtration of these fields.

It seems that the first terms (e.g. order 2 for right handed Maxwell and order 3 for right handed gravity) give the fields with non-vanishing 'charges',and the remainder give an expansion in 'multipoles' where, for example, a 2^p- pole for a right handed helicity $n/2$ field is given by

$$\phi_{\underbrace{A'\dots K'}_{n}} = \oint \sigma^{\overbrace{A\dots N}^{n+p}\overbrace{P\dots S}^{p}} \dot{y}_P^{L'} \dots \dot{y}_S^{N'} \nabla_{AA'} \dots \nabla_{NN'} \frac{ds}{(x - y(s))^2}$$

where $\sigma^{A\dots S}$ is a totally symmetric spinor function of s, the proper time along the world-line $y^a(s)$. Under conformal transformations, a 2^p-pole gets mixed with lower ordered terms, which is what one might expect given that the twistor space expansion is not a direct sum.

There are still many details to be tidied up here, and further work is in progress. I am very grateful to Michael Singer for discussions about this work.

Comment (1994). Most of this work, and some related material, appear in Bailey (1991).

References

Bailey, T.N. (1985) Proc. Roy. Soc. **A397**, 143–155.

Bailey, T.N. (1991) J. Math. Phys. **32** 1465–9.

Penrose, R. & Rindler, W. (1986) *Spinors and space-time*, Vol. Vol. 2, Cambridge University Press.

§II.2.12 **Preferred parameters on curves in conformal manifolds**

by T.N.Bailey & M.G.Eastwood (TN 31, October 1990)

What follows are some observations made while considering whether one can construct analogues in conformal differential geometry of the 'pinched curvature, injectivity radii and minimising geodesics' ideas that lead (for example) to the Sphere Theorem in Riemannian differential geometry. These considerations are at a very early stage, but have led indirectly to an interesting (and as far as we know original) result about the distribution of curvature on a closed curve in $\mathbb{R}^n$.

Let γ be a *curve* (i.e. a smoothly immersed 1-dimensional submanifold) in an n-dimensional conformal Riemannian manifold. Choose a metric in which to work and parametrise γ by an arc-length parameter t and let U^a be the unit tangent vector. The *curvature* is given by $\kappa = \sqrt{A^b A_b}$, where $A^b = U^a \nabla_a U^b$.

As observed by Cartan (for an account in this language see Bailey & Eastwood 1990), such a curve has a natural *projective structure*—i.e. a family of preferred parameters related by fractional linear transformations under $SL(2, \mathbb{R})$. The function s on γ is a preferred parameter if it obeys the ('inhomogeneous Schwarzian') equation

$$(s')^{-1}s''' - \tfrac{3}{2}(s')^{-2}(s'')^2 = \tfrac{1}{2}\kappa^2 + P$$

where 'dash' denotes differentiation with respect to t and $P = P_{ab}U^aU^b$ where P_{ab} is the usual trace-modified multiple of the Ricci tensor. The simplified form of the equation when compared with the above reference is due to our use of an arc-length parameter.

In order to study the behaviour of these parameters we substitute $\xi = s(s')^{-1/2}$ since a brief calculation shows that these ξ*-parameters* obey the linear equation

$$\xi'' + (\tfrac{1}{4}\kappa^2 + \tfrac{1}{2}P)\xi = 0.$$

We define the *index* of a curve from A to B to be the number of zeroes (excluding the initial one) that the ξ-parameter with $\xi(A) = 0, \xi'(A) = 1$ has *before* reaching B. The location of, (and hence the number of) such zeroes is conformally invariant. As a first step towards investigating the properties of this index we have considered its behaviour on closed curves in $\mathbb{R}^n$. If γ is a (geometric) circle, it is easy to check that the first zero of any ξ-parameter occurs exactly at the starting point after one full traverse of the curve. As we see below, this characterises the circles among all closed curves.

We begin with a result (perhaps of interest in its own right) about the distribution of curvature on a closed curve.

PROPOSITION 1. *Let γ be a closed curve in $\mathbb{R}^n$ of length l, parametrised by arc-length t. Let κ be the curvature of γ. Then*

$$\int_0^l \kappa^2 \sin^2\left(\frac{\pi t}{l}\right) \mathrm{d}t \geqslant \frac{2\pi^2}{l}$$

with equality if and only if γ is a circle.

The proposition is proved by expanding the coordinates as functions of t in Fourier series and performing some essentially trivial manipulations.

PROPOSITION 2. *Let γ be a closed 1-dimensional submanifold of $\mathbb{R}^n$ which is not a geometrical circle. Then any closed curve which consists of traversing γ once from some chosen starting point has index at least 1.*

Proof. Let γ be of length l and consider the eigenvalue problem

$$\left(-\frac{\mathrm{d}^2}{\mathrm{d}t^2} - \tfrac{1}{4}\kappa^2\right)\xi = \lambda\xi, \qquad \xi(0) = \xi(l) = 0$$

on the interval $[0, l]$. Then the ξ-parameter with $\xi(0) = 0, \xi'(0) = 1$ will have a zero before $t = l$ if and only if zero is greater than the least eigenvalue of this problem. Since the operator on the left-hand side is bounded below we know that the least eigenvalue is always less than

$$\int_0^l \phi(t)\left(-\frac{\mathrm{d}^2}{\mathrm{d}t^2} - \tfrac{1}{4}\kappa^2\right)\phi(t)\mathrm{d}t$$

for any function the integral of whose square over $[0, l]$ is unity.

Taking $\phi(t) = \sqrt{2/l}\sin(\pi t/l)$ we see that a sufficient condition is that

$$\int_0^l \kappa^2 \sin^2\left(\frac{\pi t}{l}\right) \mathrm{d}t \geqslant \frac{2\pi^2}{l}.$$

The result then follows from Proposition 1. □

It is unclear at this stage whether these ideas (together perhaps with a study of the 'exponential map' for conformal circles) will lead to any interesting results on conformal manifolds. It would be interesting to know (for example) whether conformal circles are in any sense index-minimising curves. As a starting point however one can (easily) prove results such as:

PROPOSITION 3. *If M is a compact Riemannian conformal manifold such that there is a metric in the conformal class with pinched sectional curvatures*

$$\frac{2(n-1)}{4(n-2)k^2 + n} \leqslant K \leqslant 1$$

and $k \geqslant 1$ is an integer, then any two points can be joined by a curve of index less than k.

The authors thank John Baez for assistance with the proof of the second proposition.

References

Bailey, T.N. & Eastwood, M.G. (1990) Proc. AMS **108**, p. 215–221.

§II.2.13 **The Fefferman-Graham Conformal Invariant** *by M.G.Eastwood* (TN 20, September 1985)

In Eastwood & Rice (1987) it is shown that, on an arbitrary holomorphic spin 4-manifold, there are natural conformally invariant differential operators:

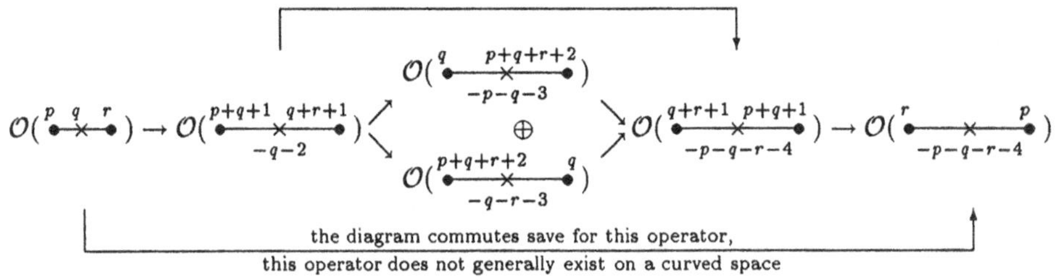

the diagram commutes save for this operator,
this operator does not generally exist on a curved space

(cf. R.J. Baston & M.G. Eastwood, §1.3.15) where

$$\mathcal{O}(\overset{p\ \ q\ \ r}{\bullet\!\!-\!\!\times\!\!-\!\!\bullet}) = \mathcal{O}_{\underbrace{(AB\ldots D)}_{p}\ \underbrace{(E'F'\ldots H')}_{r}}\{p+q+r\}.$$

As a typical example ($p=1, q=r=0$) with the notation of [3]:

$$\begin{array}{ccccccc} \overset{2\ -2\ 1}{\bullet\!\!-\!\!\times\!\!-\!\!\bullet} & = & \mathcal{O}_{(AB)C'}\{1\} & \to & \mathcal{O}_{(A'B'C')}\{-1\} & = & \overset{0\ -4\ 3}{\bullet\!\!-\!\!\times\!\!-\!\!\bullet} \\ & & \Cup & & \Cup & & \\ & & \phi_{ABC'} & \mapsto & \nabla^{A}_{(A'}\nabla^{B}_{B'}\phi_{C')AB} + \Phi^{AB}_{(A'B'}\phi_{C')AB} & & \end{array}$$

On (compactified) Minkowski space there are no others and the sequence (without the long arrow) is exact. On a curved space, except for the case $p=q=r=0$ (the deRham sequence), the composition of two consecutive operators involves (conformal) curvature. For example, composing

$$\overset{1\ \ 0\ \ 0}{\bullet\!\!-\!\!\times\!\!-\!\!\bullet} = \mathcal{O}_A\{1\} \xrightarrow{\nabla_{BB'}} \mathcal{O}_{(AB)B'} \xrightarrow{\nabla_C^{A'}} \mathcal{O}_{(ABC)} = \overset{3\ -3\ 0}{\bullet\!\!-\!\!\times\!\!-\!\!\bullet}$$

gives the anti-self-dual Weyl curvature $\phi^A \mapsto \Psi_{ABCD}\phi^D$. Since this is merely a tensor it acts more generally:

$$\overset{p\ \ q\ \ r}{\bullet\!\!-\!\!\times\!\!-\!\!\bullet} = \mathcal{O}^{\overbrace{(AB\ldots D)}^{p}\,\overbrace{(E'F'\ldots H')}^{r}}\{q\} \xrightarrow{\Psi_A{}^{EFG}} \mathcal{O}^{(BC\ldots G)(E'F'\ldots H')}\{q+3\} = \overset{p+2\ \ q-3\ \ \ r}{\bullet\!\!-\!\!\!-\!\!\times\!\!-\!\!\!-\!\!\bullet}$$

and a conformally invariant zeroth order operator is obtained even though it is not strictly speaking a composition of two in the list in Eastwood & Rice (1987). It is remarkable that the same thing seems to happen for higher order operators too provided a judicious adjustment of constants is made. Some heavy calculation is required to check this thoroughly. In this relaxed way an algebra of invariant operators may be generated by the "basic" ones (i.e. those in Eastwood & Rice 1987).

Example:

$$\begin{array}{ccc} \mathcal{O}_A\{1\} & \longrightarrow \mathcal{O}_{(AB)A'}\{1\} \longrightarrow & \mathcal{O}_{(B'C'D')}\{-1\} \\ \Cup & & \Cup \\ \phi_A & \longrightarrow & -\frac{1}{2}\tilde{\Psi}_{A'B'C'D'}\nabla^{AA'}\phi_A + (\nabla^{AA'}\tilde{\Psi}_{A'B'C'D'})\phi_A \end{array}$$

is an invariant operator obtained directly as a combination of basic ones but by altering constants it is clear that $\tilde{\Psi}_{A'B'C'D'}\nabla^{AA'}\phi_{AB\ldots D} - (w+1)(\nabla^{AA'}\tilde{\Psi}_{A'B'C'D'})\phi_{AB\ldots D}$ is conformally invariant when acting on $\phi_{AB\ldots D}$ of weight w. In particular, $\mathrm{i}[\tilde{\Psi}_{A'B'C'D'}\nabla^{AA'}\Psi_{ABCD} - \Psi_{ABCD}\nabla^{AA'}\tilde{\Psi}_{A'B'C'D'}]$ is a conformally invariant tensor (like but not equal to the du Plessis tensor). It is real on a real spacetime and gives a third order scalar invariant by taking discriminant for primed/unprimed spinor indices.

A more exotic example: Changing constants of $(p = q = 1, r = 0)$:

$$\mathcal{O}(\overset{4\ -4\ 1}{\bullet\!\!-\!\!\times\!\!-\!\!\bullet}) \to \mathcal{O}(\overset{2\ -6\ 3}{\times\!\!-\!\!\bullet\!\!-\!\!\bullet}) \to \mathcal{O}(\overset{0\ -6\ 1}{\bullet\!\!-\!\!\times\!\!-\!\!\bullet})$$

gives:

$$\begin{array}{ccccl} \mathcal{O}(\overset{4\ -4\ 0}{\bullet\!\!-\!\!\times\!\!-\!\!\bullet}) & = & \mathcal{O}_{(ABCD)} & \to & \mathcal{O}\{-6\} = \mathcal{O}(\overset{0\ -6\ 0}{\bullet\!\!-\!\!\times\!\!-\!\!\bullet}) \\ & & \Cup & & \Cup \\ & & \phi_{ABCD} & \mapsto & 2\Psi^{ABCD}\Box\phi_{ABCD} + 12(\nabla^{A'E}\Psi^{ABC}{}_E)(\nabla^D_{A'}\phi_{ABCD}) \\ & & & & \quad + (\nabla^{EE'}\Psi^{ABCD})(\nabla_{EE'}\phi_{ABCD}) \\ & & & & \quad + (2\Box\Psi^{ABCD} + 32\Lambda\Psi^{ABCD})\phi_{ABCD}, \end{array}$$

a conformally invariant operator! In particular, putting $\phi_{ABCD} = \Psi_{ABCD}$ gives:

$$\boxed{2\Box\Psi^2 + 12(\nabla^{A'E}\Psi^{ABC}{}_E)(\nabla^D_{A'}\Psi_{ABCD}) - 3(\nabla^{EE'}\Psi^{ABCD})(\nabla_{EE'}\Psi_{ABCD}) + 32\Lambda\Psi^2}$$

as a conformally invariant scalar (of weight -6). Save for adding a multiple of Ψ^3 this is the Fefferman-Graham fourth order invariant (Fefferman & Graham 1985).

Question. Do all conformally invariant operators or tensors arise in this fashion?

Thanks to Roger Penrose and Robert Bryant.

References

Eastwood, M.G. & Rice, J.W. (1987) Conformally invariant differential operators on Minkowski space and their curved analogues, *Comm. Math. Phys.* **109**, 207–228.

Fefferman, C. & Graham, C.R. (1985) Conformal invariants, in *Élie Cartan et les Mathématiques d'Aujourdui*, Astérisque, 95–116.

Penrose, R. & Rindler, W. (1984) *Spinors and space-time*, Vol. 1, Cambridge University Press.

§II.2.14 **On the Weights of Conformally Invariant Operators** *by M.G.Eastwood* (TN 24, September 1987)

Introduction: In Eastwood & Rice (1987) the *conformally invariant differential operators* on $\mathbb{M}$ (compactified complexified Minkowski space) were classified and it was shown that most admit an invariant *curved analogue* by the addition of suitable curvature correction terms ($\Box + R/6$ is the prototype). In this article, however, there is one vital point which was skipped over in the process of complexification. I am grateful to Robin Graham for drawing my attention to this omission and to Rice University for hospitality in February 1987 during which the point was cleared up as follows.

On a general (pseudo-)Riemannian manifold M one can consider rescaling the metric according to $g_{ab} \mapsto \hat{g}_{ab} = \Omega^2 g_{ab}$ for an arbitrary nowhere-vanishing function Ω known as a *conformal factor.* There are various reasons for rescaling by the *square* of a conformal factor rather than just insisting on a positive rescaling or rescaling by e^λ for some function λ, the main reason being that Ω has units of length rather than length^2. Also, in four dimensions, if M is spin then Ω rescales the spinor epsilons which is a pleasant feature: in general, and also in the complexification, rescaling the metric by Ω^2 is preferred since it corresponds to rescaling a spin structure by Ω. Another popular convention is $g_{ab} \mapsto \hat{g}_{ab} = u^{4/(n-2)} g_{ab}$ for $n = \dim M$ but this is specifically designed so that the Yamabe equation simplifies. Thus, for the purposes of this article, a *conformal density* of *weight* w is a "function" f which rescales according to $f \mapsto \hat{f} = \Omega^w f$ when $g_{ab} \mapsto \hat{g}_{ab} = \Omega^2 g_{ab}$. As usual, f may be equivalently regarded as a section of an appropriate line bundle. Similar comments apply to conformally weighted spinor fields. There is now a significant difference between the real and complex cases for on a real manifold w is an unrestricted real number whereas on a complex manifold with complex-valued Ω the weight w must be integral else Ω^w makes no sense. In Eastwood & Rice (1987) the argument for classification proceeded on the complexification $\mathbb{M}$ and so the weights were integral. It turns out that this assumption is justified but does require additional argument as follows.

Dimension 4: To deal with the real case one should attempt to classify invariant differential operators on real compactified Minkowski space $M = \mathrm{SO}(4,2)/P$. As in Eastwood & Rice (1987) the question is equivalent to the classification of homomorphisms between Verma modules induced from finite-dimensional irreducible representations of P. Since the Verma modules are purely Lie-algebraic constructs, there is no harm in complexification at this level. Thus, one searches for homomorphisms (see Baston and Eastwood (1985) for notation)

$$V(\overset{s}{\bullet}\!-\!\overset{t}{\times}\!-\!\overset{u}{\bullet}) \longrightarrow V(\overset{p}{\bullet}\!-\!\overset{q}{\times}\!-\!\overset{r}{\bullet})$$

as equivalent to invariant differential operators

$$\overset{p}{\bullet}\!-\!\overset{q}{\times}\!-\!\overset{r}{\bullet} \longrightarrow \overset{s}{\bullet}\!-\!\overset{t}{\times}\!-\!\overset{u}{\bullet}$$

except that it is no longer the case that q and t are required to be integral. More specifically, a section of $\overset{p}{\bullet}\!-\!\overset{q}{\times}\!-\!\overset{r}{\bullet}$ is a spinor field

$$\phi_{\overbrace{AB\cdots D}^{r}\overbrace{E'F'\cdots H'}^{p}} \qquad \text{(slight notational change from Baston and Eastwood 1985)}$$

of conformal weight $w = p + q + r$ (which is no longer required to be integral). This coincides with asking for a finite-dimensional irreducible representation of the Lie algebra (*not* Lie group)

$$\mathfrak{p} = \bullet\!-\!\times\!-\!\bullet = \begin{pmatrix} * & * & * & * \\ * & * & * & * \\ 0 & 0 & * & * \\ 0 & 0 & * & * \end{pmatrix}.$$

These are classified by $\overset{p}{\bullet}\!-\!\overset{q}{\times}\!-\!\overset{r}{\bullet}$ with p, $r \in \mathbb{Z}_{\geqslant 0}$ and $q \in \mathbb{C}$ in the usual way ($\overset{p}{\bullet}\!-\!\overset{q}{\bullet}\!-\!\overset{r}{\bullet}$ gives *minus* the weight of the *lowest* weight vector).

The argument for the classification is much the same as that for integral weights Eastwood & Rice (1987). Thus, the first requirement in order that there be a non-zero homomorphism between two Verma modules is that they have the same *central character* which, in this case, reduces one to looking for invariant operators between two of:

$$\overset{p+q+r+2}{\bullet}\!-\!\underset{-q-r-3}{\times}\!-\!\overset{q}{\bullet}$$

$$\overset{p}{\bullet}\!-\!\overset{q}{\times}\!-\!\overset{r}{\bullet} \qquad \overset{p+q+1}{\bullet}\!-\!\underset{-q-2}{\times}\!-\!\overset{q+r+1}{\bullet} \qquad\qquad \overset{q+r+1}{\bullet}\!-\!\underset{-p-q-r-4}{\times}\!-\!\overset{p+q+1}{\bullet} \qquad \overset{r}{\bullet}\!-\!\underset{-p-q-r-4}{\times}\!-\!\overset{p}{\bullet}$$

$$\overset{q}{\bullet}\!-\!\underset{-p-q-3}{\times}\!-\!\overset{p+q+r+2}{\bullet}$$

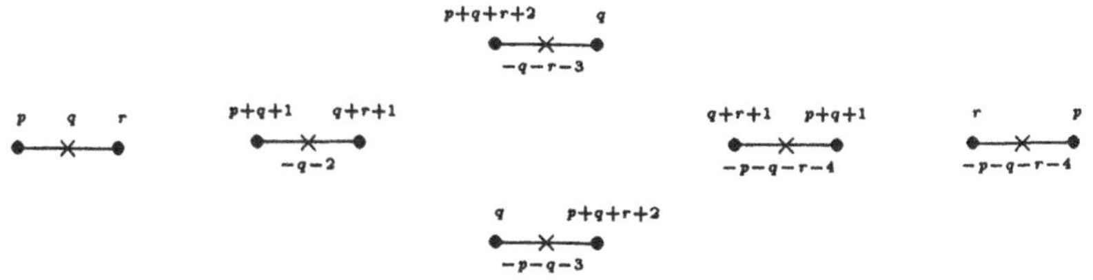

whence, the only possibility for an invariant differential operator with non-integral conformal weight is

$$\overset{p}{\bullet}\!-\!\overset{q}{\times}\!-\!\overset{r}{\bullet} \longrightarrow \overset{r}{\bullet}\!-\!\underset{-p-q-r-4}{\times}\!-\!\overset{p}{\bullet}\,.$$

There are two ways of eliminating this possibility.

Method 1 Consider the possible symbols for this differential operator.

$$\Omega^1 = \overset{1}{\bullet}\!-\!\overset{-2}{\times}\!-\!\overset{1}{\bullet}, \quad \Omega^2 = \overset{2}{\bullet}\!-\!\overset{-4}{\times}\!-\!\overset{2}{\bullet} \oplus \overset{0}{\bullet}\!-\!\overset{-2}{\times}\!-\!\overset{0}{\bullet}, \quad \Omega^3 = \overset{3}{\bullet}\!-\!\overset{-6}{\times}\!-\!\overset{3}{\bullet} \oplus \overset{1}{\bullet}\!-\!\overset{-4}{\times}\!-\!\overset{1}{\bullet},$$

$$\Omega^4 = \overset{4}{\bullet}\!-\!\overset{-8}{\times}\!-\!\overset{4}{\bullet} \oplus \overset{2}{\bullet}\!-\!\overset{-6}{\times}\!-\!\overset{2}{\bullet} \oplus \overset{0}{\bullet}\!-\!\overset{-4}{\times}\!-\!\overset{0}{\bullet} \quad \text{etc.}$$

so one is led to ask whether one can have

$$\overset{k}{\bullet}\!-\!\overset{-2n}{\times}\!-\!\overset{k}{\bullet} \otimes \overset{p}{\bullet}\!-\!\overset{q}{\times}\!-\!\overset{r}{\bullet} = \cdots \oplus \overset{r}{\bullet}\!-\!\underset{-p-q-r-4}{\times}\!-\!\overset{p}{\bullet} \oplus \cdots.$$

If so, then, amongst other restrictions,

$$\left.\begin{array}{r} k+p+2(-2n+q)+k+r = r+2(-p-q-r-4)+p \\ \text{and 2 divides } k+p+r \end{array}\right\}$$

so $2q = 2n-4-(k+p+r)$ and q must be an integer.

Method 2 By means of the *translation principle*, as in Eastwood & Rice (1987), a differential operator

$$\overset{p}{\bullet}\!-\!\overset{q}{\times}\!-\!\overset{r}{\bullet} \longrightarrow \overset{r}{\bullet}\!-\!\underset{-p-q-r-4}{\times}\!-\!\overset{p}{\bullet}$$

gives rise to operators

$$\overset{p}{\bullet}\!-\!\underset{q+1}{\times}\!-\!\overset{r-1}{\bullet} \longrightarrow \overset{r-1}{\bullet}\!-\!\underset{-p-q-r-4}{\times}\!-\!\overset{p}{\bullet} \quad \text{and} \quad \overset{p-1}{\bullet}\!-\!\overset{q}{\times}\!-\!\overset{r}{\bullet} \longrightarrow \overset{r}{\bullet}\!-\!\underset{-p-q-r-3}{\times}\!-\!\overset{p-1}{\bullet}$$

so one can reduce to the case $p = r = 0$, ask whether there is an operator

$$\overset{0}{\bullet}\!-\!\overset{q}{\times}\!-\!\overset{0}{\bullet} \longrightarrow \overset{0}{\bullet}\!-\!\overset{-q-4}{\times}\!-\!\overset{0}{\bullet}$$

and now consider possible symbols. Evidently, one is forced into

$$\overset{0}{\bullet}\!-\!\overset{-2n}{\times}\!-\!\overset{0}{\bullet} \otimes \overset{0}{\bullet}\!-\!\overset{q}{\times}\!-\!\overset{0}{\bullet} = \overset{0}{\bullet}\!-\!\overset{-q-4}{\times}\!-\!\overset{0}{\bullet}$$

so $q = n-2$ is integral (and the operator is $\Box^n$).

Notice that these arguments apply only to flat space. Presumably there is an argument directly applicable to curved space: I anticipate that operators on a curved 4-dimensional spacetime which are invariant under conformal rescaling must act on fields with integral conformal weight. As shown below, the corresponding result is false in 3-dimensions (and presumably the pattern continues with even versus odd dimensions (cf. Huygens' principle etc.)). Notice that complexification is more satisfactory in dimension 4.

Dimension 3: The same general reasoning applies to the search for invariant operators

$$\overset{q}{\times}\!\!\Rightarrow\!\!\overset{r}{\bullet} \longrightarrow \overset{t}{\times}\!\!\Rightarrow\!\!\overset{u}{\bullet}$$

where again r is non-negative integral and q is unrestricted. Central character considerations restrict the search to

$$\overset{q}{\times}\!\!\Rightarrow\!\!\overset{r}{\bullet} \qquad \overset{-q-2}{\times}\!\!\Rightarrow\!\!\overset{2q+r+2}{\bullet} \qquad \overset{-q-r-3}{\times}\!\!\Rightarrow\!\!\overset{2q+r+2}{\bullet} \qquad \overset{-q-r-3}{\times}\!\!\Rightarrow\!\!\overset{r}{\bullet}\,.$$

If q is integral one obtains, by translation, the deRham sequence

$$\begin{array}{ccccccc} \overset{0}{\times}\!\!\Rightarrow\!\!\overset{0}{\bullet} & \longrightarrow & \overset{-2}{\times}\!\!\Rightarrow\!\!\overset{2}{\bullet} & \longrightarrow & \overset{-3}{\times}\!\!\Rightarrow\!\!\overset{2}{\bullet} & \longrightarrow & \overset{-3}{\times}\!\!\Rightarrow\!\!\overset{0}{\bullet} \\ \| & & \| & & \| & & \| \\ \Omega^0 & \xrightarrow{d} & \Omega^1 & \xrightarrow{d} & \Omega^2 & \xrightarrow{d} & \Omega^3 \end{array}$$

and its siblings (the Bernstein-Gelfand-Gelfand resolutions). If q is not integral then there is the possibility of an invariant operator

$$\overset{q}{\times}\!\!\Rightarrow\!\!\overset{r}{\bullet} \longrightarrow \overset{-q-r-3}{\times}\!\!\Rightarrow\!\!\overset{r}{\bullet}\,.$$

To investigate this case (and also to show that there are no other operators in the deRham case) one can start with symbol considerations:

$$\Omega^1 = \overset{-2}{\times}\!\!\Rightarrow\!\!\overset{2}{\bullet}, \quad \odot^2\Omega^1 = \overset{-4}{\times}\!\!\Rightarrow\!\!\overset{4}{\bullet} \oplus \overset{-2}{\times}\!\!\Rightarrow\!\!\overset{0}{\bullet}, \quad \odot^3\Omega^1 = \overset{-6}{\times}\!\!\Rightarrow\!\!\overset{6}{\bullet} \oplus \overset{-4}{\times}\!\!\Rightarrow\!\!\overset{2}{\bullet},$$
$$\odot^4\Omega^1 = \overset{-8}{\times}\!\!\Rightarrow\!\!\overset{8}{\bullet} \oplus \overset{-6}{\times}\!\!\Rightarrow\!\!\overset{4}{\bullet} \oplus \overset{-4}{\times}\!\!\Rightarrow\!\!\overset{0}{\bullet}, \text{ etc.}$$

so one asks for

$$\overset{-2n}{\times}\!\!\Rightarrow\!\!\overset{2k}{\bullet} \otimes \overset{q}{\times}\!\!\Rightarrow\!\!\overset{r}{\bullet} = \cdots \oplus \overset{-q-r-3}{\times}\!\!\Rightarrow\!\!\overset{r}{\bullet} \oplus \cdots$$

whence $2(-2n+q)+2k+r = 2(-q-r-3)+r$ so $2q = 2n-k-r-3$. Thus, $2q$ is integral but q itself need not be. By applying the translation functor (tensor with $\overset{0}{\bullet}\!\!\Rightarrow\!\!\overset{1}{\bullet} = \overset{0}{\times}\!\!\Rightarrow\!\!\overset{1}{\bullet} + \overset{-1}{\times}\!\!\Rightarrow\!\!\overset{1}{\bullet}$) one is reduced to deciding on the existence of a differential operator $\overset{-1/2}{\times}\!\!\Rightarrow\!\!\overset{0}{\bullet} \longrightarrow \overset{-5/2}{\times}\!\!\Rightarrow\!\!\overset{0}{\bullet}$.

The multiplication table for $SO(5,\mathbb{C})$ is (entries give $[a,b]$):

a \ b	H	h	X	x_1	x_2	x_3	Y	y_1	y_2	y_3
H	0	0	$-2X$	$2x_1$	0	$-2x_3$	$2Y$	$-2y_1$	0	$2y_3$
h	0	0	X	$-2x_1$	$-x_2$	0	$-Y$	$2y_1$	y_2	0
X	$2X$	$-X$	0	$-x_2$	$-x_3$	0	$-H$	0	$2y_1$	$2y_2$
x_1	$-2x_1$	$2x_1$	x_2	0	0	0	0	$-h$	$-Y$	0
x_2	0	x_2	x_3	0	0	0	$2x_1$	$-X$	$-H-2h$	$-2Y$
x_3	$2x_3$	0	0	0	0	0	$2x_2$	0	$-2X$	$-4H-4h$
Y	$-2Y$	Y	H	0	$-2x_1$	$-2x_2$	0	y_2	y_3	0
y_1	$2y_1$	$-2y_1$	0	h	X	0	$-y_2$	0	0	0
y_2	0	$-y_2$	$-2y_1$	Y	$H+2h$	$2X$	$-y_3$	0	0	0
y_3	$-2y_3$	0	$-2y_2$	0	$2Y$	$4H+4h$	0	0	0	0

The Lie subalgebra $s\ell(2,\mathbb{C}) \cong \langle h, x_1, y_1\rangle$ corresponds to this node	$\longrightarrow$ $\bullet\!\!\Rightarrow\!\!\bullet$ $\longleftarrow$	The Lie subalgebra $s\ell(2,\mathbb{C}) \cong \langle H, X, Y\rangle$ corresponds to this node

$p = \times\!\!\Rightarrow\!\!\bullet$ is therefore generated by everything *except* y_1, y_2, y_3 *namely* $H, h, X, x_1, x_2, x_3, Y$

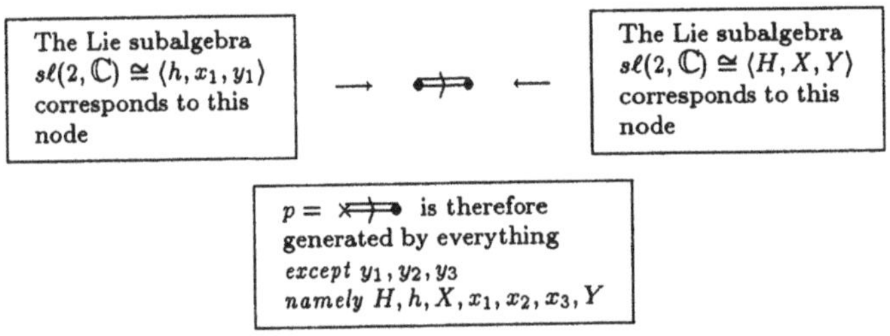

The possible differential operator $\overset{-1/2}{\times}\!\!\Rightarrow\!\!\overset{0}{\bullet} \longrightarrow \overset{-5/2}{\times}\!\!\Rightarrow\!\!\overset{0}{\bullet}$ is second order by symbol considerations: $\odot^2\Omega^1 = \overset{-4}{\times}\!\!\Rightarrow\!\!\overset{4}{\bullet} \oplus \overset{-2}{\times}\!\!\Rightarrow\!\!\overset{0}{\bullet}$ and $\overset{-2}{\times}\!\!\Rightarrow\!\!\overset{0}{\bullet} \otimes \overset{-1/2}{\times}\!\!\Rightarrow\!\!\overset{0}{\bullet} = \overset{-5/2}{\times}\!\!\Rightarrow\!\!\overset{0}{\bullet}$. By Poincaré-Birkhoff-Witt $V(\overset{-1/2}{\times}\!\!\Rightarrow\!\!\overset{0}{\bullet}) = \mathbb{C}[y_1, y_2, y_3]\alpha$ where α is *highest* (annihilated by X and x_1) with weight given by $h\alpha = -\frac{1}{2}\,\alpha$ and $H\alpha = 0$. One easily checks that $(y_2^2 - 2y_1, y_3)\alpha \equiv \beta$ satisfies $h\beta = -\frac{5}{2}\,\alpha$, $H\beta = 0$, and $X\beta = 0$ so β corresponds to the proposed symbol and lifts to an invariant operator iff $x_1\beta = 0$. But $x_1\beta = x_1(y_2^2 - 2y_1y_3)\alpha = (Yy_2 + y_2x_1y_2 - 2hy_3 - 2y_1x_1y_3)\alpha = (-y_3 + y_2Y + y_2Y + y_2^2x_1 - 2y_3h - 2y_1y_3x_1)\alpha = -y_3\alpha + y_3\alpha = 0$. The operator in question is, of course, the conformally invariant Laplacian in dimension 3. There is also an invariant first order operator $\overset{-3/2}{\times}\!\!\Rightarrow\!\!\overset{1}{\bullet} \longrightarrow \overset{-5/2}{\times}\!\!\Rightarrow\!\!\overset{1}{\bullet}$ (if α is the highest weight vector in $V(\overset{-3/2}{\times}\!\!\Rightarrow\!\!\overset{1}{\bullet})$ then $(y_2 + 2y_1Y)\alpha$ is also maximal). These operators

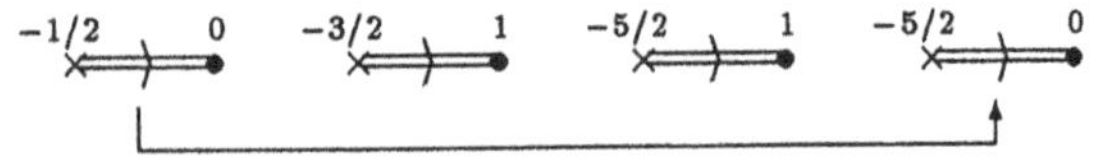

and translates together with the Bernstein-Gelfand-Gelfand operators comprise a complete list of invariant operators between bundles of non-singular character. These operators all have curved analogues either by a curved translation principle Eastwood & Rice (1987) or (better) by direct application of the Cartan connection, Baston (1985). The half integral weights which occur in this example suggest an analogue for G_2:

Case of $\times\!\!\Lleftarrow\!\!\bullet$ **:**  Central character considerations restrict attention to

q r
−q−2 3q+r+3
−q−r−3 3q+2r+4
−2q−r−4 3q+2r+4
−2q−r−4 3q+r+3
−q−r−3 r

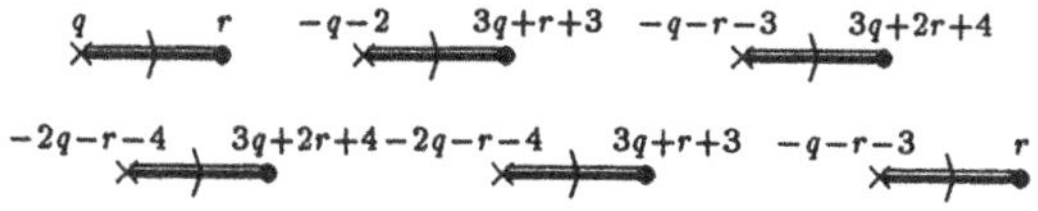

and symbol considerations force $6q \in \mathbb{Z}$. The multiplication table for G_2 is:

	H	h	X	x_1	x_2	x_3	x_4	x_5	Y	y_1	y_2	y_3	y_4	y_5
H	0	0	$-2X$	$3x_1$	x_2	$-x_3$	$-3x_4$	0	$2Y$	$-3y_1$	$-y_2$	y_3	$3y_4$	0
	h	0	X	$-2x_1$	$-x_2$	0	x_4	x_5	$-Y$	$2y_1$	y_2	0	$-y_4$	y_5
		X	0	$-x_2$	$-x_3$	$-x_4$	0	0	$-H$	0	$3y_1$	$4y_2$	$3y_3$	0
			x_1	0	0	0	x_5	0	0	$-h$	$-Y$	0	0	$-y_4$
				x_2	0	$-x_5$	0	0	$3x_1$	$-X$	$-H-3h$	$-4Y$	0	$3y_3$
					x_3	0	0	0	$4x_2$	0	$-4X$	$-8H-12h$	$-12Y$	$-12y_2$
						x_4	0	0	$3x_3$	0	0	$-12X$	$-36H-36h$	$36y_1$
							x_5	0	0	$-x_4$	$3x_3$	$-12x_2$	$36x_1$	$-36H-72h$
								Y	0	y_2	y_3	y_4	0	0
									y_1	0	0	0	$-y_5$	0
										y_2	0	y_5	0	0
											y_3	0	0	0
												y_4	0	0
													y_5	0

This table can be computed from the root diagram.

It is now elementary to check that, with obvious notation, $(y_2 + 3y_1Y)\alpha$ provides an invariant differential operator $\overset{-4/3}{\times}\!\!\Lleftarrow\!\!\overset{1}{\bullet} \longrightarrow \overset{-7/3}{\times}\!\!\Lleftarrow\!\!\overset{2}{\bullet}$ 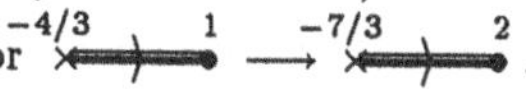. However, although

$$\left(9y_1^2y_4^2 - 18y_1y_2y_3y_4 + 6y_1y_3^3 + 8y_2^3y_4 - 3y_2^2y_3^2 - 27y_1y_4y_5 + 3y_2y_3y_5\right)\alpha$$

provides a symbol for a differential operator $\overset{-1/2}{\times}\!\!\Lleftarrow\!\!\overset{0}{\bullet} \longrightarrow \overset{-5/2}{\times}\!\!\Lleftarrow\!\!\overset{0}{\bullet}$ , it does *not* lift and, by translation it follows that there are no differential operators with $\frac{1}{2}$-integral weights.

References

Baston, R.J. and Eastwood, M.G. (1985) The Penrose transform for complex homogeneous spaces, TN 20 and §I.3.15.

Baston, R.J. (1985) Conformally invariant differential operators for curved space-time, TN 20.

Baston, R.J. (1985) *The algebraic construction of invariant differential operators*, D. Phil. Thesis (Oxford University).

Eastwood, M.G. & Rice, J. (1987) Conformally invariant differential operators on Minkowski space and their curved analogues, Comm. Math. Phys. **109**, 207–228, and Erratum, Comm. Math. Phys., **144** (1992), 213.

§II.2.15 Tensor products of Verma modules and conformally invariant tensors
by R.J.Baston (TN 24, September 1987)

Let $P = \bullet\!-\!\!\times\!\!-\!\bullet$ be the conformal complex Poincaré group in $SL(4,\mathbb{C})$. If $G = SL(4,\mathbb{C})$ then the flag variety $G/P = Gr_2(\mathbb{C}^4) = \mathbb{CM}$. At least locally (and globally, modulo topological obstructions) on any 4-dim$_{\mathbb{C}}$ conformal manifold there is an unique P-principal bundle $\mathcal{G} \to \mathcal{M}$ together with a $\mathfrak{g}$ = Lie algebra $sl(4,\mathbb{C})$-valued one-form on $\mathcal{G}$, ω giving an isomorphism $T_g\mathcal{G} \xrightarrow{\simeq} \mathfrak{g}$ (so if $\mathcal{M}$ is conformally flat, $\mathcal{G}$ is covered by G and the pull back of ω is the Maurer-Cartan form). ω has the property that $\omega^{-1} : \mathfrak{g} \to \Gamma(T\mathcal{G})$ restricts to a homomorphism of Lie algebras on $\mathfrak{p}$ = Lie Alg. of P and for $u \in \mathfrak{p}$, $v \in \mathfrak{g}$, $\omega^{-1}[u,v] = [\omega^{-1}u, \omega^{-1}v]$. The failure of ω^{-1} to a be homomorphism of algebras is thus determined on a complement to $\mathfrak{p}$ in $\mathfrak{g}$: set $\mathfrak{g} = \mathfrak{p} \oplus \mathfrak{u}_-$ and take $x, v \in \mathfrak{u}_-$. Then $\Omega(\omega^{-1}x, \omega^{-1}v) = [x,v] - \omega[\omega^{-1}x, \omega^{-1}v]$ is the curvature of ω. All of this is well known in twistor theory as the *local twistor transport* and its curvature. (See Penrose & Rindler 1986.) The same sort of picture holds in dims$_{\mathbb{C}} \geqslant 3$. (In dim 2, the conformal group is ∞-dim, which is what all the fuss is about). If V is any representation of P, one has a homogeneous sheaf of holomorphic sections of the vector bundle $\mathcal{G} \times_P V$ induced by V. So $\overset{1\ \ 0\ \ 0}{\bullet\!-\!\!\times\!\!-\!\bullet} = \mathcal{O}^{A'}$ etc. (see Baston & Eastwood 1985). From the flat case, we know there is a good class of linear differential invariants of the conformal structure of $\mathcal{M}$ (see Eastwood and Rice 1987, Baston 1985, Baston & Eastwood 1991): the calculation of these depends on ω and the identification of $J^\infty(V)$ (the formal jet bundle of $\mathcal{G} \times_P V$) as a homogeneous bundle, induced by the (G-finite) dual of a *Verma module* $\mathcal{M}_{\mathfrak{p}}(V^*) = U(\mathfrak{g}) \otimes_{U(\mathfrak{p})} V^*$. The differential invariants we want correspond to the translation invariant differential operators on $\mathcal{G} \times_P V$. These are classified by $\mathfrak{p}$-module homomorphisms $\mathcal{M}_p(V^*) \leftarrow W^*$ (giving a differential operator on homogeneous sheaves $\mathcal{O}(V^*) \to \mathcal{O}(W^*)$). Frobenius reciprocity says such an operator is equivalent to a $\mathfrak{g}$-module homomorphism of Verma modules $\mathcal{M}_{\mathfrak{p}}(V^*) \leftarrow \mathcal{M}_{\mathfrak{p}}(W^*)$.

RECAP. *homomorphisms of Verma modules* $\Longleftrightarrow$ *distinguished differential invariants of conformal structure of* $\mathcal{M}$.

For precise details, see Baston and Eastwood (1991).

REMARK. In particular, almost any conformally invariant operator on $\mathbb{M}$ (e.g. $\Box$) gives rise to one on $\mathcal{M}$ (e.g. $\Box + R/6$). One may even calculate the curvature terms explicitly.

The invariant operators on G/P fit into certain Bernstein-Gelfand-Gelfand resolutions (generalized de Rham resolutions). On $\mathcal{M}$ these fail to be complexes ($D^2 \neq 0$) by virtue of curvature (i.e. ω^{-1} is not a homomorphism on all of $\mathfrak{g}$). Their composition gives further differential invariants: thus $\overset{1\ \ 0\ \ 0}{\bullet\!-\!\!\times\!\!-\!\bullet} \xrightarrow{d_1} \overset{2\ -2\ \ 1}{\bullet\!-\!\!\times\!\!-\!\bullet} \xrightarrow{d_2} \overset{3\ -3\ \ 0}{\bullet\!-\!\!\times\!\!-\!\bullet}$ by $\pi^{A'} \xrightarrow{d_1} \nabla^A_{(A'}\pi_{B')}$ and $\phi^A_{A'B'} \to \nabla^A_{(A'}\phi_{B'C')}$ so that $d_2 \circ d_1 : \pi^{A'} \longrightarrow \nabla^A_{(A'}\nabla_{|A|B'}\pi_{C')} \propto \tilde{\psi}_{A'B'C'D'}\pi^{D'}$ (see Eastwood 1985).

It has been conjectured that taking all possible combinations of this kind, in various linear combinations and repeatedly yields all differential invariants of the conformal structure (see Eastwood 1985). To begin to try to prove such a result, one can examine the known results more closely. For instance

$$E_{abc} \equiv \tilde{\psi}_{A'B'C'D'}\nabla^{AA'}\psi_{ABCD} - \psi_{ABCD}\nabla^{AA'}\tilde{\psi}_{A'B'C'D'} \tag{1}$$

is invariant (and extremely useful). It is clear that this ought to be thought of as some kind of anti-symmetrization "$\alpha \cdot \beta - \beta \cdot \alpha$". Now both $\tilde{\psi}$ and ψ, as sections of $\overset{4\ -4\ 0}{\bullet\!-\!\times\!-\!\bullet}$ and $\overset{0\ -4\ 4}{\bullet\!-\!\times\!-\!\bullet}$, induce sections of the corresponding jet sheaves. Over each point of $\mathcal{M}$ this gives elements of dual Verma modules, $\mathcal{M}(\overset{4\ -4\ 0}{\bullet\!-\!\times\!-\!\bullet})^*$ and $\mathcal{M}(\overset{0\ -4\ 4}{\bullet\!-\!\times\!-\!\bullet})^*$. Just as in the finite dimensional case, we may take tensor products of these modules and decompose (hopefully into dual Verma modules) so obtaining fresh spinorial invariants, like (1). Indeed, an explicit calculation readily checks that (1) comes this way.

RECAP.[1] *Decompose tensor products to compute conformal invariants.*

CONJECTURE. *all come this way*

The general theory of such tensor products must be complicated. But, using the Penrose transform, one can consider a special case: the inclusion $P \subset G$, with "transpose" determines a p^+ at infinity in $\mathbb{M}$ and so a line $\mathbb{L}$ in $\mathbb{PT}$ and its annihilator $\mathbb{L}^*$ in $\mathbb{PT}^*$. Then as $\mathfrak{g}$-modules

$$\begin{aligned} & & & L_1 \\ & & & \parallel \\ H^1(\overset{1\ 0\ 1}{\times\!-\!\bullet\!-\!\bullet})(\text{on } \mathbb{PT}-\mathbb{L}) &\overset{\mathcal{P}}{\cong} \text{unique irreducible submodule of } \mathcal{M}(\overset{0\ -4\ 4}{\bullet\!-\!\times\!-\!\bullet})^* &=& L(\overset{0\ -4\ 4}{\bullet\!-\!\times\!-\!\bullet}) \\ H^1(\overset{1\ 0\ 1}{\bullet\!-\!\bullet\!-\!\times})(\text{on } \mathbb{PT}^*-\mathbb{L}^*) &\cong \text{unique irreducible submodule of } \mathcal{M}(\overset{4\ -4\ 0}{\bullet\!-\!\times\!-\!\bullet})^* &=& L(\overset{4\ -4\ 0}{\bullet\!-\!\times\!-\!\bullet}) \\ & & & \parallel \\ & & & L_2 \end{aligned}$$

(sheaf $\overset{1\ 0\ 1}{\times\!-\!\bullet\!-\!\bullet} \cong \Theta =$ holomorphic tangent bundle so this concerns linearized deformations). Under $\mathfrak{p}$ (*not* under $\mathfrak{g}$) one has a splitting (differential) $\mathcal{M}(\overset{0\ -4\ 4}{\bullet\!-\!\times\!-\!\bullet})^* \to L(\overset{0\ -4\ 4}{\bullet\!-\!\times\!-\!\bullet})$ (etc.). So we try to decompose $L^1 \otimes L^2$ under $\mathfrak{g}$ (strictly, we should do it under $\mathfrak{p}$ but, since we hope to encounter only $\mathfrak{g}$-irreducibles, this is irrelevant). Compute (using Künneth):

$$L^1 \otimes L^2 \cong H^2((\mathbb{P}-\mathbb{L}) \times (\mathbb{P}^*-\mathbb{L}^*), \overset{1\ 0\ 1}{\times\!-\!\bullet\!-\!\bullet} \otimes \overset{1\ 0\ 1}{\bullet\!-\!\bullet\!-\!\times}) \cong H^4_{\mathbb{L}\times\mathbb{L}^*}(\mathbb{P}\times\mathbb{P}^*, \overset{1\ 0\ 1}{\times\!-\!\bullet\!-\!\bullet} \otimes \overset{1\ 0\ 1}{\bullet\!-\!\bullet\!-\!\times}). \quad \circledast$$

(Mayer Vietoris + relative cohomology sequence). The map is just *cross product*. But then multiplication by $\frac{1}{(z\cdot\omega)^k}$ gives maps

②

$$\frac{1}{(z\cdot\omega)^k} : H^3_{\mathbb{L}\times\mathbb{L}^*}(\mathbb{A}, (\overset{1\ 0\ 1}{\times\!-\!\bullet\!-\!\bullet} \otimes \overset{1\ 0\ 1}{\bullet\!-\!\bullet\!-\!\times})|_{\mathbb{A}} \otimes \overset{k\ 0\ k}{\times\!-\!\bullet\!-\!\times}) \longrightarrow L^1 \otimes L^2 \quad (k \geqslant 1)$$

[1] Something quite close to this has in fact turned out to be true: the reader should consult Eastwood, Graham and Gover

whose images span $L^1 \times L^2$ (standard theory). The groups ② are easy to calculate, using $(\overset{1\ 0\ 1}{\times\!-\!\bullet\!-\!\bullet} \otimes \overset{1\ 0\ 1}{\bullet\!-\!\bullet\!-\!\times})\,|_{\mathbb{A}} \cong \overset{2\ 0\ 2}{\times\!-\!\bullet\!-\!\times} + (\overset{2\ 1\ 0}{\times\!-\!\bullet\!-\!\times} \oplus \overset{0\ 1\ 2}{\times\!-\!\bullet\!-\!\times}) + (\overset{0\ 2\ 0}{\times\!-\!\bullet\!-\!\times} \oplus \overset{1\ 0\ 1}{\times\!-\!\bullet\!-\!\times})$. If λ is one of these weights, tensored by $\overset{k\ 0\ k}{\times\!-\!\bullet\!-\!\times}$ $(k \geqslant 1)$ then one has a contribution $L(\sigma_2\sigma_1\sigma_3 \cdot \lambda)$ in ② (See Baston & Eastwood 1985 for notation, Weyl group etc.)

So one gets invariants as sections of $\overset{3+k\ \ -8-2k\ \ 3+k}{\bullet\!-\!-\!\times\!-\!-\!\bullet} \cong \mathcal{O}_{\underbrace{(A'\ldots D')}_{3+k}\underbrace{(A\ldots D)}_{k+k}}[-2]$ (from $\overset{2\ 0\ 2}{\times\!-\!\bullet\!-\!\times}$ and the other constituents give the following, after some pure thought along the lines of exact sequences:

(i)	$\overset{2\ 1\ 0}{\times\!-\!\bullet\!-\!\times} \rightsquigarrow \overset{k+2\ \ -2k-8\ \ k+4}{\bullet\!-\!-\!\times\!-\!-\!\bullet}$	$\ni \tilde{\psi}_{A'B'C'D'}\nabla^{A'}_{(E}\psi_{ABCD)} - \psi_{(ABCD}\nabla^{A'}_{E)}\tilde{\psi}_{A'B'C'D'}$
(ii)	$\overset{0\ 1\ 2}{\times\!-\!\bullet\!-\!\times} \rightsquigarrow \overset{k+4\ \ -2k-8\ \ k+2}{\bullet\!-\!-\!\times\!-\!-\!\bullet}$	$\ni \tilde{\psi}_{(A'B'C'D'}\nabla^{A}_{E')}\psi_{ABCD} - \psi_{ABCD}\nabla^{A}_{(E'}\tilde{\psi}_{A'B'C'D')}$
(iii)	$\overset{1\ 0\ 1}{\times\!-\!\bullet\!-\!\times} \rightsquigarrow \overset{k+2\ \ -2k-6\ \ k+2}{\bullet\!-\!-\!\times\!-\!-\!\bullet}$	$\ni E_{abc} = \tilde{\psi}_{A'B'C'D'}\nabla^{AA'}\psi_{ABCD} - \psi_{ABCD}\nabla^{AA'}\tilde{\psi}_{A'B'C'D'}$
(iv)	$\overset{0\ 2\ 0}{\times\!-\!\bullet\!-\!\times} \rightsquigarrow \overset{k+3\ \ -6-2k\ \ k+3}{\bullet\!-\!-\!\times\!-\!-\!\bullet}$	$\ni \psi_{ABCD}\tilde{\psi}_{A'B'C'D'}$
	$(k \geqslant 1)$	case $k = 1$ examples

Notice the "generalizations", at (i) and (ii) of E_{abc} (which are easily and explicitly checked) (Penrose points out that E_{abc} occurs in Dighton's thesis). All the tensors obtained in this way are subject to invariant differential operators, constructed as at the start of this note. Iff the Bach tensor vanishes, they should be automatically in the kernel of this operator, since then each lies in an irreducible subsheaf of the jet sheaf. (Explicit calculations verify this for E_{abc}).

Remark. A linearized contact deformation of $\mathbb{A}$ lies in $H^1(\mathbb{A}, \overset{1\ 0\ 1}{\times\!-\!\bullet\!-\!\times})$; cf. Mason & Baston (1987). The usual obstruction theory and representation theory implies that this extends to

$$H^1(\mathbb{P}, \overset{1\ 0\ 1}{\times\!-\!\bullet\!-\!\bullet}) \oplus H^1(\mathbb{P}^*, \overset{1\ 0\ 1}{\bullet\!-\!\bullet\!-\!\times}) \subset H^1(\mathbb{P} \times \mathbb{P}^*, \Theta)$$

iff Bach tensor vanishes (use argument at ⊛, previous page). Then the commutator (as in Kodaira-Spencer deformation theory) $[v, w]$, $(v \in H^1(\mathbb{P}, \overset{1\ 0\ 1}{\times\!-\!\bullet\!-\!\bullet})$, $w \in H^1(\mathbb{P}^*, \overset{1\ 0\ 1}{\bullet\!-\!\bullet\!-\!\times}))$ lies in $H^2(\mathbb{P} \times \mathbb{P}^*, \Theta)$. Restriction to formal neighbourhoods of $\mathbb{A}$ gives one elements of

$$H^2(\mathbb{A}, \overset{1\ 0\ 1}{\times\!-\!\bullet\!-\!\times} \otimes \overset{-k'\ 0\ -k'}{\times\!-\!\bullet\!-\!\times}) \qquad k' \geqslant s;\ \text{set } k = t' - s$$

$$\wr\|$$

$$\ker(\overset{k+2\ \ -2k-6\ \ k+2}{\bullet\!-\!-\!\times\!-\!-\!\bullet} \xrightarrow{\nabla^{AA'}} \overset{k+1\ \ -2k-6\ \ k+1}{\bullet\!-\!-\!\times\!-\!-\!\bullet})$$

But this must be a constituent in $L^1 \otimes L^2$ and $[v, \omega] \mapsto$ projection of $[v] \otimes [\omega]$. So the obstructions are indentified as the invariants of (iii). The other invariants admit similar geometric descriptions.

References

Bailey, Eastwood, Graham (1992) Preprint, to appear in Annals of Mathematics.

Baston, R.J. (1985) The Algebraic construction of invariant differential operators, D. Phil. Thesis (Oxford University).

Baston, R.J. & Eastwood, M.G. (1985) The Penrose transform for complex homogeneous spaces, TN 20 and §I.3.15.

Baston, R.J. & Eastwood, M.G. (1991), in *Twistors in Mathematics and Physics*, (T.N.Bailey & R.J.Baston eds.), CUP.

Eastwood, M.G. & Rice, J.W. (1987) Conformally invariant differential operators on Minkowski space and their curved analogues, *Comm. Math. Phys.* **109**, 207–228.

Eastwood, M.G. (1985) The Fefferman Graham conformal invariant, TN 20 and §II.2.13.

Mason, L.J. & Baston, R.J. (1987) Conformal gravity, the Einstein equations and spaces of complex null geodesics, *Classical and Quantum Gravity* **4**, 815–826.

Penrose, R. & Rindler, W. (1986) *Spinors and Space Time*, Vol. II, Cambridge University Press.

§II.2.16 Structure of the jet bundle for manifolds with conformal or projective structure

by A.R.Gover (TN 29, November 1989)

Let G be a complex semi-simple Lie group and $P < G$ a Lie subgroup. To each P-module E there corresponds a homogeneous vector bundle $\mathbf{E}$, over G/P, which is just $G \times E$ modulo the equivalence relation $(g, e) \sim (gp, p^{-1}e)$ for $g \in G$, $p \in P$ and $e \in E$. We write

$$\boldsymbol{E} = G \times_P E$$

It is an interesting exercise (see Eastwood & Rice 1987) to check that the bundle of infinite jets, $\boldsymbol{JE}$, associated with this homogeneous bundle is itself homogeneous:

$$\boldsymbol{JE} = G \times_P JE$$

where JE may be obtained as the fibre of $\boldsymbol{JE}$ over the identity coset $P \in G/P$.

It is interesting and useful to see the extent to which this goes through in case G/P is replaced by a more general structure. So let $\mathcal{M}$ be an arbitrary complex holomorphic manifold equipped with a P-principal bundle $\mathcal{G}$. Then $\mathcal{M}$ is the fibrewise quotient $\mathcal{G}/P$ and we have an appropriate generalisation of the above situation. Corresponding to the homogeneous bundles above are *semi-homogeneous* bundles which are constructed in exactly the same way: If E a P-module then we have $\boldsymbol{E} := \mathcal{G} \times_P E$.

A section f of $\mathbf{E}$ corresponds to a function F,

$$F : \mathcal{G} \to E,$$

such that

$$pF(gp) = F(g) \qquad \text{for } g \in \mathcal{G} \text{ and } p \in P.$$

Such functions F will be said to be *semi-homogeneous.* The space of such E-valued semi-homogeneous functions is itself a P-module in the obvious way. This has a P-submodule of semi-homogeneous functions which vanish to order $k+1$ on gP for $x = gP$ an arbitrary point of $\mathcal{M}$. The quotient P-module shall be denoted $J^k_x E$. Points of the bundle of k-jets associated to $\mathbf{E}$, over x, correspond to points in $gP \times J^k_x E$ modulo the equivalence relation $(g, F) \sim (gp, p^{-1}F)$ for $p \in P$ and $F \in J^k_x E$. Thus we see that the bundle of k-jets, $\boldsymbol{J}^k\mathbf{E}$, has an underlying P-structure as in the homogeneous case above. However in this more general setting the inducing P-module may vary from point to point of $\mathcal{M}$. We could write

$$\boldsymbol{J}^k\mathbf{E} = \mathcal{G} \times_P J^k_{gP} E$$

to describe this. With the same notation the bundle of infinite jets is given by

$$\boldsymbol{J}\mathbf{E} = \mathcal{G} \times_P J_{gP} E,$$

where, for each $gP \in \mathcal{G}/P$, $J_{gP}E$ is the projective limit over k of the $J^k_{gP}E$.

The dual version of this proceeds as follows. Let $\mathcal{D}$ denote the space of differential operators from E-valued functions on $\mathcal{G}$ to $\mathbb{C}$-valued functions on $\mathcal{G}$. When restricted to act on semi-homogeneous functions, F, $\mathcal{D}$ gains a P-module structure: For $D \in \mathcal{D}$, pD is defined by

$$[pD]F(g) := DF(\tilde{g})\big|_{\tilde{g}=gp} = Dp^{-1}F(\tilde{g}p^{-1})$$

where $g \in \mathcal{G}$, $p \in P$ and on the extreme right hand side $F(\tilde{g}p^{-1}) = F(g)$ is to be regarded as a function of $\tilde{g}$. $\mathcal{D}$ has a P-submodule of operators which act as zero on semi-homogeneous functions. Denote by $\tilde{X}(E)$ the quotient P-module. Let $\mathcal{O}_{\mathcal{M}}$ denote the $\mathbb{C}$-valued semi-homogeneous functions on $\mathcal{G}$, i.e. the functions constant on each fibre gP, $g \in \mathcal{G}$. For some fixed $x \in \mathcal{M}$, let $\mathcal{I}_x < \mathcal{O}_{\mathcal{M}}$ be the subspace of functions which vanish over x. Denote by $\mathcal{I}_x \cdot \tilde{X}(E)$ the P-submodule of $\tilde{X}(E)$ which consists of elements of $\tilde{X}(E)$ left multiplied by functions in $\mathcal{I}_x$. Since $\tilde{X}(E)$ is naturally a $\mathcal{O}_{\mathcal{M}}$-module it is clear that $\mathcal{I}_x \cdot \tilde{X}(E)$ is a P-submodule of $\tilde{X}(E)$. Once again we can form the quotient module which we shall denote by $\tilde{V}_x(E)$. This P-module is filtered naturally by order of the operators involved; write $\tilde{V}_{k,x}(E)$ to denote the submodule consisting of operators of order no greater than k. Since each element of $\tilde{V}_{k,x}(E)$ determines a map

$$J^k_x E \to \mathbb{C}$$

and $\tilde{X}(E)$ consists of all non-trivial operators on E-valued semi-homogeneous functions it is at once clear that $\tilde{V}_{k,x}(E)$ is precisely the vector dual of $J^k_x E$. It is easily checked that it is also dual as a P-module. Evidently then,

$$(\boldsymbol{J}^k \boldsymbol{E})^* = \mathcal{G} \times_P \tilde{V}_{k,gP}(E)$$

with notation understood to be as above.

While this and the dual version first mentioned provide a description of the jet bundle the situation is less than ideal. Even at the level of k-jets, since the inducing module is point dependent there is little scope for reducing the problem of finding differential operators of order $\leqslant k$ to a finite dimensional one. Nevertheless without more structure this is probably as far as one can go. However in many instances such structure is readily available ...

For example if $\mathcal{M}^n$ has a conformal (or projective) structure then one obtains a principal P-bundle, $\mathcal{G}$, where P is a particular parabolic subgroup of Spin$(n+2)$ (SL$(n+1)$ respectively). Moreover the bundle $\mathcal{G}$ comes equipped with a canonical notion of horizontality called the normal conformal (resp. projective) Cartan connection. We shall see that in either of these cases the jet bundle is almost as simple to describe as in the homogeneous case. The Cartan connection (which will always refer to the *normal* version) is usually described by a 1-form ϑ which satisfies (where $\mathfrak{p}$ and $\mathfrak{g}$ are the Lie algebras of P and G respectively):

(i) $\vartheta_q : T_q\mathcal{G} \longrightarrow \mathfrak{g}$ is a vector space isomorphism $\forall q \in \mathcal{G}$.

(ii) $\vartheta(X^*_q) = X$ if X^* is the Killing field corresponding to $X \in \mathfrak{p}$.

(iii) $R^*_p\vartheta = \mathrm{Ad}(p^{-1})\vartheta$, where R_p describes the right action of $p \in P$ on $\mathcal{G}$.

as well as some curvature conditions. Note that if we write $\mathfrak{g}^* := \vartheta^{-1}(\mathfrak{g})$ then, regarding the vector fields in $\mathfrak{g}^*$ as differential operators, (iii) is equivalent to

$$[XY]^* = [X^*, Y^*]$$

for arbitrary $X \in \mathfrak{p}$ and $Y \in \mathfrak{g}$. We can, in the obvious way, extend ϑ^{-1} to act on the tensor algebra, $\otimes\mathfrak{g} := \bigoplus_{k=0}^{\infty} \otimes^k \mathfrak{g}$. The result of this is a space of special differential operators on $\mathcal{G}$ which will be dented by $\mathcal{U}(\mathfrak{g}^*)$. There is a natural filtration of $\mathcal{U}(\mathfrak{g}^*)$ induced from the grading of the tensor algebra $\otimes\mathfrak{g}$; i.e. $\mathcal{U}_k(\mathfrak{g}^*)$ is the image of $\bigoplus_{i=0}^{k} \otimes^i \mathfrak{g}$. We note that $\mathcal{U}(\mathfrak{g}^*)$ is strictly contained in $\mathcal{D}$, in fact $\mathcal{U}_k(\mathfrak{g}^*)$ is finite dimensional.

The left $\mathcal{U}(\mathfrak{g}^*)$-module

$$\mathcal{U}(\mathfrak{g}^*) \otimes E^*$$

may be thought of as a special class of differential operators from E-valued functions on $\mathcal{G}$ to $\mathbb{C}$-valued functions on $\mathcal{G}$. Taken as operators restricted to semi-homogeneous functions, we may consider the action of P (as described above for all of $\mathcal{D}$) on this space. Now we may regard this $\mathcal{U}(\mathfrak{g}^*)$-module as a $\mathfrak{p}^*$-module (or equivalently a $\mathfrak{p}$-module) by restriction and it is readily verified that this agrees precisely with the P-action (at least treating the elements of $\mathcal{U}(\mathfrak{g}^*) \otimes E^*$ as differential operators on semi-homogeneous functions). Thus $\mathcal{U}(\mathfrak{g}^*) \otimes E^*$ is closed under this P-action and so, given property (i) of ϑ, is an ideal candidate to replace $\mathcal{D}$.

$\mathcal{U}(\mathfrak{g}^*) \otimes E^*$ has a P-submodule of operators which annihilate all semi-homogeneous functions. Let $X(E)$ be the quotient and $\mathcal{O}_{\mathcal{M}} \cdot X(E)$ consist of elements of $X(E)$ left multiplied by functions from $\mathcal{O}_{\mathcal{M}}$. Then $\mathcal{O}_{\mathcal{M}} \cdot X(E)$ is also a P-module and, for any fixed $x \in \mathcal{M}$, has $\mathcal{I}_x \cdot X(E)$ as a P-submodule. Again we form the quotient and denote the resulting P-module by $V_x(E)$. With similar reasoning to that used in the $\tilde{V}_x(E)$ case it is not difficult to see that $V_x(E) = (J_x E)^*$ and that

$$\mathcal{G} \times_P V_{gP} E \equiv (\mathbf{JE})^*.$$

Also in this construction the inducing P-module varies on $\mathcal{M}$. Thus at first glance it would seem that we are no better off than with the construction that began with $\mathcal{D}$. In fact, however, we now have a considerably more rigid structure as consideration at the level of k-jets reveals.

Write $\mathcal{D}_k < \mathcal{D}$ to mean the subspace of differential operators of order $\leqslant k$. Corresponding to this $\tilde{X}(E)$ will inherit a filtration, by $\tilde{X}_k(E)$ say. In the approach that begins with all differential operators, this is the key P-module leading to the construction of the dual k-jet bundle. The problem is that $\tilde{X}_k(E)$ is infinite dimensional and we know nothing about its structure. If a conformal or projective structure is present then corresponding to this one has $X_k(E)$, where the filtration of $X(E)$ by the $X_k(E)$ arises from the filtration of $\mathcal{U}(\mathfrak{g}^*)$ by $\mathcal{U}_k(\mathfrak{g}^*)$. Now, in contrast to $\tilde{X}_k(E)$, $X_k(E)$ is finite dimensional. In fact $X_k(E)$ looks just like certain Verma modules which arise in the homogeneous case with some modification due to the curvature of the Cartan connection. Thus, although the structure of $V_{k,x}(E)$ varies over $\mathcal{M}$, the variation involved is a relatively minor detail involving the actual value of the curvature at each point. The important point, however, is that we have a natural bundle epimorphism from a finite dimensional semi-homogeneous bundle onto $(\mathbf{J}^k\mathbf{E})^*$:

$$\mathcal{G} \times_P X_k(E) \to (\mathbf{J}^k\mathbf{E})^*.$$

There is an immediate application of this result. Suppose there is a P-module monomorphism

$$i : H^* \to X_k(E).$$

Then this induces an invariant homomorphism of the corresponding semi-homogeneous bundles:

$$\mathcal{G} \times_P H^* \to \mathcal{G} \times_P X_k(E)$$

and thus a vector bundle homomorphism,

$$\mathcal{G} \times_P H^* \to (\mathbf{J}^k\mathbf{E})^*.$$

Dually then, we have a bundle homomorphism

$$J^k\boldsymbol{E} \to \boldsymbol{H},$$

that is, a differential operator; here $\boldsymbol{H}$ is of course $\mathcal{G} \times_P H$. Beginning with irreducible modules H, finding injections such as i above is straightforward (in principle at least) and just involves finding certain vectors in $X(E)$ which are annihilated by a special subalgebra of $\mathfrak{p}$. (These are called *maximal* vectors.) It is thus easy to see that for irreduible H and any k there are a finite number of differential operators which arise in this fashion. Indeed beginning also with E irreducible, the resulting operators are, in a real sense, analogues of the *invariant* operators in the homogeneous case (as in Baston 1985 and in Eastwood & Rice 1987) or composites thereof.

Examples of applications of these ideas can be found in Gover (1989).

References

Baston, R. J. (1985) *The Algebraic Construction of Invariant Operators*, D. Phil. thesis (Oxford University).

Eastwood, M. G. and Rice, J.W. (1987) Conformally invariant differential operators on Minkowski space and their curved analogues, Commun. Math. Phys. **109**, p. 207–228.

Gover, A. R. (1989) *A Geometrical Construction of Conformally Invariant Differential Operators*, D. Phil. thesis (Oxford University).

§II.2.17 **Exceptional invariants** *by A.R.Gover* ($\mathbb{TN}$ 26, March 1988)

Regard $\mathbf{P}^n$ as a homogeneous space for $\mathrm{SL}(n+1,\mathbf{R})$ and, for a particular homogeneous bundle over $\mathbf{P}^n$, consider the problem of constructing all density valued differential invariants on the bundle which are polynomial in the jets. We will call such objects *projective invariants*. An example for $n = 1$ is given by the formula,

$$wf\nabla\nabla f - (w-1)\nabla f\nabla f$$

where f has weight w. Here ∇, (often written $\eth$ and called edth) is a local flat affine connection. Of course $\mathbf{P}$ does not have a unique such connection but rather a family of them related by transformation formulae, $\nabla \mapsto \nabla f + w\Upsilon f$ where Υ satisfies $\nabla\Upsilon = \Upsilon^2$ (see e.g. Bailey, Eastwood & Gover

(1992) for the corresponding formulae on $\mathbf{P}^n$). The point is that this differential operator is invariant under these transformations. (An analogous, and more familiar, problem is to find conformally invariant differential equations for flat conformal structures. The standard model for the latter is S^n as a homogeneous space for $\mathrm{SO}(n+1,1)$. Of course for $n=1$ these are the *same* problem.)

Recall that a function f of weight w on $\mathbf{P}^n$ corresponds to a function on $\mathbf{R}^{n+1}$ which is homogeneous of degree w, i.e. $f(\lambda X^A) = \lambda^w f(X^A)$, where X^A are the standard coordinates on $\mathbf{R}^{n+1}$. So a good trick for proliferating many projective invariants on $\mathbf{P}^n$ is simply to write down affine invariants on $\mathbf{R}^{n+1}$ and then regard these as invariants on $\mathbf{P}^n$ by simply insisting that f be homogeneous of some weight. For example if $n=1$ then the following is an affine invariant on $\mathbf{R}^{n+1}$,

$$\epsilon^{AC}\epsilon^{BD}\partial_A\partial_B f \partial_C\partial_D f$$

where $\partial_A := \partial/\partial X^A$. The standard representation theory of $\mathrm{SL}(n+1,\mathbf{R})$, due to Weyl and others, tells us scalar valued affine invariants on $\mathbf{R}^{n+1}$ are always linear combinations of such complete contractions. If, in the last formula, we now restrict f to be homogeneous of weight w then we obtain a projective invariant. In fact for $w \neq 1$ this is precisely the invariant mentioned earlier. Invariants which arise this way are called *Weyl invariants.*

Since it is evidently possible to list all Weyl invariants, it is interesting to ask if all projective invariants are Weyl. It turns out Gover (1994a) that if the weight w of f is non-integral or negative integral then all invariants are Weyl. However for the remaining cases with w non-negative integral, to which we now restrict our attention, we shall see that it is easy to write down some invariants which are not Weyl. Such will be called *exceptional* invariants. Even for the simple case of densities on $\mathbf{P}^n$ it is not known how to sort out the Weyl invariants from the exceptional invariants. However, there is a simpler, yet very important, problem on which much progress has been made. For each weight w there is a linear invariant differential operator $\mathcal{O}(w) \to \mathcal{O}_{\underbrace{(ab\cdots d)}_{w+1}}(w)$; in terms of a local affine connection ∇_a on $\mathbf{P}^n$, this is given by $f \mapsto \nabla_a\nabla_b\cdots\nabla_d f$. This operator splits the jet bundle. So instead of looking for invariants on the jets of $\mathcal{O}(w)$ one can look for invariants on the"slightly smaller" space which, at a particular point of $\mathbf{P}^n$, is the jets of $\mathcal{O}(w)$ modulo the kernel of this operator. There are analogous conformal and CR versions of this latter problem too Eastwood & Graham (1991),Fefferman & Graham (1985) and they are geometrical equivalents of some difficult algebraic problems first posed and discussed in Fefferman (1979). On $\mathbf{P}^n$, I have completely solved this problem Gover (1994a). It turns out that, in this case, there are non-vanishing exceptional invariants. For example, if $w=1$, then $\nabla_a\nabla_b f$ is invariant and therefore when $n=2$ one can construct the projective invariant

$$\epsilon^{ac}\epsilon^{bd}\nabla_a\nabla_b f\nabla_c\nabla_d f.$$

Since its homogeniety with respect to f is just two it cannot be a Weyl invariant (the construction of which requires that a ϵ^{ABC} be used). The general situation is well characterised by this example; in n dimensions the exceptional invariants are always constructed by a contraction of $w+1$ $\epsilon^{ab\cdots d}$'s into an n-fold juxtaposition of the linear operator with itself.

By adapting the methods in Gover (1994a) and invoking some new tricks Bailey, Eastwood and Graham (1994) were able to solve corresponding problems related to CR and conformal S^n geometries. There are no exceptional invariants in the CR case but for the conformal case there are. Here again it turns out that operators which are homogeneous of degree n in the argument density f can be exceptional while all others are Weyl. However, in Bailey, Eastwood & Graham (1994) the authors posed the question of whether the exceptionals were, as in the projective case, constructed purely from juxtapositions of the linear invariant $\underbrace{\nabla_{(a}\nabla_b\cdots\nabla_{d)_0}}_{w+1}f$, where w is the (non-negative integral) weight of f.

More recently I was investigating the same question for invariants of vector fields on $\mathbf{P}^n$ and discovered a means of generating exceptionals which are rather more vile. Here is a simple example. Consider the problem of constructing invariants of the module which is jets of vectors v^a of weight 0, at some point modulo the kernel of the linear invariant differential operator. In this case this operator is trace-free$(\nabla_a\nabla_b v^c)$ and is given in terms of $\mathbf{R}^{n+1}$ objects by $v^C_{AB} := \partial_A\partial_B v^C$, where v^C satisfies (the divergence free "gauge" condition) $\partial_C v^C = 0$. Note that, as well as being trace free, v^C_{AB} is annihilated upon contraction with X^A, since v^A is homogeneous of degree 1 with respect to X^A. Now let β_A be a covector in R^{n+1} which satisfies $X^A\beta_A = 1$. So β_A is homogeneous of degree -1 and is only defined up to transformations $\beta_A \mapsto \hat{\beta}_A = \beta_A + \Upsilon_A$ where $X^A\Upsilon_A = 0$. Consider now the object

$$\partial_E\partial_F(v^E_{AB}v^F_{CD}\epsilon^{ACI}\epsilon^{BDJ}\beta_I\beta_J).$$

This is clearly an invariant provided it is independent of the choice of β_A. Indeed it is independent of β_A, and so an invariant, because $\epsilon^{ACI}\hat{\beta}_I = \epsilon^{ACI}\beta_I + X^{[A}\gamma^{C]}$ for some γ^C. When expanded out it is given by the formula

$$\epsilon^{cd}\epsilon^{ef}(\nabla_c\nabla_e\nabla_a v^b\nabla_d\nabla_f\nabla_b v^a + 2\nabla_c\nabla_e\nabla_b\nabla_a v^a\nabla_d\nabla_f v^b + \nabla_c\nabla_e\nabla_a v^a\nabla_d\nabla_f\nabla_b v^b)$$

and so is clearly non-zero. (Thanks to Michael Eastwood for helping check this expansion.) Furthermore this is certainly an exceptional invariant since it is homogeneous of degree just two with respect to v^A and is an example which is not simply a juxtaposition of the linear invariant with itself. Fortunately it turns out Gover (1994b) that, for vectors on $\mathbf{P}^n$, all exceptional invariants can be constructed by a generalisation of the method used for this example. So the exceptional invariants can now be listed as readily as Weyl invariants. These methods work for many other similar modules and structures. For example Toby Bailey and I,Bailey & Gover (1994), have shown that an analogue of the above works for the conformal case.

These results have applications for curved geometries. In Bailey, Eastwood & Gover (1992) a "tractor calculus" is developed which includes curved analogues of the operators and tensors which are described above for the homogeneous (or "flat") case. For example there is an operator D_A to generalise ∂_A and a tensor $\epsilon^{AB\cdots E}$ which generalises the dual volume form on $\mathbf{R}^{n+1}$. Thus one can immediately construct analogues of the above exceptional operators for the corresponding curved geometries. In Bailey, Eastwood & Gover (1992) a new invariant for four dimensional conformal geometries is constructed in this spirit. In fact the main motivation for the above work and its various generalisations is the problem of listing all such invariants of the curved geometries themselves (rather than invariants of vector bundles over these geometries). It turns out that the latter problem can often be reduced to a problem, on the corresponding homogeneous geometry, of the sort discussed above (see e.g. Bailey, Eastwood & Graham (1994),Fefferman (1979)) A new approach to the projective case is discussed in Gover (1993).

References

Bailey, T.N., Eastwood, M.G. & Gover, A.R. (1992) Thomas's structure bundle for conformal, projective and related structures. To appear in: Rocky Mtn. J. Math.

Bailey, T.N., Eastwood, M.G. & Graham, C.R. (1994) Invariant theory for conformal and CR geometry. To appear in: Annals of Math.

Bailey, T.N. & Gover, A.R. (1994) Exceptional invariants in the parabolic invariant theory of conformal geometry. To appear in: Proc. Amer. Math. Soc.

Eastwood, M.G. & Graham, C.R. (1991) Invariants of conformal densities. Duke Math. Jour. **63**, 633–671.

Fefferman, C. (1979) Parabolic invariant theory in complex analysis. Adv. in Math. **31**, 131–262.

Fefferman, C. & Graham, C.R. (1985) Conformal invariants. In: *Élie Cartan et les Mathématiques d'Aujourdui.* Astérisque, 95–116.

Gover, A.R. (1994a) Invariants on projective space. J. Amer. Math. Soc. **7**, 145–158.

Gover, A.R. (1994b) Invariant theory for a parabolic subgroup of $\mathrm{SL}(n+1,\mathbf{R})$. To appear in: Proc. Amer. Math. Soc.

Gover, A.R. (1993) The Invariants of Projective Geometries, *preprint*.

§II.2.18 **The conformal Einstein equations** *by L.J.Mason & R.J.Baston* (TN 22, September 1986)

In twistor related problems in general relativity it is often useful to know when there exists a metric in a given conformal class which is vacuum (possibly with cosmological constant). If such a metric exists it is then useful to be able to characterise it in a conformally invariant fashion. Here we describe a set of conditions for a space-time to be conformal to vacuum emerged which completes an approach due to Kozameh, Newman & Tod (1985):

PROPOSITION. *If* $I = \Psi_{ABCD}\Psi^{ABCD} \neq 0 \neq \tilde{\Psi}_{A'B'C'D'}\tilde{\Psi}^{A'B'C'D} = \tilde{I}$ *then:*

$$(1) \qquad B_{ab} = (\nabla^C_{A'}\nabla^D_{B'} + \Phi^{CD}_{A'B'})\Psi_{ABCD} = 0$$

$$(2) \qquad E_{abc} = \Psi_{ABCD}\nabla^{DD'}\tilde{\Psi}_{A'B'C'D'} - \tilde{\Psi}_{A'B'C'D'}\nabla^{DD'}\Psi_{ABCD} = 0$$

iff the space-time is conformal to a solution of the Einstein equations. (These equations are conformally invariant.)

Sketch of Proof. The Einstein equations are $\nabla^A_{A'}\Psi_{ABCD} = 0$ and its conjugate. Under rescaling $g_{ab} \to \hat{g}_{ab} = \Omega^2 g_{ab}$ these become: $\hat{\nabla}^A_{A'}\Psi_{ABCD} = \Upsilon^A_{A'}\Psi_{ABCD}$, where $\Upsilon_a = \nabla_a \ln\Omega$. To obtain this from above, multiply (2) by $\tilde{\Psi}^{A'B'C'}{}_{E'}$, then, using $\tilde{\Psi}^{A'B'C'}{}_{E'}\tilde{\Psi}_{A'B'C'D'} = \frac{1}{2}\tilde{I}\epsilon_{D'E'}$, we get:

$$\nabla^A_{A'}\Psi_{ABCD} = \nu^A_{A'}\Psi_{ABCD} \qquad \text{where } \nu_{AA'} = \frac{2}{\tilde{I}}\tilde{\Psi}_{A'}{}^{B'C'D'}\nabla^{E'}_A\tilde{\Psi}_{B'C'D'E'}.$$

After some manipulation, (1) $\Rightarrow \nu_a$ is a gradient; $\nu_a = \nabla_a \ln\Omega$. Rescaling the metric by Ω^2 provides the metric in the conformal class satisfying Einstein's equations. (Thanks to K.P. Tod for the proof.) □

These equations have a particularly striking from when written in terms of the local twistor connection (see Penrose & Rindler 1986 for derivation). Let Ω^+, Ω^-, be the self-dual, anti-self-dual parts of the curvature respectively, and D the covariant exterior derivation. Then:

(1)	$B_{ab} = 0 \iff D\Omega^+ = 0$	(or $D\Omega^- = 0$)
(2)	$E_{abc} = 0 \iff [\Omega^+, \Omega^-] = 0$	([,] on Lie algebra indices, no wedge on form indices.)

Conformal invariance of the local twistor connection is broken by introducing an infinity twistor, $I_{\alpha\beta}$, $I_{\alpha\beta} = I_{[\alpha\beta]}$ etc. We have:

LEMMA. *There exists $I_{\alpha\beta}$ such that $DI_{\alpha\beta} = 0$ iff the space-time is conformal to Einstein. $I_{\alpha\beta}$ is simple iff the cosmological constant vanishes. Knowledge of $I_{\alpha\beta}$ is sufficient to recover the Einstein metric.*

The proof is obtained by straightforward manipulation.

Relationship to formal neighbourhoods of ambitwistor space. Afficionados will recognize (1) & (2) as the obstructions to extending the Ward transform of the local twistor bundle on ambitwistor space to the third and fourth formal neighbourhoods respectively. (1) and (2) were originally obtained as the obstructions to constructing the fifth and sixth formal neighbourhoods respectively for second order perturbations to the conformal structure of $\mathbb{M}$. This leads to:

CONJECTURE. *Existence of the* $(n+2)^{th}$ *formal neighbourhood of an ambitwistor space* $\Longleftrightarrow$ *The Ward transform of the local twistor bundle extends to* n^{th} *order.*

The Ward transform of $\mathbb{T}_{[\alpha\beta]}$ (the bundle of $I_{\alpha\beta}$'s) is closely related to LeBrun's Einstein bundle, LeBrun (1985). $I_{\alpha\beta}$ s.t. $DI_{\alpha\beta} = 0$ provides a holomorphic section of this bundle via the Ward transform. A proof of the conjecture would therefore unite LeBrun's approach with the formal neighbourhood approach as well as proving the general extendability result.

References

Kozameh, C.N., Newman, E.T. & Tod, K.P. (1985) Conformal Einstein spaces, Gen. Rel. Grav. **17**, p. 343–352.

LeBrun, C.R. (1985) Ambitwistors and the Einstein equations, Class. & Quant. Grav. **2**, 555–563.

Penrose, R. & Rindler, W. (1986) *Spinors and Space-Time, vol. II: Spinor and Twistor Methods in Space-Time Geometry*, C.U.P.

§II.2.19 **Self-dual manifolds need not be locally conformal to Einstein** *by T.N.Bailey & M.G.Eastwood* ($\mathbb{TN}$ 31, October 1990)

There are topological obstructions to the existence of Einstein metrics in four dimensions which imply, for example, that $\mathbb{CP}_2\#\mathbb{CP}_2\#\mathbb{CP}_2\#\mathbb{CP}_2$ does not admit such a metric. On the other hand, $\mathbb{CP}_2\#\cdots\#\mathbb{CP}_2$ always admits a self-dual metric (as shown abstractly by Donaldson & Friedman 1989 and explicitly by LeBrun 1991). The purpose of this note is to observe that there are also local obstructions to the existence of an Einstein metric within a given self-dual conformal class. We shall discuss the obstruction

$$K_{abc} = C^{efgh}C_{efgh}\nabla^d C_{abcd} - 4C^{efgh}C_{abcd}\nabla^d C_{efgd}$$

of Kozameh, Newman & Tod (1985) where C_{abcd} is the Weyl tensor. This tensor is easily shown to be conformally invariant and the contracted Bianchi identity shows that it vanishes in the case of an Einstein metric. If C_{abcd} is self-dual, then it may be written as $K_{abc} = 4K_{A'B'C'C}\varepsilon_{AB}$ where

$$K_{A'B'C'C} = \widetilde{\Psi}^2 \nabla_C^{D'} \widetilde{\Psi}_{A'B'C'D'} - 2\widetilde{\Psi}^{E'F'G'H'} \widetilde{\Psi}_{A'B'C'H'} \nabla_C^{D'} \widetilde{\Psi}_{E'F'G'D'}$$

(and $\widetilde{X}^2 = \widetilde{X}^{A'B'C'D'} \widetilde{X}_{A'B'C'D'}$).

In Baston & Mason (1987) two tensors were identified

$$E_{abc} = \Psi_{ABCD} \nabla^{DD'} \widetilde{\Psi}_{A'B'C'D'} - \widetilde{\Psi}_{A'B'C'D'} \nabla^{DD'} \Psi_{ABCD}$$
$$B_{ab} = (\nabla_{A'}^{C} \nabla_{B'}^{D} + \Phi_{A'B'}^{CD})\, \Psi_{ABCD}$$

whose vanishing in the case of algebraically general Weyl curvature is necessary and sufficient for the existence of an Einstein scale. These tensors evidently vanish for a self-dual metric. It is, therefore, enlightening to notice that

$$2\Psi_C{}^{EFG} \left(\delta_{A'}^{E'} \delta_{B'}^{F'} \delta_{C'}^{G'} \widetilde{\Psi}^2 - 2\widetilde{\Psi}^{E'F'G'H'} \widetilde{\Psi}_{A'B'C'H'} \right) E_{efg} = \Psi^2 K_{A'B'C'C}$$

Thus, if E_{abc} vanishes and the Weyl curvature is algebraically general (whence, in particular, Ψ^2 is nowhere vanishing), then K_{abc} also vanishes.

We claim, however, that there are self-dual metrics for which K_{abc} is non-zero and hence a genuine obstruction to the existence of an Einstein scale. In fact, the metric

$$(dw + 2xy\,dy - x\,dz)^2 + 2y^2 z\,(dx^2 + dy^2) + y^2\,dz^2$$

is self-dual but K_{abc} is nowhere vanishing—for example,

$$K \lrcorner \left(\frac{\partial}{\partial y} \otimes \frac{\partial}{\partial x} \otimes \frac{\partial}{\partial x} \right) = \frac{9}{8y^7 z^4}$$

and

$$K \lrcorner \left(\frac{\partial}{\partial y} \otimes \frac{\partial}{\partial x} \otimes \frac{\partial}{\partial y} \right) = \frac{9x(y^2 - 2z)}{4y^8 z^5} \,.$$

The explicit computation of K_{abc} is far from easy. These results were obtained via a computer program (written in 'maple' and available upon request: it computes K_{abc} along with other differential geometric fauna for any explicit metric and can also check whether the metric is self-dual).

Of course, the metric above was constructed to be self-dual—notice that $\partial/\partial w$ is a Killing vector so the metric arises from an Einstein-Weyl space together with a generalized monopole as described by Jones & Tod (1985). In fact, the Einstein-Weyl space in question is hyperbolic 3-space. This is also the basis of LeBrun's explicit metrics (LeBrun 1991) on $\mathbb{CP}_2\# \cdots \#\mathbb{CP}_2$. The generalized monopoles that he employs, however, are derived from Green's functions for the hyperbolic Laplacian.

Whilst extremely natural, they are computationally more difficult and we have not yet succeeded in completing the calculation of K_{abc} in this case. We suspect that it will turn out to be non-zero.[1]

The tensor K_{abc} is evidently an obstruction to rescaling the metric so that $\widehat{\nabla}^d \widehat{C}_{abcd} = 0$, a so-called '$C$ metric'. Indeed, its deviation in Kozameh, Newman & Tod (1985) is from

$$\nabla^d C_{abcd} + \Upsilon^d C_{abcd} = 0, \qquad (\star)$$

noting that if $C^2 \equiv C^{abcd}C_{abcd} \neq 0$, then $\Upsilon^h = -4C^{efgh}\nabla^d C_{efgd}/C^2$. Substituting back into $(\star)$ gives K_{abc}/C^2. In the algebraically general case Υ^h is automatically closed (Kozameh, Newman & Tod 1985). We presume this is an extra condition in the self-dual case.

We thank Claude LeBrun for many helpful communications.

References

Baston, R.J. & Mason, L.J. (1987) Conformal gravity, the Einstein equations and spaces of complex null geodesics, *Class. Quan. Grav.* **4**, 815–826.

Donaldson, S.K. & Friedman, R. (1989) Connected sums of self-dual manifolds and deformations of singular spaces, *Nonlinearity* **2**, 197–239.

Jones, P.E. & Tod, K.P. (1985) Minitwistor spaces and Einstein-Weyl spaces, *Class. Quan. Grav.* **2**, 565–577.

Kozameh, C., Newman, E.T. & Tod, K.P. (1985) Conformal Einstein spaces, *Gen. Rel. Grav.* **17**, 343–352.

LeBrun, C.R. (1991) Explicit self-dual metrics on $\mathbb{CP}_2 \# \cdots \# \mathbb{CP}_2$, *J. Diff. Geom.* **34**, 223–253.

Pedersen, H. and Tod, P. (1991) Class. Quan. Grav. **8**, 751–760.

[1] See Pedersen & Tod (1991) for further results on whether LeBrun's metrics are conformally Einstein.

Chapter 3

Aspects of general relativity

§II.3.1 **Introduction** *by L.P.Hughston & L.J.Mason*

The most profound connections between twistor theory and general relativity arise in connection with the relation between *curved* twistor spaces and curved space-time. These ideas are explored at length in volume III. But there are other connections with general relativity, of varying degrees of significance, and some of these are described in the articles appearing in this chapter.

In Penrose (1967), the very first paper written on twistor theory, there is an interesting section on the metrical structure of Minkowski space, where the concept of what we now call the *infinity twistor* is introduced, and it is indicated in passing that by use of a different type of infinity twistor it is possible to represent within the twistor framework global properties of the de Sitter space-time and the Einstein static universe. The de Sitter universe is obtained by use of an infinity twistor that is real but not simple; whereas the Einstein universe is obtained by use of an infinity twistor that is simple but not real. In Minkowski space the infinity twistor is both real and simple. The matter is not pursued further in Penrose (1967) but is here in §§II.3.2-3 in articles by Penrose and Hurd treating the general class of Freidman-Robertson-Walker cosmological models. For further details see Penrose & Rindler (1986) §9.5, and Hurd (1985).

The reason why twistor theory can be applied directly to these cosmological models, but not in general (one might say *yet*) to real curved space-times, is on account of the local conformally flatness of these models. But it is worth bearing in mind the very important role these cosmological models play in astrophysical arguments. Moreover, if the *zero initial Weyl curvature* hypothesis (Penrose 1979) on the nature of the initial cosmological singularity turns out to be correct, then it may be that cosmological applications of twistor theory will come into play in some kind of *asymptotic* sense, even if the space-time as a whole has non-trivial Weyl curvature.

Since the conformal structure of space-time undoubtedly *does* differ from that of Minkowski space, the role of conventional *flat* twistor theory is necessarily circumscribed. But it is surprising how here and there certain 'outcrops' of twistor-related structures arise in a fully curved space-time context. An important class of such structures arises in connection with the study of type $\{2,2\}$ space-times. These are space-times for which the Weyl spinor is doubly degenerate, and they include the most important examples of *vacuum* solutions of Einstein's equations, namely the Schwarzschild solution and the Kerr solution. The unexpected link with twistor theory was established by Walker & Penrose (1970) who showed that the Kerr solution admits a valence two *Killing spinor*, i.e. a non-trivial solution of the *twistor equation*.

In flat space-time this equation is intimately related to the momentum and angular momentum structure of a massive relativistic particle (i.e. to the conservation laws of the Poincaré group), but in curved space-time in general there are no solutions—integrability conditions involving curvature terms obstruct the existence of solutions. Walker & Penrose (1970) showed how the existence of this Killing spinor explained the existence of a hitherto mysterious fourth integral of the equations of motion for geodesic orbits in the Kerr solution, discovered by B. Carter. In a series of subsequent investigations (Hughston, Penrose, Sommers & Walker 1972, Hughston & Sommers 1973a,b, Sommers 1973a,b, Floyd 1973, Dietz & Rudiger 1981, and Jeffryes 1984, to list but a few) these results were extended to the charged Kerr solution and a number of other results were established interconnecting the Kerr family of gravitational fields and various twistor-related ideas.

Some of the articles in this chapter take this line of enquiry a few steps further. In §II.3.4 Jeffryes looks at Killing spinors of valence greater than two, and establishes algebraic relations relating such spinors to the Weyl spinor. In §II.3.5 and §II.3.6 the investigation of charged particle orbits in the Kerr family is taken up again, and some unusual features of the Kerr and Kerr-Newman geometries are examined in connection with properties of the special anti-symmetric *Killing-Yano* type tensor associated with such gravitational fields. (Cf. Floyd 1973, Dietz & Rudiger 1981, Penrose & Rindler 1986, pp. 107-111). It is shown that for a charged particle moving under the influence of the standard coulomb-like *test* electromagnetic field in the Kerr background, there exists an absolutely conserved vectorial quantity.

Another connection with flat space twistors arises from the observation due to L.Witten that the vacuum equations for space-times with two commuting Killing vectors can effectively be reduced to the self-dual Yang-Mills equations with a two-dimensional symmetry group. Thus, according to the philosophy espoused in chapter 1, these equations are integrable and a reduction of R.S.Ward's correspondence between solutions of the equations and certain holomorphic bundles on flat twistor space applies. In Ward (1983) an ansatz for the twistor description of a large number of such solutions was given and analyzed. In Woodhouse & Mason (1988) the Ward construction was reduced systematically to give a reduced twistor correspondence between solutions of the equations and holomorphic vector bundles on a *non-Hausdorff* Riemann surface. In that article, building on Woodhouse (1987), it was shown that the hidden symmetries of the stationary axisymmetric Einstein equations (the Geroch group) have a simple and direct interpretation on twistor space. For further details of the Ward transform see §I.4.0 and for a discussion of the structure of the reduced twistor space see the appendix to §I.6.1. For applications to cylindrical waves see Woodhouse (1989). The articles by Fletcher in this chapter, §§II.3.7-9, are concerned with understanding Killing horizons and singularities in space-time in terms of the structure of the bundle on the reduced twistor space. These are studied in the context of the Kerr solution in §§II.3.7-8, and a technique is given in §II.3.9 to understand Killing horizons more generally. For more on this work, see Fletcher & Woodhouse (1990).

The articles in §§II.3.10-11 by Penrose and Ashtekar are concerned with a geometric prooof of the positive energy theorem using the global behaviour of null geodesics in space-times with energy-momentum tensors satisfying the weak energy condition. The argument relies on showing that when the ADM energy is negative or zero there exists a null geodesic without conjugate points. This then contradicts either null geodesic completeness, the weak energy condition or the genericity condition. In §II.3.10 it is argued that for negative mass at space-like infinity, it should be possible to arrive at a contradiction if one assumes that every null geodesic has a conjugate point. In §II.3.11, it is shown that the behaviour of null geodesics near space-like infinity is indeed sensitive to the ADM energy momentum.

In §II.3.12, V.Thomas explains a programme, based on the exact sets and characteristic hypersurface initial data ideas of Penrose (see Penrose & Rindler 1984, and Penrose 1960, 1963, 1980), for which the aim is to calculate the full power series of the Weyl tensor at a point in terms of its symmetrized derivatives (symmetrized in the sense of the spinor indices). The symmetrized derivatives are in effect the gravitational field's null datum, the freely specifiable characteristic initial data for the field defined on the light cone through the point in question. A realization of the programme would give an alternate approach to the approximate numerical solution of Einstein's equations, in which one calculates the Taylor series at a point to some finite order, rather than solving the field equations using finite difference methods.

References

Carter, B. (1968) Global structure of the Kerr family of gravitational fields, Phys, Rev. **174**, 1559.

Dietz, W., & Rudiger, R. (1980,81) Space-times admitting Killing-Yano tensors, I, II. Proc. Roy. Soc. **A375**, 361-78, and **A381**, 315-22.

Fletcher, J. & Woodhouse, N.M.J. (1990) Twistor characterization of stationary axisymmetric solutions of Einstein's equations, in *Twistors in Mathematics and Physics*, (T.N.Bailey & R.J.Baston eds.), LMS Lecture Notes Series **156**, CUP.

Floyd, R. (1973) The dynamics of Kerr Fields, Ph.D. thesis, Birkbeck College, University of London.

Hughston, L.P., Penrose, R., Sommers, P.D. & Walker, M. (1972) On a quadratic first integral for the charged particle orbits in the charged Kerr solution, Comm. Math. Phys. **27**, 303-308.

Hughston, L.P., & Sommers, P.D. (1973a) Spacetimes with Killing tensors, Comm. Math. Phys. **32**, 147-152.

Hughston, L.P., & Sommers, P.D. (1973b) The symmetries of Kerr black holes, Comm. Math. Phys.

33, 129-133.

Hurd, T.R. (1985) The projective geometry of simple cosmological models, Proc. Roy. Soc. Lond. **A397**, 233-43.

Jeffryes, B.P. (1984) Space-times with two-index Killing spinors, Proc. Roy. Soc. **A392**, 323-341.

Penrose, R. (1960) A spinor approach to general relativity, Ann. Phys. **10**, 171-201.

Penrose, R. (1967) Twistor Algebra, J. Math. Phys., **8**, 345-66.

Penrose, R. (1979) Singularities and time-asymmetry, in General Relativity: an Einstein centenary survey (S.W.Hawking & W.Israel, eds) CUP.

Penrose, R. (1963), Null hypersurface initial data for classical fields of arbitrary spin and for general relativity, in Aerospace Research Laboratories Report 63-56, (P.G.Bergmann ed.) reprinted in Gen. Rel. Grav. **12**, pp. 225-264 (1980).

Penrose, R. & Rindler, W. (1984) *Spinors and space-time*, vol. I, CUP.

Sommers, P.D. (1973a) Killing tensors and type $\{2,2\}$ space-times, Ph.D. thesis, Department of Physics, University of Texas.

Sommers, P.D. (1973b) On Killing tensors and constants of motion, J. Math. Phys. **14**, 787-790.

Ward, R.S.(1983) Stationary axisymmetric space-times a new approach, Gen. Rel. Grav., **15**, no. 2, 105-9.

Witten, L. (1979) Phys. Rev. D, **19**, pp. 718-720.

Woodhouse, N.M.J. (1987) Twistor description of the symmetries of the Einstein equations for stationary axisymmetric space-times, Class. Quant. Grav., **4**, 799-814.

Woodhouse, N.M.J. & Mason, L.J. (1988) The Geroch group and non-Hausdorff twistor spaces, Nonlinearity, **1**, 73.

Woodhouse, N.M.J. (1989) Cylindrically gravitational waves, Class. Quant. Grav., **6**, 933-943.

§II.3.2 **Twistors for Cosmological Models** *by R. Penrose* (𝕋ℕ 12, June 1981)

The Friedmann-Robertson-Walker [FRW] cosmological models are all conformally flat (as follows at once from their spatial isotropy, since the Weyl curvature, with its principal null directions, must vanish). Thus twistors (of the standard 'flat' variety) can be directly used to study their properties. These models differ from Minkowski space in that they have a different scale factor determining the metric, but also their global properties are different in the sense that they are conformal to different portions of the Einstein cylinder $\mathcal{E}$ ($\cong S^3 \times \mathbb{R}$). The space $\mathcal{E}$ is conformally the universal covering space of $\mathbb{M}$ (compactified Minkowski space), while $\mathbb{M}$ itself may be regarded as a real quadric (signature $++----$) in $\mathbb{RP}^5$.

Let us first consider de Sitter space $\mathcal{M}$ and compare it with Minkowski space $\mathbb{M}^I$. The boundary of $\mathcal{M}$ (considered a conformal subset of $\mathcal{E}$) is a pair of spacelike hypersurfaces $\tilde{\mathscr{I}}$ and $\mathscr{I}$ which get identified with one another when $\mathcal{E}$ is wrapped back up to the quadric $\mathbb{M}$. This identified hypersurface turns out to be a hyperplane section of $\mathbb{M}$ in the projective space $\mathbb{RP}^5$. Removing this hypersurface from $\mathbb{M}$ we get a space conformal to $\mathcal{M}$. This is very similar to the case of $\mathbb{M}^I$, the only difference being that for $\mathbb{M}^I$, the hyperplane in $\mathbb{RP}^5$ touches $\mathbb{M}$, the contact point being i (or I). A hyperplane R in $\mathbb{RP}^5$ can be represented twistorially by a skew twistor $R_{\alpha\beta}$ which is 'real' in the sense that its twistor complex conjugate $\overline{R}^{\alpha\beta}$ and its dual $R^{\alpha\beta}$ are equal. We may also think of $R^{\alpha\beta}$ as representing a point in $\mathbb{RP}^5$, this point being the pole of R with respect to the quadric $\mathbb{M}$, but the point and hyperplane R represent equivalent information. The point lies on $\mathbb{M}$ iff its polar hyperplane R touches $\mathbb{M}$ iff the twistor $R_{\alpha\beta}$ is simple. Thus de Sitter space has an 'infinity twistor' $I^{\alpha\beta}$, with dual $I_{\alpha\beta}$, which is *real* in the above sense and which is *not simple*. In fact, for *de Sitter* space we always have

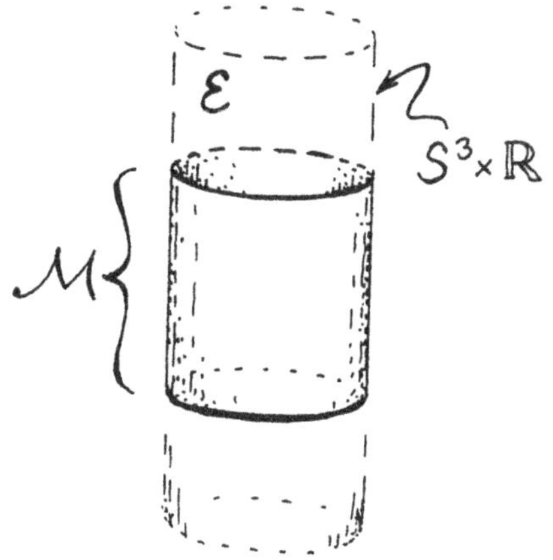

$$I_{\alpha\beta}I^{\alpha\beta} > 0,$$

while for *anti-de Sitter* space we should have

$$I_{\alpha\beta}I^{\alpha\beta} < 0,$$

the corresponding hypersurface $\mathscr{I}$ being now timelike. We can take the radius of the de Sitter pseudosphere to be

$$(I_{\alpha\beta}I^{\alpha\beta})^{-1/2}.$$

The de Sitter [anti-de Sitter] group is the subgroup of the twistor group $SU(2,2)$ which leaves $I^{\alpha\beta}$ invariant.

The particular FRW models just considered have the uncharacteristic property that they do *not* single out a unique 1-parameter family of hypersurfaces of homogeneity. In all other cases such a family *is* singled out, and they are determined in $\mathbb{RP}^5$ as the intersections of $\mathbb{M}$ with the hyperplanes of a pencil, i.e. by the family of skew twistors

$$\lambda I_{\alpha\beta} + \mu \tilde{I}_{\alpha\beta} \qquad (1)$$

given by varying the ratio $\lambda : \mu$, where $I_{\alpha\beta}$ and $\tilde{I}_{\alpha\beta}$ are fixed non-proportional skew twistors. The members of the family (1) are to be all real (in the above twistor sense). One way of achieving this is to make $I_{\alpha\beta}$ and $\tilde{I}_{\alpha\beta}$ both real and to take $\lambda, \mu \in \mathbb{R}$. Another is to take $\tilde{I}_{\alpha\beta} = \bar{I}_{\alpha\beta}$ (= complex conjugate of dual of $I_{\alpha\beta}$) and $\lambda = \bar{\mu}$.

There are three allowed possibilities for the pencil (1): the axis X of the pencil (= the 3-plane common to all members of the pencil) may (i) miss $\mathbb{M}$ altogether, (ii) touch $\mathbb{M}$, or (iii) meet $\mathbb{M}$ in an S^2.

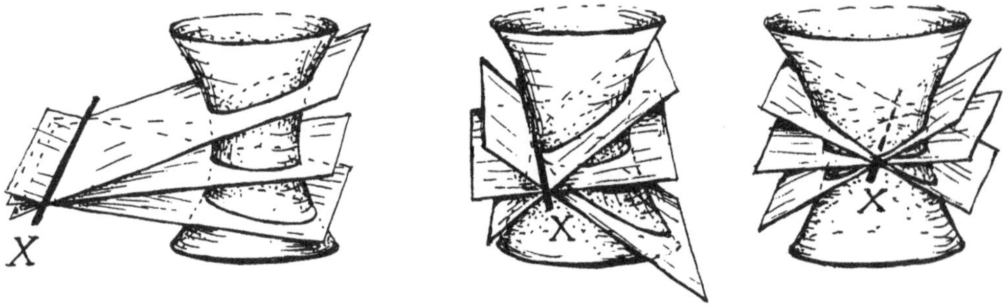

In case (i) the hypersurfaces of homogeneity are all S^3's ($k = 1$); in case (ii) they are all flat ($k = 0$); and in case (iii) the spacelike members of the family are all Lobachevsky (hyperbolic) spaces. There are also some disallowed cases in which the sections of $\mathbb{M}$ by the members of the pencil are all non-spacelike. In case (iii), there are two members of the pencil which touch $\mathbb{M}$ and we can, if desired, choose the corresponding members of the family (1) to be $I_{\alpha\beta}$ and $\tilde{I}_{\alpha\beta}$ themselves. In this case $I_{\alpha\beta}$ and $\tilde{I}_{\alpha\beta}$ will both be simple. They may or may not represent infinity for the FRW model, and they only could do so (for reasonable matter) if the cosmological constant vanishes. For non-vanishing cosmological constant we may prefer to choose $I_{\alpha\beta}$ (and/or $\tilde{I}_{\alpha\beta}$) actually to represent infinity. In that case we get $I_{\alpha\beta}I^{\alpha\beta}$ having the same sign as the cosmological constant (because $\mathscr{I}$ is spacelike for positive, null for zero, and timelike for negative cosmological constant). In case (i), we find that the members of the family which touch $\mathbb{M}$ would be conjugate complex, so if we choose these to be given by $I_{\alpha\beta}$ and $\tilde{I}_{\alpha\beta}$ we get $I_{\alpha\beta}$ complex and simple, and we can take $\tilde{I}_{\alpha\beta} = \bar{I}_{\alpha\beta}$. In case (ii) we find that only one member of the pencil touches $\mathbb{M}$ (giving (1) simple). In each one of these cases the symmetry group (leaving the homogeneity surfaces invariant) is the subgroup of $SU(2,2)$ leaving $I_{\alpha\beta}$ *and* $\tilde{I}_{\alpha\beta}$ invariant.

In models where there is a big bang (and/or big crunch) there will be a member of the pencil which corresponds to this. Thus we shall have a *bang twistor* $B_{\alpha\beta}$ (and/or a *crunch twistor* $C_{\alpha\beta}$) which is skew, twistor-real, and non-simple with

$$B_{\alpha\beta}B^{\alpha\beta} > 0 \quad (C_{\alpha\beta}C^{\alpha\beta} > 0)$$

since the bang (and/or crunch) is spacelike. If there is both a bang and a crunch there is the possibility that $B_{\alpha\beta}$ and $C_{\alpha\beta}$ turn out to be proportional—which would be the case with $k = +1$ if an observer 'just sees all the way around the universe' only at the moment that he encounters the crunch. (This is the case conformally identical with the de Sitter case, only here the boundary surfaces $\tilde{\mathscr{I}}$ and $\mathscr{I}$ at infinity are replaced by the singular boundaries of the bang and crunch, respectively). But in the general case, $B_{\alpha\beta}$ and $C_{\alpha\beta}$ would be non-proportional and could, if desired, be used in place of $I_{\alpha\beta}$ and $\tilde{I}_{\alpha\beta}$ in defining the pencil (1). An alternative—and this might be felt to be more appropriate in the case of a (say $k = -1$) model which starts from a bang and then expands to infinity—would be to use $B_{\alpha\beta}$ and $I_{\alpha\beta}$ (assuming they are non-porportional, as they would always be if $k = -1$).

Consider $k = \pm 1$ and make the choice (described earlier) where $I_{\alpha\beta}$ and $\tilde{I}_{\alpha\beta}$ are both simple ($\tilde{I}_{\alpha\beta} = \bar{I}_{\alpha\beta}$ if $k = +1$; $I_{\alpha\beta}$, $\tilde{I}_{\alpha\beta}$ both real if $k = -1$). We can scale things, assuming there is a bang, so that

$$B_{\alpha\beta} = \frac{1}{2}(I_{\alpha\beta} + \tilde{I}_{\alpha\beta}).$$

A point P of the space-time lies on the hypersurface given by (1) iff $(\lambda I_{\alpha\beta} + \mu \tilde{I}_{\alpha\beta})P^{\alpha\beta} = 0$ where $P^{\alpha\beta}$ is skew, simple and twistor-real, representing P. We can take this as

$$(\tau^{-1} I_{\alpha\beta} + \tau \tilde{I}_{\alpha\beta})P^{\alpha\beta} = 0$$

where $\tau = e^T$ for $k = -1$, and $\tau = e^{iT}$ for $k = +1$, with

$$T = \int_0 \frac{dt}{R(t)}.$$

Here t and $R = R(t)$ are the standard time and universe radius parameters for

$$ds^2 = dt^2 - R^2 d\Sigma^2$$

and $d\Sigma^2$ = squared metric of unit S^3 ($k = +1$) or unit Lobachevsky space ($k = -1$).

Editor's comment (1994). For further details see Penrose & Rindler (1986) *Spinors and space-time* Vol.2, CUP, §9.5 and Hurd, T.R. (1985) The projective geometry of simple cosmological models, Proc. Roy. Soc. **A397**, 233-43. See also §II.3.3.

§II.3.3 **Cosmological Models in $\mathbb{P}^5$** *by T.R. Hurd* (TN 16, August 1983)

In §II.3.2 in this volume and Penrose & Rindler (1986) a remarkably simple picture of the general Friedmann-Robertson-Walker [FRW] cosmological models is described. The point was that being conformally flat these models can be represented as conformal subsets of the Einstein cylinder, or isometrically as sections through a 5-dimensional cone in $\mathbb{R}^6$, or in terms of a quadric hypersurface Ω in $\mathbb{P}^5$.

In the $\mathbb{P}^5$ picture, the general geometry is characterized by the choice of a 3-plane $\mathscr{P}$ and the associated pencil of 4-planes which contain $\mathscr{P}$. This pencil foliates the space-time by a family of space-like hypersurfaces, the surfaces of constant 'cosmological time'. The three distinct ways the pencil can slice up the space-time, given by

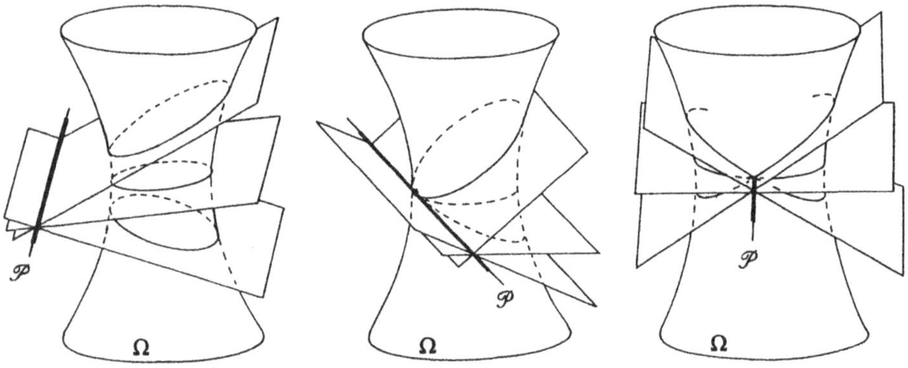

$\mathscr{P} \cap \Omega = \emptyset$ $\qquad$ $\mathscr{P} \cap \Omega =$ one point $\qquad$ $\mathscr{P} \cap \Omega = S^2$

correspond to the $k = +1, 0, -1$ models respectively. The metric of a general FRW model is now characterized by calculating the conformal factor $\Omega(X)$ (constant on each slice) which relates it to the flat metric. This gives a homogeneity $+1$ section $\tilde{I}(X) = \Omega(X) I_{\alpha\beta} X^{\alpha\beta}$ on $\mathbb{P}^5$ which can be thought of as giving the space-time as the section of the 5-cone in $\mathbb{R}^6$ with $\tilde{I}(X) = 2$.

I wish to point out how, with a slight change of point of view, this fits neatly into the general $\mathbb{P}^5$ framework developed by Hughston & Hurd (1983). As usual, take homogeneous coordinates X^i, $i = 0, 1, ..., 5$ on $\mathbb{P}^5$ and define the quadric hypersurface Ω by the equation $\Omega_{ij} X^i X^j = 0$ where Ω_{ij} has signature $++----$. The conformal structure of flat space is given by the null separation condition $X^i Y^j \Omega_{ij} := X \cdot Y = 0$ for points $X, Y \in \Omega$. Ω_{ij} acts as a conformal metric 'tensor': to obtain the general conformally flat metric tensor on Ω we adjust the conformal weight of Ω_{ij} to -2 by introducing a homogeneity $+1$ section $J(X)$. That is,

$$g_{ij} = J^{-2} \Omega_{ij},$$

or in line element form

$$ds^2 = J^{-2}\Omega_{ij}\, dX^i\, dX^j.$$

The curvature of such a metric on Ω can be calculated to give:

$$R_{ij} = -2J^{-1}J_{ij} + \left\{3J^{-2}J_kJ^k - J^{-1}J_{kl}\Omega^{kl}\right\}\Omega_{ij}$$

$$R = 12J_kJ^k - 6JJ_{kl}\Omega^{kl} = 6J^3\nabla\cdot\nabla(J^{-1})$$

$$\text{and} \quad G_{ij} = -2J^{-1}J_{ij} + \left\{-3J^{-2}J_kJ^k + 2J^{-1}J_{kl}\Omega^{kl}\right\}\Omega_{ij}(1)$$

where $J_{kl} := \nabla_k\nabla_l J$ and $J_k := \nabla_k J$. The metric of flat space for example derives from $J(X) = 2I_iX^i$ where $I_i = I_{\alpha\beta}$ is the 'infinity twistor'. More generally, the zeroes of J define $\mathscr{I}$ while the singularities of J are curvature singularities.

For a FRW space-time, we choose a pencil $\mathscr{P}$ of 4-planes $\alpha A_i + \beta B_i$, $\alpha, \beta \in \mathbb{C}$ which is real in the sense that $\overline{A}, \overline{B} \in \mathscr{P}$. The intersection of each real 4-plane C_i of this pencil with Ω is a 3-dimensional hypersurface which is space-like, null, or time-like if $C\cdot C$ is > 0, $= 0$ or < 0 respectively. Of the possible real pencils, only the three cases illustrated above provide space-like 3-surfaces. When $k = \pm 1$, we can without loss of generality choose A_i and B_i to be tangent to Ω, i.e. $A^2 = B^2 = 0$. Then the reality and normalization conditions

$$A = \overline{B}; \quad A\cdot B = +1/2 \qquad (\text{for } k = +1)$$
$$A = \overline{A}, \quad B = \overline{B}; \quad A\cdot B = -1/2 \qquad (\text{for } k = -1)$$

can be imposed. (The pencil for a $k = 0$ model is spanned by real planes C_i and D_i with $C\cdot C = C\cdot D = 0$ and $D\cdot D = +1$. This case must be dealt with separately).

The assumption of spatial isotropy restricts the general form of $J(X)$ for a $k = \pm 1$ FRW model:

$$J(X) = \sqrt{ab}\, S(\tfrac{b}{a})$$

where $a := 2(A\cdot X)$, $b := 2(B\cdot X)$ and S is a free function of one variable. For a J of this form the combinations appearing in (1) can be worked out

$$J_kJ^k = S^2\left[-1 + 4\left(\frac{b}{a}\right)^2\left(\frac{S'}{S}\right)^2\right]$$

$$J^{-1}J_{ij} = 4\left[\left(\frac{b}{a}\right)^2\frac{S''}{S} + \left(\frac{b}{a}\right)\frac{S'}{S} - \frac{1}{4}\right]\left(\frac{B_i}{b} - \frac{A_i}{a}\right)\left(\frac{B_j}{b} - \frac{A_j}{a}\right)$$
$$= J(J_{kl}\Omega^{kl})V_iV_j$$

where V_i is the unit time-like vector normal to the constant time surfaces, and

$$J(J_{kl}\Omega^{kl}) = 4S^2\left[\left(\frac{b}{a}\right)^2\frac{S''}{S} + \left(\frac{b}{a}\right)\frac{S'}{S} - \frac{1}{4}\right].$$

The stress tensor T_{ij} for a FRW model has the general form $T_{ij} = -p(X)g_{ij} + (\mu+p)V_iV_j$ where the scalar fields p and μ are known as the pressure and energy density. Comparison with G_{ij} via the Einstein equation then leads to the following equations:

$$p(X) = 3J_kJ^k - 2JJ_{kl}\Omega^{kl} = S^2\left[-1 + 12\left(\frac{b}{a}\right)^2\left(\frac{S'}{S}\right)^2 - 8\left(\frac{b}{a}\right)^2\frac{S''}{S} - 8\left(\frac{b}{a}\right)\frac{S'}{S}\right]$$

$$\mu(X) = -3J_kJ^k = S^2\left[3 - 12\left(\frac{b}{a}\right)^2\left(\frac{S'}{S}\right)^2\right].$$

Particular models are derived from an equation of state such as $p = 0$ (dust models), $p = -\mu/3$ (radiation models) etc., which gives a differential equation which can be solved for S and hence leads to $J(X)$.

With less work, one can find $J(X)$ for one's favourite model by choosing coordinates in the following way. For $k = -1$ models, write

$$X^i(\eta, \chi, \theta, \phi) = e^\eta A^i + e^{-\eta}B^i + Q^i(\chi, \theta, \phi)$$

where

$$A^i = (1/2, 1/2; 0, 0, 0, 0),$$

$$B^i = (-1/2, 1/2; 0, 0, 0, 0),$$

and

$$Q^i = (0, 0; \cosh\chi, \sinh\chi\cos\theta, \sinh\chi\sin\theta\cos\phi, \sinh\chi\sin\theta\sin\phi)$$

satisfy $(Q\cdot A) = (Q\cdot B) = 0$ and $A\cdot B = -\frac{1}{2}$. Now $a = 2(A\cdot X) = -e^{-\eta}$ and $b = 2(B\cdot X) = -e^{\eta}$. If we put $\Omega_{ij} = diag(1, -1; 1, -1, -1, -1)$ then $(X\cdot X) = -1 + Q^iQ_i = 0$ and the points with $\eta =$ constant form space-like 3-surfaces. Now

$$dX^i = e^\eta A^i d\eta - e^{-\eta}B^i d\eta + dQ^i$$

$$\Omega_{ij}\, dQ^i\, dQ^j = -d\chi^2 - \sinh^2\chi\big[d\theta^2 + \sin^2\theta\, d\phi^2\big] := d\Sigma^2$$

and $J(X) = S(e^{2\eta})$. Therefore

$$ds^2 = J^{-2}\Omega_{ij}\, dX^i dX^j$$

$$= \big[S(e^{2\eta})\big]^2\{d\eta^2 - d\Sigma^2\}$$

Having put the metric in this standard form, one can read off S and hence $J(X)$ for any cosmological model one wants. For one example, consider the *open Friedmann dust model.* This is the $k = -1$ solution with equation of state $p = 0$ and has

$$S(x) = (\sqrt{x} + \frac{1}{\sqrt{x}} - 2)^{-1};$$

$$J(X) = 2(A\cdot X)(B\cdot X)\big(\sqrt{A\cdot X} - \sqrt{B\cdot X}\big)^{-2}.$$

The model has

$\mathscr{I}^+ =$ null cone $\{X : A \cdot X = 0\}$

$\mathscr{I}^- =$ null cone $\{X : B \cdot X = 0\}$

'big bang' = hypersurface

$\{X : (A - B) \cdot X = 0\}$

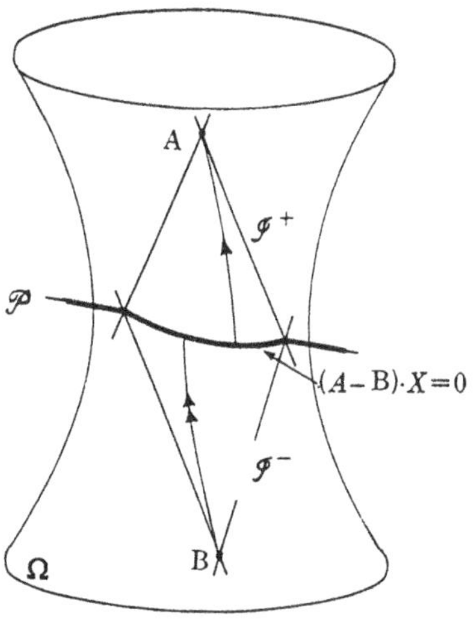

One can now use the general $\mathbb{P}^5$ framework for tensor, spinor and twistor analysis on FRW models. For example, one could study anisotropic perturbations $g_{ij} = J^{-2}\Omega_{ij} + \epsilon N_{ij}$ in the early universe (such a metric has a first order Weyl tensor $C_{ijkl} = J^{-2}\nabla_{[i}\nabla_{[k}J^2 N_{j]l]}$). Another direction where this framework might be useful is in the study of more general asymptotic behaviour than asymptotic flatness.

Thanks are due to A.D.Helfer and L.P.Hughston.

References

Penrose, R. & Rindler, W. (1986) *Spinors and Space-Time, vol. II: Spinor and Twistor Methods in Space-Time Geometry*, CUP.

Hughston, L.P. & Hurd, T.R. (1983) A $\mathbb{CP}^5$ calculus for space-time fields, Phys. Rep. **100**, p. 273–326.

Hurd, T.R. (1985) The projective geometry of simple cosmological models, Proc. Roy. Soc. Lond. **A397**, p. 233–243.

§II.3.4 **Curved Space Twistors and GHP** *by B.P.Jeffryes* (TN 15, January 1983)

As is well known, if $\omega^{A\ldots B}$ is a solution to the n index twistor equation,

$$\nabla^{A'(A}\omega^{B\ldots C)} = 0 \tag{1}$$

then an additional consistency condition may be found by applying $\nabla^{D}_{A'}$ and symmetrizing over unprimed indices (Penrose 1975), namely

$$\Psi_F^{(ABC}\omega^{D\ldots E)F} = 0. \tag{2}$$

It is clear that if ω is null then Ψ_{ABCD} is also null and they share the same principal null direction (henceforth abbreviated to pnd). It is not immediately clear what the restriction on Ψ_{ABCD} is if ω is of a more general type. The conditions may be simply found by taking components of (1) in the GHP formalism (Geroch, Held & Penrose 1973). Those not familiar with the formalism should read what follows in conjunction with the above reference.

THEOREM. *If $\omega^{A\ldots B}$ is a solution to the n-index twistor equation then Ψ_{ABCD} is proportional to $(\omega_{A\ldots B})^{4/n}$.*

Proof. If we choose $\{\iota^A, o_A\}$ as a spinor dyad normalized by $\iota^A o_A = 1$, then if $\omega^{AB\ldots CD}$ is a solution to (1) we write

$$\begin{aligned}
\omega_0 &= o^A \ldots o^D \omega_{A\ldots D} && \text{type } \{n, 0\} \\
\omega_1 &= \iota^A o^B \ldots o^D \omega_{A\ldots D} && \text{type } \{n-2, 0\} \\
&\ \vdots && \\
\omega_{n-1} &= \iota^A \ldots \iota^C o^D \omega_{A\ldots D} && \text{type } \{2-n, 0\} \\
\omega_n &= \iota^A \ldots \iota^D \omega_{A\ldots D} && \text{type } \{-n, 0\}\ .
\end{aligned}$$

First we consider the case when ω is null. We set o_A as its pnd. ω_n is then the only non-zero component and (1) becomes

$$\kappa\omega_n = 0 \qquad \sigma\omega_n = 0 \tag{3}$$

$$(\text{þ} + n\rho)\omega_n = 0 \qquad (\eth + n\tau)\omega_n = 0 \tag{4}$$

$$\eth'\omega_n = 0 \qquad \text{þ}'\omega_n = 0. \tag{5}$$

Equation (3) implies $\sigma = \kappa = 0$. By GHP (2.23) therefore $\Psi_0 = 0$. Applying the commutator $[\text{þ}', \eth']$ to ω_n and using GHP (2.31) gives

$$0 = n\Psi_3;$$

applying $[\text{þ}, \eth]$ and using GHP (2.21) and (2.24) we have

$$0 = 3n\Psi_1.$$

$[\text{þ}', \text{þ}]$ and $[\eth', \eth]$ give us

$$\begin{aligned} \text{þ}'\rho &= -\tau\bar{\tau} + \Psi_2 + \Phi_{11} - \Lambda \\ \eth'\tau &= -\rho\bar{\rho}' - \Psi_2 + \Phi_{11} - \Lambda. \end{aligned}$$

Substituting into GHP (2.26) gives $\Psi_2 = 0$; thus Ψ_4 is the only non-vanishing component.

If ω is not null, then it has at least two pnds. We choose two of these as our dyad. Thus $\omega_0 = \omega_n = 0$. (1) becomes

$$\begin{array}{cccccc} \kappa\omega_1 & & & & & = 0 \\ (\text{þ} + \rho)\omega_1 & - & n\,\kappa\omega_2 & & & = 0 \\ (\eth' + (n-1)\tau')\omega_1 - & & n(\text{þ} + 2\rho)\omega_2 & + \ldots & & = 0 \\ \tau'\omega_1 & & -\,n(\eth' + (n-2)\tau')\omega_2 + & \ldots & & = 0 \\ & - & n\,\sigma'\omega_2 & + \ldots + & \ldots & = 0 \\ & & & \vdots & \vdots & \end{array} \tag{6}$$

and similarly if σ, $\eth$, $\text{þ}'$, κ' are substituted for κ, þ, $\eth'$, σ' respectively. Firstly we have $\sigma = \kappa = \sigma' = \kappa' = 0$. Thus $\Psi_0 = \Psi_4 = 0$. Applying $[\text{þ}, \eth]$ to the non-vanishing ω_m with the lowest value of m gives

$$(n - 4m)\Psi_1 = 0. \tag{7}$$

Similarly applying $[\text{þ}', \eth']$ to the non-vanishing ω_{n-k} with the lowest value of k implies

$$(n - 4k)\Psi_3 = 0. \tag{8}$$

First we consider the case where n is not a multiple of 4. Then $\Psi_1 = \Psi_3 = 0$, and thus Ψ_{ABCD} has only 2 pnds. Since all pnds of ω are also pnds of Ψ_{ABCD}, ω has only one non-vanishing component, whose derivatives are given by

$$\begin{aligned} (\text{þ} + m\rho)\omega_m &= (\eth + m\tau)\omega_m = 0 \\ (\text{þ}' + (n-m)\rho')\omega_m &= (\eth' + (n-m)\tau')\omega_m = 0. \end{aligned} \tag{9}$$

Applying $[\text{þ}', \text{þ}]$ and $[\eth, \eth']$ to ω_m and using GHP (2.26) we find

$$(n - 2m)\Psi_2 = 0 \tag{10}$$

To recap: if n is not a multiple of 4 then ω is either a power of $\omega_0\, o_A$ or of $\omega_1\, o_{(A}\iota_{B)}$ and the Weyl spinor is given respectively by $\Psi_4\, o_A o_B o_C o_D$ and $6\Psi_2\, o_{(A} o_B \iota_C \iota_{D)}$.

If $n = 4m$ then we seek to show that ω is a power of a 4-twistor which is proportional to Ψ_{ABCD}. ω can have a maximum of 4 pnds since every pnd of ω is also one of Ψ_{ABCD}. We examine the cases according to the type of Ψ_{ABCD}.

$\Psi_{ABCD} = o_{(A}\iota_B\alpha_C\beta_{D)}$. Since for no choice of pnd of ω do either Ψ_1 or Ψ_3 vanish each pnd must by (7) and (8) have multiplicity $n/4$.

$\Psi_{ABCD} = o_{(A}o_B\iota_C\alpha_{D)}$. If our dyad is $\{\iota^A, o_A\}$ then (8) tells us ι_A has multiplicity $n/4$; similarly if $\{\alpha^A, o_A\}$ is the dyad α_A has multiplicity $n/4$. Hence o_A has multiplicity $n/2$.

$\Psi_{ABCD} = o_{(A}o_B\iota_C\iota_{D)}$. ω has only one term. By (10) we see that o_A and ι_A both have multiplicity $n/2$.

$\Psi_{ABCD} = o_{(A}o_Bo_C\iota_{D)}$. By (8) ι_A has multiplicity $n/4$, hence o_A has multiplicity $3n/4$. □

References

Geroch, R., Held, A. & Penrose, R. (1973) A space-time calculus based on pairs of null directions, J. Math. Phys. **14**, pp. 874–881.

Penrose, R. (1975) Twistor theory, its aims and achievements, in *Quantum Gravity: an Oxford Symposium* (C.J.Isham, R.Penrose, & D.W.Sciama, eds), OUP p. 370.

§II.3.5 A Note on Conserved Vectorial Quantities associated with the Kerr Solution *by L.P.Hughston* (TN 24, September 1987)

1. Introduction. Chandrasekhar (1983) points out that whereas the equations governing gravitational perturbations of the Kerr solution can be decoupled and separated, and hence solved, despite their byzantine complexity, the corresponding equations governing the combined gravitational-electromagnetic pertubations of the Kerr-Newman solution cannot. Although a definitive proof to this effect is lacking, numerous efforts to separate the equations have been unsuccessful.

The separability in the one case and apparent unseparability in the other cannot simply be an accident, since we know in the case of the Kerr solution that the separability of various systems of partial differential equations is intimately linked with the existence of a Killing tensor for that space-time, and hence ultimately to the existence of a solution of the *twistor equation* $\nabla_{A'(A}\chi_{BC)} = 0$

(Walker & Penrose 1970). In a general way therefore we conclude that there must very likely be some feature of the Kerr geometry in connection with the twistor field χ_{AB} that does not carry through to the Kerr-Newman geometry—and that this feature, or rather lack thereof, is responsible for the non-separability of the perturbation equations in the Kerr-Newman case. In this note I shall examine an unusual feature of just this sort which arises in connection with the *Killing-Yano tensor* associated with the Kerr and Kerr-Newman space-times.

2. The Killing-Yano Tensor. Let us begin by fixing a few conventions. In the case of both the Kerr solution (K) and the Kerr-Newman solution (K-N) there exists a solution χ_{AB} of the twistor equation $\nabla_{A'(A}\chi_{BC)} = 0$. The freedom of a constant factor is fixed by requiring that the vector $\xi_{A'A}$, defined by

$$\nabla_{A'A}\chi_{BC} = \xi_{A'(B}\varepsilon_{C)A} \tag{1}$$

or equivalently by

$$\nabla^{A'A}\chi_{AB} = \tfrac{3}{2}\xi_B^{A'}, \tag{2}$$

which is necessarily a Killing vector (cf. Hughston & Sommers 1973, Penrose & Rindler 1986, p. 108), should be future-pointing and asymptotically of unit norm. Then the skew tensor L_{ab} defined by

$$L_{ab} = \mathrm{i}\chi_{AB}\varepsilon_{A'B'} - \mathrm{i}\bar{\chi}_{A'B'}\varepsilon_{AB} \tag{3}$$

satisfies the Killing-Yano equation $\nabla_{(a}L_{b)c} = 0$: it is the trace-free part of this equation that is equivalent to the twistor equation, whereas the vanishing divergence $\nabla^a L_{ab} = 0$ is implied by the reality of ξ_a (cf. Hughston & Tod 1990, p. 150.)

In both K and K-N an absolutely conserved (i.e. parallelly propagated) vectorial quantity can be constructed for *geodesic* orbits by use of L_{ab} (Floyd 1973; cf. also Penrose & Rindler 1986, p. 110). Thus if $U^a\nabla_a U^b = 0$ then the vector H_a defined by $H_a = L_{ab}U^b$ is parallelly propagated: $U^a\nabla_a H_b = 0$, which follows as a simple consequence of the Killing-Yano relation.

3. Charged Particle Orbits. Here the situation is rather more complicated, for in the case of K-N we consider a charged particle moving under the influence of the associated electromagnetic field of the K-N solution—whereas in the case of K we consider the motion of a charged particle in a special 'test' electromagnetic field of the Kerr type with its principal directions aligned with those of the gravitational field.

In each case, as with geodesic orbits, we can construct a Carter-type fourth integral of the motion (Hughston *et al* 1972) given by $Q = H_a H^a$ with H_a defined as above, where U^a is tangent to the orbits in question.

On the other hand the situation as regards conserved *vectorial* quantities for charged particle orbits is more subtle. Floyd (1973) shows that in K-N although a conserved (i.e. parallelly propagated) vector cannot apparently be constructed for charged particle orbits nevertheless a vectorial quantity can indeed be found which satisfies a natural *modified* propagation law. Suppose F_{ab} is the electromagnetic field associated with K-N, and let the equations of motion for a charged particle in that background be

$$m\,U^a\nabla_a U^b = \varepsilon\,F^{bc}U_c\,, \tag{4}$$

where m is the mass and ε is the charge of the particle. The vector $H_a = L_{ab}U^b$, defined as before, satisfies according to Floyd the following propagation equation:

$$m\,U^a\nabla_a H^b = \varepsilon\,F^{bc}H_c\,. \tag{5}$$

This arises from the fact that the tensor $\theta_{ab} = L_{ac}F^c_{\ b}$ is *automatically symmetric*, a circumstance deriving from the fact that the spinor ϕ_{AB} associated with F_{ab} is *proportional* to χ_{AB} (Hughston *et al* 1972).

In the case of K the vector H^a is likewise 'Lorentz propagated', i.e. according to equation (5), if here we interpret F^{ab} as the special test field as described above (with a specified charge).

4. Conserved Vectorial Quantities for Charged Particle Orbits. Now what about the prospects for constructing an absolutely *absolutely* conserved vector in the case of charged particle motion? In this connection it is helpful to recall a simple result (Hughston *et al* 1972) concerning the *energy* integral for charged particle motion:

LEMMA. *Let F_{ab} satisfy the Maxwell equation $\nabla_{[a}F_{bc]} = 0$, and suppose U^a satisfies the Lorentz equation (4). Suppose that ξ^a is a Killing vector and that the Lie derivation of F_{ab} with respect to ξ^a vanishes: $\mathcal{L}_\xi F_{ab} = 0$. Then $F_{ab}\xi^b = \nabla_a\phi$ for some scalar field ϕ, and hence $I = mU^a\xi_a + \varepsilon\phi$ is a constant of motion for the charged particle $U^a\nabla_a I = 0$.*

The important point to note is the existence, under the given assumptions, of a scalar ϕ such that $F_{ab}\xi^b = \nabla_a\phi$, which is essentially the potential energy of the charged particle. By analogy with this situation we are led in the present case to consider a *vectorial* integral of the form

$$G_a = L_{ab}U^b + M_a. \tag{6}$$

In order for G_a to be *absolutely* conserved, i.e. $U^a\nabla_a G_b = 0$, we must by (4) have

$$\varepsilon\,L_{bc}F^c_{\ a} + m\nabla_a M_b = 0 \tag{7}$$

and hence $\nabla_{[a}M_{b]} = 0$ since the other term is symmetric. Thus if G_a is to be conserved there must exist a scalar Φ such that $M_a = \nabla_a\Phi$ and hence that $\varepsilon\,L_{bc}F^c_{\ a} + m\nabla_a\nabla_b\Phi = 0$.

Remarkably, in the case of the *Kerr* solution, where F_{ab} is a test field (of charge q), such a scalar Φ *does* exist, and is given by the following formula:

$$\Phi = -\frac{1}{2}\,\mathrm{i}\,\frac{\varepsilon q}{mM}\left(\chi_{AB}\chi^{AB} - \overline{\chi}_{A'B'}\overline{\chi}^{A'B'}\right). \tag{8}$$

Thus for a charged particle moving under the influence of a test electromagnetic field F^{ab} in the Kerr background there exists an absolutely conserved vectorial integral G_a, a feature of the Kerr geometry which insofar as I am aware has not hitherto been pointed out. (Here M is the mass parameter of K.)

In the case of the K-N geometry, on the other hand, the construction breaks down, and it is apparently *not* possible to construct a conserved vectorial quantity for charged particle motion, at least along the lines suggested above.

Interestingly enough the terms which arise in the K-N case that 'obstruct' the existence of a conserved vector quantity are essentially the *same* expressions as those arising in the combined gravitational-electromagnetic perturbations studied by Chandrasekhar which render the equations non-separable there (cf. p. 583 in his book). By the same token however we might therefore expect the vector G_a to play a role in the Kerr case in the decoupling and separability of Maxwell's equations in the Kerr background, though this is far from being obviously the case.

Gratitude is expressed to Ian Gatenby for helpful discussions and for checking through details of some of the calculations.

References

Chandrasekhar, S. (1983) *The Mathematical Theory of Black Holes* Oxford University Press.

Floyd, R. (1973) *The Dynamics of Kerr Fields* Ph.D. Thesis London University.

Hughston, L.P., Penrose, R., Sommers, P. & Walker, M. (1972) On a Quadratic First Integral for the Charged Particle Orbits in the Charged Kerr Solution. *Comm. Math. Phys.* **27** pp. 303–308.

Hughston, L.P. & Sommers, P. (1973) The Symmetries of Kerr Black Holes *Comm. Math. Phys.* **33** pp. 129–133.

Hughston, L.P. & Tod, K.P. (1990) *An Introduction to General Relativity*, Cambridge University Press.

Penrose, R. & Rindler, W. (1986) *Spinors and Space-Time, vol. II: Spinor and Twistor Methods in Space-Time Geometry*, Cambridge University Press.

Walker, M. & Penrose, R. (1970) On Quadratic First Integrals of the Geodesic Equations for Type $\{2,2\}$ Space-Times, Comm. Math. Phys. **18**, pp. 265–274.

§II.3.6 Further Remarks on Conserved Vectorial Quantities associated with the Kerr solution *by L.P.Hughston* (TN 25, September 1987)

Here I would like to summarize some of the technical details that lead to the conclusions drawn in the previous article II.3.5.

LEMMA 1. *Suppose χ_{AB} satisfies $\nabla_{A'(A}\chi_{BC)} = 0$ and thus $\nabla_{A'A}\chi_{BC} = \xi_{A'(B}\varepsilon_{C)A}$ for some $\xi_{A'A}$. Then the divergence of ξ_a vanishes. Moreover if $R_{ab} - \frac{1}{4}Rg_{ab} = 0$ then ξ_a is a Killing vector.*

More generally, if R_{ab} is determined by the stress-energy of an electromagnetic field for which the principal spinors are aligned with those of χ_{AB} (as in the Kerr-Newman solution) then ξ_a is a Killing vector.

Proof. See Hughston & Sommers (1973) or Penrose & Rindler (1986), Proposition 6.7.17, p. 108. □

In the case of Kerr (K) and Kerr-Newman (K-N) the Killing vector thus constructed can be chosen (by suitable adjustment of the disposable constant factor) to be *real*, and in the region where it is time-like to be future-pointing with unit norm. The constant factor in χ_{AB} can be fixed by these conventions.

LEMMA 2. *Let ξ_a be a real Killing vector constructed as above and let $f_{ab} = \nabla_{[a}\xi_{b]}$. If $f_{ab} = \phi_{AB}\varepsilon_{A'B'} + \bar{\phi}_{A'B'}$ varepsilon$_{AB}$ where*

$$\phi_{AB} = \tfrac{1}{2}\nabla_{A'(A}\xi^{A'}_{B)}$$

then the spinor ϕ_{AB} is given by

$$\phi_{AB} = \psi_{ABCD}\chi^{CD}$$

where ψ_{ABCD} is the Weyl spinor.

Proof. Apply $\nabla^{A'}_{D}$ to $\nabla_{A'A}\chi_{BC} = \xi_{A'(B}\varepsilon_{C)A}$ and symmetrize over A and D, then use a Ricci identity. □

Note that the 'charge integral' associated with f_{ab} gives the Komar mass. In the case of the Kerr solution the Komar mass agrees with the charge integral of $\psi_{ABCD}\chi^{CD}$, the spin 1 field obtained by lowering the spin of ψ_{ABCD} by use of χ^{AB}. Thus the 'charge' associated with ϕ_{AB} as defined above is M. An alternative formula for ϕ_{AB} is given by $\phi_{AB} = \Box\chi_{AB}$.

Now let us write $L_{ab} = \mathrm{i}\chi_{AB}\varepsilon_{A'B'} - \mathrm{i}\overline{\chi}_{A'B'}\varepsilon_{AB}$ for the Killing-Yano tensor, so $\nabla_{(a}L_{b)c} = 0$ as a consequence of the twistor equation and the reality of ξ_a. We are interested in the motion of a particle of charge ε under the influence of a field F_{ab} of charge q defined by $F_{ab} = qM^{-1}f_{ab}$ with f_{ab} as defined earlier. The equation of motion is $mU^a\nabla_a U^b = \varepsilon\, F^{bc}U_c$.

LEMMA 3. *The vector $G_a = L_{ab}U^b + M_a$ is a constant of the motion (i.e. is parallelly propagated) if and only if there exists a scalar Φ such that $M_a = \nabla_a\Phi$ and*

$$\varepsilon\, L_{bc}F^c{}_a + m\,\nabla_a\nabla_b\Phi = 0$$

where m is the mass of the particle.

Proof.

$$\begin{aligned} U^a\nabla_a G_b &= U^a\nabla_a(L_{bc}U^c + M_b) \\ &= U^aU^c\nabla_a L_{bc} + L_{bc}U^a\nabla_a U^c + U^a\nabla_a M_b \\ &= m^{-1}\varepsilon\, L_{bc}F^c{}_aU^a + U^a\nabla_a M_b, \end{aligned}$$

which vanishes for all U^a if and only if $\varepsilon L_{bc}F^c{}_a + m\nabla_a M_b$ vanishes. However $L_{bc}F^c{}_a$ is automatically symmetric (on account of the proportionality of ϕ_{AB} and χ_{AB}) and thus $\nabla_{[a}M_{b]} = 0$. □

LEMMA 4. *Suppose $\nabla_{AA'}\chi_{BC} = \xi_{A'(B}\varepsilon_{C)A}$. Then the vector N_a defined by $N_{A'A} = \xi^B_{A'}\chi_{AB}$ is curl-free.*

Proof. We have $\chi^{BC}[\nabla_{A'A}\chi_{BC} - \xi_{A'(B}\varepsilon_{C)A}] = 0$, and therefore

$$\tfrac{1}{2}\nabla_{A'A}\chi^2 = \xi_{A'B}\chi^B{}_A,$$

where $\chi^2 = \chi_{AB}\chi^{AB}$. □

THEOREM. *There exists a scalar Φ such that $\varepsilon\, L_{bc}F^c{}_a + m\nabla_a\nabla_b\Phi = 0$.*

Proof. Let L_{ab} be the Killing-Yano tensor and ξ^a the associated Killing vector, which for the Kerr solution is real. Then

$$\begin{aligned} L_{ab}\xi^a &= \mathrm{i}\,\chi_{AB}\varepsilon_{A'B'}\xi^{A'A} - \mathrm{i}\,\overline{\chi}_{A'B'}\varepsilon_{AB}\xi^{A'A} \\ &= \mathrm{i}\,\chi_{AB}\xi^A_{B'} - \mathrm{i}\,\overline{\chi}_{A'B'}\xi^{A'}_B \\ &= -\tfrac{1}{2}\mathrm{i}\,\nabla_b(\chi^2 - \overline{\chi}^2) \end{aligned}$$

by Lemma 4. Thus by differentiation we obtain

$$\nabla_a(L_{cb}\xi^c) = -\tfrac{1}{2}\mathrm{i}\,\nabla_a\nabla_b(\chi^2 - \overline{\chi}^2)$$

and therefore

$$(\nabla_a L_{cb})\xi^c + L_{cb} f_a{}^c = -\tfrac{1}{2}\mathrm{i}\,\nabla_a\nabla_b(\chi^2 - \bar{\chi}^2)$$

where $\nabla_a\xi_b = f_{ab}$. But $\xi^c\nabla_a L_{cb} = -\xi^c\nabla_c L_{ab}$ and $\xi^c\nabla_c L_{ab} = 0$ since $\xi^c\nabla_c\chi_{AB} = 0$, which in turn follows from the relation $\nabla_{A'A}\chi_{BC} = \xi_{A'(B}\varepsilon_{C)A}$. Therefore

$$L_{bc}F^c{}_a = -\frac{1}{2}\mathrm{i}qM^{-1}\nabla_a\nabla_b(\chi^2 - \overline{\chi}^2)$$

and the scalar Φ is given by

$$\Phi = -\frac{1}{2}\mathrm{i}\,\frac{\varepsilon q}{mM}(\chi^2 - \overline{\chi}^2),$$

as given in §II.3.5. □

Thanks are due to Ian Gatenby and Roger Floyd.

§II.3.7 Non-Hausdorff Twistor Spaces for Kerr and Schwarzschild *by J.Fletcher* (TN 27, December 1988)

[**Introduction** (1994). Einstein's equations for a stationary, axisymmetric space-time are equivalent to a form of the anti-self-dual Yang-Mills equations on flat space known as Yang's equation (Witten, 1979). This means that twistor methods can be used to construct solutions from bundles over appropriate portions of $\mathbb{CP}^3$, factored out by the lifts of the actions of the two space-time Killing vectors.

This article and articles §II.3.8, §II.3.9 concern the relationship between the geometry of the vector bundles and the structure of the space-times which they generate. This work was begun by Ward (1980) and continued by Woodhouse (1987), Woodhouse & Mason (1988) and Woodhouse (1989). Further details of both the topics coverd in these articles and a twistor approach to the black hole uniqueness theorem can be found in Fletcher (1990) and Fletcher and Woodhouse (1990).]

It is well-known that Einstein's equations for stationary, axisymmetric vacuum space-times can be reduced to a form of the rank 2 anti-self-dual Yang-Mills equations by the introduction of Weyl coordinates (see, for example, Witten 1979). Woodhouse & Mason (1988) showed that if these are solved locally by using the usual Ward transform, the holomorphic vector bundle over ordinary twistor space is actually the pull-back of a bundle E over a non-Hausdorff 'reduced' space.

Weyl co-ordinates are notorious for concealing the interesting parts of space-time geometry. For example, in the Schwarzschild solution, they only represent $R \geqslant 2m$, and the horizon is a part of the symmetry axis. (I shall use R to denote the radial co-ordinate in Kerr and Schwarzschild.) Since all the information about the analytic continuation of the manifold is contained in the exterior part of the metric, one might expect to find it in the twistor description; and this is in fact possible. The Kerr and Schwarzschild solutions have reduced twistor spaces which consist of the Riemann spheres S_0 and S_1 which are identified except at three pairs of points; these points are the points at infinity and $w = \pm(m^2 - a^2)^{\frac{1}{2}}$ where w is a co-ordinate on the spheres and $a = 0$ in the Schwarzschild case. The bundle can be described in a standard form which consists of first restricting it to each sphere and then giving the patching matrix P between them. In each case $E|_{S_0}$ is $L_1 \oplus L_0$; $E|_{S_1}$ is $L_{-1} \oplus L_0$ (where L_1 is the tautological bundle, L_0 the trivial bundle and L_{-1} the hyperplane section bundle); and

$$P = \frac{1}{w^2 - m^2 + a^2}\begin{pmatrix} (w+m)^2 + a^2 & 2am \\ 2am & (w-m)^2 + a^2 \end{pmatrix}.$$

(again, $a = 0$ for Schwarzschild). This description is unique if we demand that P be real (in the sense $\overline{P(w)} = P(\overline{w})$) and symmetric, and provided we know which points belong to which sphere in the reduced twistor space. We can, however, obtain a different patching matrix from the same bundle by changing our minds about which of the double points belongs to each of S_0 and S_1, and then putting the bundle in standard form over the two new spheres.

We therefore have four different possibilities. If we take S_0 and S_1 to be labelled by the points at infinity, then as well as the original description we can swap the points at $+b$, at $-b$ or at both, where $b = (m^2 - a^2)^{\frac{1}{2}}$. In order to see what this means in terms of the space-time, we have to introduce the idea of the patching matrix's being 'adapted' to one part of the axis in the (z, r)-plane. (Here z and r are the co-ordinates on the space of the orbits of the Killing vectors in the space-time; $r = 0$ represents the symmetry axis.) This simply means that we can find the metric on the space of Killing vectors on this part of the axis by taking the limit as $r \to 0$ of its value in a neighbourhood of it. The patching matrix P given above is adapted to $z > b$, which corresponds to one half of the axis of symmetry in the space-time, outside the horizon. If we interchange the points at $w = b$, we get a matrix adapted to $-b < z < b$, which is the (outer) horizon; and if we interchange the points at both $w = b$ and $w = -b$, we have a patching matrix adapted to the other half of the axis, where $z < -b$.

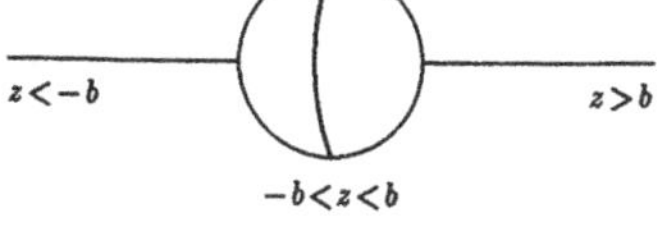

In each case, for both Kerr and Schwarzschild, the bundle E restricted to S_0 is $L_1 \oplus L_0$ and the metric on the space of Killing vectors can be extended analytically to the axis or horizon. Moreover, in the region where $-b < z < b$, we can continue this metric to the region where it is negative definite. To do this, we use the same construction as before (see Woodhouse & Mason 1988), but take values of r which are purely imaginary. Thus an orbit of the Killing vectors, which is represented by the pair of points $w = z + ir$ and $w = z - ir$, now corresponds to a pair of points on the real axis in the reduced twistor space. By taking r to be both positive and negative (when it is real) and $Im(r)$ to be both positive and negative (when r is imaginary), we can construct the usual cross-over at $R = m + b$.

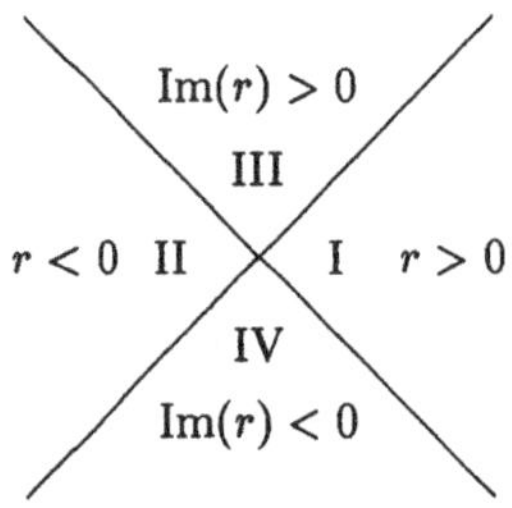

We can now choose either to identify regions I and II, and regions III and IV, or to put in the orbit $(z, r) = (b, 0)$ which corresponds to the cross-over itself. It can be shown that regularity of the metric (on the space of Killing vectors) at this point depends on the singularity structure of the patching matrix at $w = b$.

If we swap the points $w = -b$ in the Kerr solution, we find that we get another patching matrix Q of the kind that produces a cross-over (that is, Q is negative definite on the real axis in the w-plane); and that E restricted to S_0 is still $L_1 \oplus L_0$. If we take Q to be adapted to $-b < z < b$, then we can construct a similar picture to the one above; and the patching matrix adapted to $z > b$ turns out to be the inverse of the original matrix P. Since this can be obtained from P by replacing m with $-m$, the exterior region is now a negative mass Kerr solution. This of course must contain the ring singularity; the conjecture is that this is represented by the pull-back of E to the fibre of (Euclidean) twistor space above the appropriate points being non-trivial.

To obtain the Penrose diagram for the Kerr solution (see, for example, Hawking and Ellis 1973, p. 165) we have to identify region III for the '$+b$' crossover with region IV for the '$-b$' crossover. This can be done by considering the effect on the patching matrices of a reflexion of the (z, r) plane in the line $z = 0$. We also have to do the conformal rescaling which allows us to adjoin $\mathscr{I}$ to the space-time; it is at the moment less clear how the possibility of doing this is shown up by the twistor picture.

What is clear, however, is the difference between the Kerr and Schwarzschild solutions. For the latter, interchanging the points at $w = +m$ gives the same cross-over picture as for Kerr; but when we make the switch at $w = -m$, we find that the bundle E restricted to the new S_0 becomes $L_2 \oplus L_{-1}$. It is straightforward to see that this leads to a pole in the metric on the space of the Killing vectors as $r \to 0$; this is of course the usual curvature singularity at $R = 0$.

Many thanks to N.M.J.Woodhouse.

References

Fletcher, J. (1990) D.Phil. thesis, Oxford.

Fletcher, J. & Woodhouse, N.M.J. (1990) in *Twistors in Mathematics and physics*, (T.N. Bailey & R.J.Baston eds), LMS Lecture Notes Series **156**, CUP.

Hawking, S.W. & Ellis, G.F.R. (1973) *'The large scale structure of space-time'*, CUP.

Ward, R.S. (1983) Stationary axisymmetric space-times a new approach, Gen. Rel. Grav. **15**, 105-9.

Witten, L. (1979) Phys. Rev. D **19**, pp. 718–720.

Woodhouse, N.M.J. (1987) Twistor description of the symmetries of Einstein's equations for stationary axisymmetric space-times, Class. Quant. Grav. **4**, 799-814.

Woodhouse, N.M.J. (1989) Cylindrical gravitational waves, Class. Quant. Grav. **6**, 933-43.

Woodhouse, N.M.J. & Mason, L.J. (1988); Nonlinearity **1**, pp. 73–114.

§II.3.8 **More on the Twistor Description of the Kerr Solution** *by J.Fletcher* ($\mathbb{TN}$ 28, March 1989)

In my article in II.3.7, I outlined the relationship between the non-Hausdorff twistor spaces arising from Woodhouse & Mason's construction (1988) and the geometry of the Kerr and Schwarzschild solutions. The purpose of this note is to expand one or two points that arose there.

Recall that the space of orbits of the two Killing vectors in the Kerr solution corresponds to the space of quadratic maps $p : X \to R_U$, where X is a copy of $\mathbb{CP}^1$ (with coordinate q) and R_U is the reduced twistor space consisting of the Riemann spheres (coordinate w) which are identified everywhere except for the pairs of points at infinity and at $w = \pm b$. In order to determine a map p, we need to know first the values of w for which the discriminant of the equation $w = p(q)$ vanishes, and then which point of X is mapped to each of the pair of points at both $w = +b$ and at $w = -b$. If we write p in the form

$$p(q) = \tfrac{1}{2}r(q^{-1} - q) + z,$$

then the two values of w are $z + ir$ and $z - ir$, where z and r are the usual Weyl coordinates; and p is determined by these and the choice of one of four possible treatments of the double points. Orbits of the Killing vectors outside the outer horizon or inside the inner one are given by real z and real, positive r, and therefore correspond to pairs of complex conjugates in the w-plane. Orbits between the two, on the other hand, have z again real but r purely imaginary; moreover z and r are constrained so that the points $z \pm ir$ lie between $+b$ and $-b$ on the real axis.

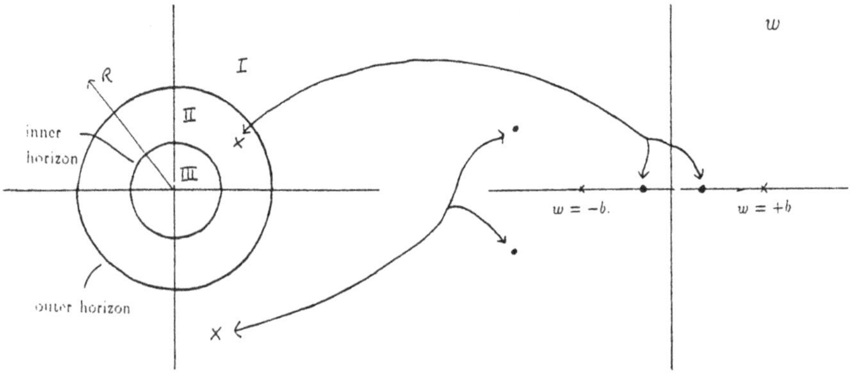

(left hand picture: cf. Hawking & Ellis p. 166)

There are, however, some values of z and r for which we cannot evaluate the metric directly by following the Ward splitting procedure. As Woodhouse & Mason (1988) showed in their paper, the method works provided the points $w = z + ir$ and $w = z - ir$ are distinct, and are both places where the two w-spheres are identified.

In the ordinary outside region I, therefore, the only problems can occur as $r \to 0$; and in their paper Woodhouse & Mason found the conditions on the bundle over R_U for the metric to be well-behaved on the axis of the Weyl coordinates (which corresponds to either an axis or a horizon in the space-time). Similarly, in region III, where the manifold can be continued analytically out to $\mathscr{I}$, there are the two parts of the axis and the horizon, and, in addition, the ring singularity. In my previous article, I mentioned the conjecture that this might correspond to a map p for which the pull-back of the bundle over R_U to one over X is non-trivial. This does in fact turn out to be the case, and it can be shown that when $z = 0$ and $r = a$, the pulled-back bundle is $L_2 \oplus L_{-2}$.

By contrast, in region II we can have values of z and r such that one of the pair $z \pm ir$ coincides with one of $\pm b$, but the other one does not. This divides region II into four, as follows:

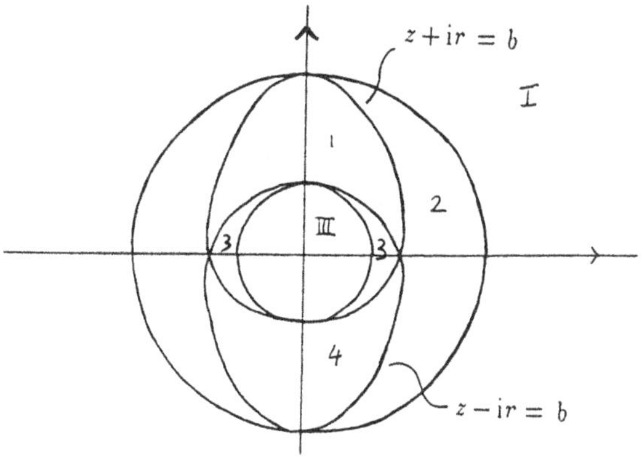

(Volumes 2 and 3 are of course connected since there is rotational symetry about the z-axis.)

and these four volumes corresponds precisely to the four different maps p that exist for each pair (z, r), and thus to the four different possible treatments of the double points $\pm b$. This means that each pair of points on the real axis in the w-plane between $+b$ and $-b$ represents four orbits of the two Killing vectors in the space-time; and if we consider the analytic continuation of the space-time (putting in the point at the $R = m + b$ cross-over)

where regions I′ and II′ are isometric to I and II in the usual way, we actually have eight orbits for each pair. (In I and II, we define r by $r = -\frac{1}{2}\mathrm{i}(w_1 - w_2)$; in I′ and II′, we take $r = +\frac{1}{2}\mathrm{i}(w_1 - w_2)$.)

This raises various questions. If each pair of points (w_1, w_2) on the real axis between $+b$ and $-b$ corresponds to four orbits in the space-time, why is the same not also true of each pair of complex conjugates, or even for each general pair of points, in the w-plane? Secondly, how can we tell that the space-time is in fact regular across the hypersurfaces where one of the points coincides with $+b$ or $-b$; and what do these surfaces mean geometrically?

The answer to the first question is that in the complexification of the Kerr solution each (ordered) pair of points does represent four Killing vector orbits, but not all of these intersect the real slice which is the space-time. Thus there are four real orbits for pairs in $(-b, +b)$ on the real axis, two real orbits (one in region I and one in region III) for pairs of distinct complex conjugates, and none otherwise. Trying to find another real orbit for the pairs of conjugates would be equivalent to interpreting the axis in volume 1 of the region II as a horizon; and this would be incompatible with regularity at the orbit $w_1 = +b = w_2$. Outside the outer horizon we are forced to think of $r = 0$ as an axis since it is the space-like Killing vector which vanishes on it. If $r = 0$ were a horizon then J, the metric on the space of orbits, would change signature to $(+, +)$ across it.

We can also use the analyticity of the complexification to see that J is well-behaved across the boundaries between volumes 1, 2, 3 and 4. By considering small variations of z into the complex, we can move (z, r) from one volume to another without $z \pm ir$ coinciding with $\pm b$, but with the explicit effect of changing the treatment of the double points by the corresponding maps $p : X \to R_U$. This is made clear by the behaviour of the open sets covering X on whose overlaps the pull-back of the bundle over R_U is described by the pull-backs of the patching matrices in the standard form I described before.

Finally, the boundaries themselves in fact represent the light-cones of the two points where the axis and the (outer) cross-over intersect.

References

Hawking, S.W. & Ellis, G.F.R. (1973) *The large scale structure of space-time*, (Cambridge University Press, Cambridge).

Woodhouse, N.M.J. & Mason, L.J. (1988) Nonlinearity **1**, pp. 73–114.

§II.3.9 **An alternative form of the Ernst potential** *by J.Fletcher* (TN 29, November 1989)

One of the strengths of the twistor description of the stationary axisymmetric solutions of Einstein's equations is the light it sheds on the relation between the metric and its Ernst potential. Recall that to represent a solution we have a bundle E over a reduced twistor space R_U which consists of two Riemann spheres S_0 and S_1 (with coordinate w) identified over the set U. The usual Ward construction gives us a solution of Yang's equation, usually denoted by J, which is the metric on the space of Killing vectors. If we impose the conditions that J be regular on the symmetry axis $r = 0$ and satisfy $\det J = -r^2$ then $E|_{S_0} = L_1 \oplus L_0$ and $E|_{S_1} = L_{-1} \oplus L_0$ where L_1 is the tautological bundle, L_{-1} is the hyperplane section bundle and L_0 is the trivial bundle. We can describe E by means of patching matrices $P_{\alpha\beta}$ defined on the overlaps of a collection of open sets U_α which cover R_U. If we take $U_0 \subset S_0$ and $U_1 \subset S_1$ to be neighbourhoods of $w = \infty$ not containing $w = 0$, and $U_2 \subset S_0$ and $U_3 \subset S_1$ to be neighbourhoods of $w = 0$ not containing $w = \infty$, then

$$P_{02} = \begin{pmatrix} 2w & 0 \\ 0 & 1 \end{pmatrix}, \quad P_{13} = \begin{pmatrix} (2w)^{-1} & 0 \\ 0 & 1 \end{pmatrix}$$

and E is completely specified by one of the patching matrices between the spheres, for example by P_{23} which I shall denote by P.

We can construct another solution $\imath(J)$ from the related bundle $\imath(E)$. To obtain $\imath(E)$, we take the same cover U_α and the same patching matrix P between the two spheres, but replace both P_{02} and P_{13} by the identity. Thus the restrictions $\imath(E)|_{S_0}$ and $\imath(E)|_{S_1}$ are both trivial; and we can think of the operations $\imath$ and $\imath^{-1}$ as untwisting and twisting the bundle round the points at infinity. We can write the corresponding matrix in the form

$$\imath(J) = \begin{pmatrix} f - \frac{\psi^2}{f} & \frac{\psi}{f} \\ \frac{\psi}{f} & \frac{1}{f} \end{pmatrix}$$

then the usual Ernst potential $\mathcal{E}$ is given by $\mathcal{E} = f + i\psi$. Woodhouse & Mason (1988) used this as the starting point for encoding the Geroch group in the twistor picture.

Instead of twisting and untwisting about $w = \infty$, however, we can do the same about $w = 0$. In other words, given our orginal bundle E, we can define a new one $\hat{E}$ which is also trivial over each of the spheres S_0 and S_1, but for which the patching matrix between them is P_{01} instead of P_{23}. Since P_{01} is given by

$$P_{01} = \begin{pmatrix} 2w & 0 \\ 0 & 1 \end{pmatrix} P \begin{pmatrix} 2w & 0 \\ 0 & 1 \end{pmatrix},$$

it has determinant equal to $(2w)^2$ and it is actually more convenient to use $P' = (1/2w)P_{01}$ to define the new bundle E'. I explained in my article in II.3.7 what it means for a patching matrix to be *adapted* to a certain part of the axis $r = 0$; if we suppose that P is adapted to an interval of the form $(0, a)$ then replacing P_{01} with P' corresponds to dividing the corresponding matrix J by u^2 to obtain J', where (in terms of the usual Weyl coordinates (z, r)), u and v are given by $r = uv$ and $z = \frac{1}{2}(u^2 - v^2)$.

Note that there is a certain amount of freedom in the construction of J'. With a particular choice of the two Killing vectors (X_1, X_2) in the orginal space-time (which must be arranged such that X_1 vanishes or is null on $r = 0$ in the interval $(0, a)$), we can still transform P by $P \to BPC$, where

$$B = \begin{pmatrix} 1 & 0 \\ b & 1 \end{pmatrix} \quad \text{and} \quad C = \begin{pmatrix} 1 & c \\ 0 & 1 \end{pmatrix}$$

for constants b and c. This will alter J' while leaving J the same. Going in the other direction, however, there is a unique P' for each J' and the only choice that can arise occurs if the twistor space $R_{U'}$ is glued down at $w = 0$; we then have to decide to which of the spheres S_0 and S_1 to assign each of the points at $w = 0$ in R_U.

There is a direct method for passing between J' and J which is analogous to changing from Ernst potential to metric. If we choose a Ward splitting $\{K'_\alpha\}$ for the patching matrices $P'_{\alpha\beta}$ descibing E' such that

$$K'_2 = \begin{pmatrix} * & 0 \\ * & 1 \end{pmatrix} \text{ at } \lambda = -(v/u) \quad \text{and} \quad K'_3 = \begin{pmatrix} 1 & * \\ 0 & * \end{pmatrix} \text{ at } \lambda = u/v$$

(where λ is the coordinate on the $\mathbb{CP}^1$ in $\mathbb{PT}$ corresponding to the orbit (u,v)) then, provided we have chosen S_0 and S_1 such that $w(u/v) \in S_0$ and $w(-v/u) \in S_1$, it follows that

$$J = H' \begin{pmatrix} -v^2 & 0 \\ 0 & u^2 \end{pmatrix} (\hat{H}')^{-1}$$

where $H' = K_0'(0)$ and $\hat{H}' = K_1'(\infty)$ (and so $J' = H'(\hat{H}')^{-1}$).

This choice of K_2' and K_3' corresponds to a choice of complex structure Ψ (in the Dolbeault version of the Ward construction: see Woodhouse & Mason 1988) such that

$$\begin{aligned} \Psi \begin{pmatrix} 1 \\ 0 \end{pmatrix} &= 0 \qquad \text{at } \lambda = u/v \\ \Psi \begin{pmatrix} 0 \\ 1 \end{pmatrix} &= 0 \qquad \text{at } \lambda = -v/u. \end{aligned} \tag{1}$$

This is because, on the pull-back of the set U_α to $\mathbb{PT}$, $\Psi = K_\alpha^{-1}\partial K_\alpha$.

It is straightforward to show (from the definition of Ψ) that (1) implies we have chosen $\hat{H}' = (s_1\, s_2)$ and $H' = J'\hat{H}'$, where s_1 and s_2 are solutions of the equations

$$\partial_r s + \frac{1}{1+\lambda^2}((J')^{-1}\partial_r J' - \lambda(J')^{-1}\partial_z J')s = 0 \tag{2}$$

$$\partial_z s + \frac{1}{1+\lambda^2}((\lambda(J')^{-1}\partial_r J' + (J')^{-1}\partial_z J')s = 0 \tag{3}$$

with $\lambda = u/v$ and $\lambda = -v/u$ respectively. To go in the other direction, we replace J' with $u^{-2}J$ and choose $H = (s_2\, s_1)$.

There are two dimensions of freedom in the choice of s_1 and s_2 in each case; and in fact defining $\hat{H}'$ in this way only implies that K_2' and K_3' are of the form

$$K_2' = \begin{pmatrix} * & \alpha \\ * & \beta \end{pmatrix} \quad \text{and} \quad K_3' = \begin{pmatrix} \gamma & * \\ \delta & * \end{pmatrix}$$

where α and β are constant on $\lambda = u/v$, and γ and δ are constant on $\lambda = -v/u$. Defining $\hat{H}$ in a similar way gives a similar form for K_2 and K_3. In both cases, the behaviour of the splitting matrices on the two surfaces in $\mathbb{PT}$ is due to the fact that the only holomorphic functions on $\mathbb{PT}$ which are invariant under the lifts of the two Killing vectors are those which are functions of w alone, where w is related to λ by the equation

$$w = \frac{r}{2}(\lambda^{-1} - \lambda) + z \tag{4}$$

When we go in the 'twisting' direction (that is to say, from J' to J) we can actually fix s_1 and s_2, and thus J, completely by considering the behaviour of K_2' and K_3' as $v \to 0$. This corresponds to the unique choice of P' in this case. On the other hand, the freedom that we have in the other direction also corresponds precisely to the freedom in the choice of patching matrix P.

Finally, a brief remark on the point of all this. It turns out that if a space-time has both a symmetry axis and a Killing horizon and is regular at the point where they intersect, then the patching matrix P has a simple pole at the point in the reduced twistor space which corresponds to the intersection and which we can assume to be at $w = 0$ (see Fletcher & Woodhouse 1990). It is straightforward to show, however, that the 'untwisted' patching matrix, P', is well-behaved on the real axis near $w = 0$, and slightly less straightforward to show that its entries are actually holomorphic in a neighbourhood of this point.

Thanks to N.M.J.Woodhouse and to Fletcher & Partners.

References

Woodhouse, N.M.J. & Mason, L.J. (1988) Nonlinearity **1**, pp. 73–114.

Fletcher, J. & Woodhouse, N.M.J. (1990) in *Twistors in Mathematics and Physics*, (T.N. Bailey & R.J.Baston eds), LMS Lecture Notes series **156**, Cambridge University Press.

§II.3.10 **Light Rays near i^0: A new Mass-Positivity Theorem** *by R.Penrose* (TN 30, June 1990)

As was emphasized by Helmut Friedrich in a recent survey talk [given at the E.T. Newman birthday conference, Pittsburgh, May 1990], one of the major problems obstructing a proper understanding of asymptotically flat space-times is the lack of a completely satisfactory geometrical framework for analysing i^0 (the conformally singular point at spacelike infinity). Perhaps twistor theory can make a significant contribution here. Whereas partial understandings have come through the work of Ashtekar-Hanson, Geroch, Beig, Sachs, Sommers and many others, there remains a distinct conceptual awkwardness, and as yet there is no elegant geometrical description to match—and to unite with—that of $\mathscr{I}^{\pm}$. The 'spi' construction of Ashtekar-Hansen (1978) seems to come closest and there are indeed some relations to twistor theory. In Minkowski space $\mathbb{M}$, the points of spi can be identified with timelike hyperplanes (i.e. timelike 3-quadrics through i) and these are represented by (proportionality classes of) skew twistor $S^{\alpha\beta}$ subject to $S^{\alpha\beta}I_{\alpha\beta} = 0$ and the "reality condition" $\bar{S}_{\alpha\beta}I^{\alpha\beta} = I_{\alpha\beta}S^{\gamma\beta}$. Such objects find their place in the *moment sequence* (see Penrose & Rindler 1986, pp 94, 96 and §II.4.2-3 in this volume). The points of $\mathscr{I}$ arise when $S^{\alpha\beta}$ is *simple* (and so has the form $Z^{[\alpha}I^{\beta]\gamma}W_{\gamma}$).

I shall postpone, until a later date, a detailed discussion of the geometry involved here. In any case, we cannot expect to capture the structure of i^0 that incorporates the total mass-momentum and angular momentum of the system using merely Minkowski-space twistors. Possible schemes for involving more general types of twistor would be to use 2-surface twistors in the manner of Shaw, or perhaps to use hypersurface twistors, e.g. for a hyperboloidal timelike hypersurface that approximates $i^0 \cup \mathscr{I}^+ \cup \mathscr{I}^-$.

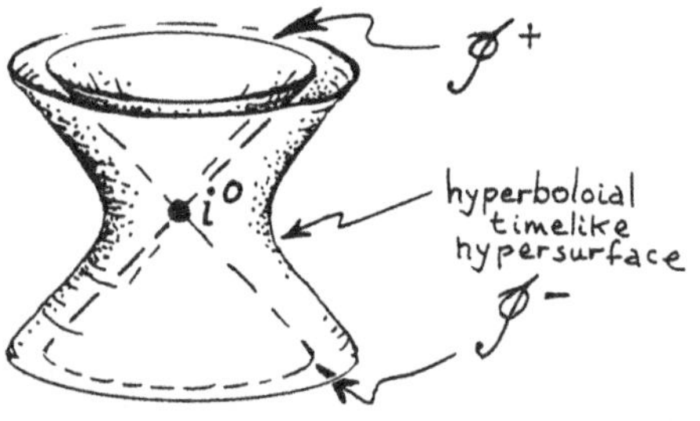

Of more immediate physical relevance would be to study the behaviour of light rays near i^0. There is a clear connection with twistor theory here, but for the moment I shall just show how this study can be used to give a new and perhaps comparatively simple proof of total energy positivity in general relativity (cf. Schoen & Yau (1979) and Witten(1981)):

THEOREM. *If $\mathcal{M}$ is asymptotically simple and satisfies the strong null convergence condition, then it cannot have a well-defined negative mass at i^0.*

N.B. By the *'strong' null convergence condition* (SNCC) I shall mean that every endless null geodesic contains a pair of conjugate points. This is a consequence of the three properties:

(1) $R_{ab}n^a n^b \leqslant 0$ for all null vectors n^a (i.e. with Einstein's equations, and my sign conventions, $T_{ab}n^a n^b \geqslant 0$),

(2) null geodesic completeness, and

(3) the genericity condition: $k_{[c}R_{a]bc[d}k_{f]}n^b n^c \neq 0$ somewhere along each null geodesic (n^a the tangent). [see Hawking & Penrose, Proc. Roy. Soc. (Lond.) **A314** (1970) 529.]

Outline of proof. The idea is to assume that the mass at i^0 *is* negative and that SNCC holds, and then to derive a contradiction. Let $a \in \mathscr{I}^+$ and consider $I^-(a)$. (All sets are in $\bar{\mathcal{M}} = \mathcal{M} \cup \mathscr{I}^+ \cup \mathscr{I}^-$; here $I^-(a)$ is to include its limit points on $\mathscr{I}^-$ (interior limit points only, so that $I^-(a)$ remains open); "$\partial I^-(a)$" is supposed to be composed of the points of $\partial I^-(a)$ in $\mathcal{M}$, together with the limit points thereof in $\bar{\mathcal{M}}$.) Let us see what would happen if it could be shown that some segment of a generator of $\mathscr{I}^-$, from a point $c \in \mathscr{I}^-$ to i^0 has a neighbourhood in $\bar{\mathcal{M}}$ that does not meet $I^-(a)$, as follows if any $c \in \mathscr{I}^-$ has a neighbourhood not meeting $I^-(a)$.

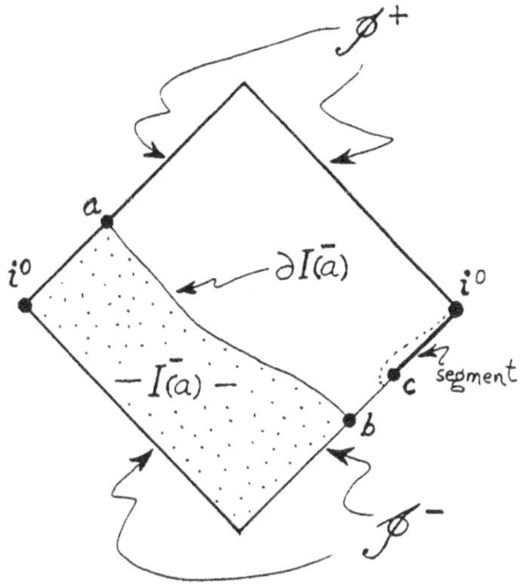

In that case $\partial I^-(a)$ would meet $\mathscr{I}^-$ transversally in some (non-vacuous) set $\mathcal{B}$ $(= \partial I^-(a) \cap \mathscr{I}^-)$. Let $b \in \mathcal{B}$. By standard theorems (since $\mathcal{M}$ is globally hyperbolic, as follows from its asymptotic simplicity) there must be a null geodesic from b to a, lying on $\partial I^-(a)$, with no pair of conjugate points between b and a, which contradicts SNCC.

In fact there is *no* such segment (or point c), but what *does* happen, with negative mass at i^0, is almost as effective. Let us examine the generators of the past light cone $\mathcal{A}$ of a (we have $\partial I^-(a) \subset \mathcal{A}$), particularly in the neighbourhood of that generator α of $\mathscr{I}^-$ that is 'diametrically opposite' to a (i.e. α extends through i^0 to become the generator of $\mathscr{I}^+$ containing a).

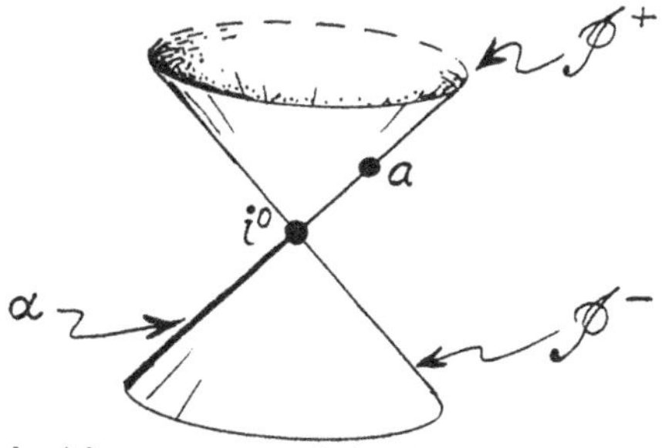

We shall be mainly concerned with generators of $\mathcal{A}$ "close" to that particular generator of $\mathscr{I}^+$ on which a lies (i.e. α extended), and with their intersection with $\mathscr{I}^-$. To assist us in picturing this situation, consider first the case of *positive* mass at i^0, to see why there is no conflict with SNCC. Not only are the light rays deflected inwards as they pass the source, but there is also a time-delay that behaves logarithmically in the impact parameter (distance of "closest approach"). This has no natural zero, so the larger the impact parameter, the larger the values of advanced time that will eventually be encountered.

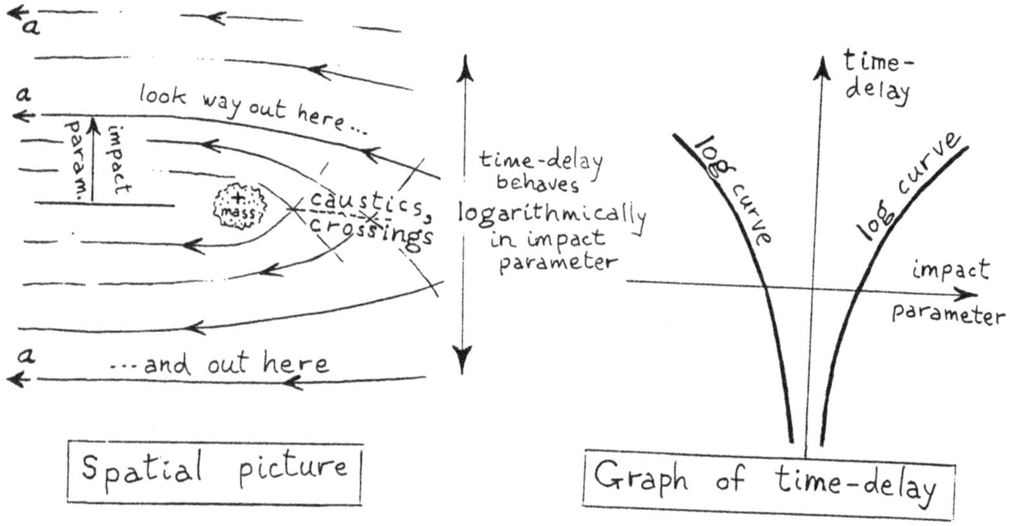

This phenomenon was described in detail in Penrose (1980). We find that the whole of $\mathscr{I}^-$ is contained in $I^-(a)$. (Hence $\partial I^-(a) \cap \mathscr{I}^- = \emptyset$, so there is no contradiction with SNCC.) To see how this comes about, examine $\mathcal{A}$ and $\partial I^-(a)$ near α. We find

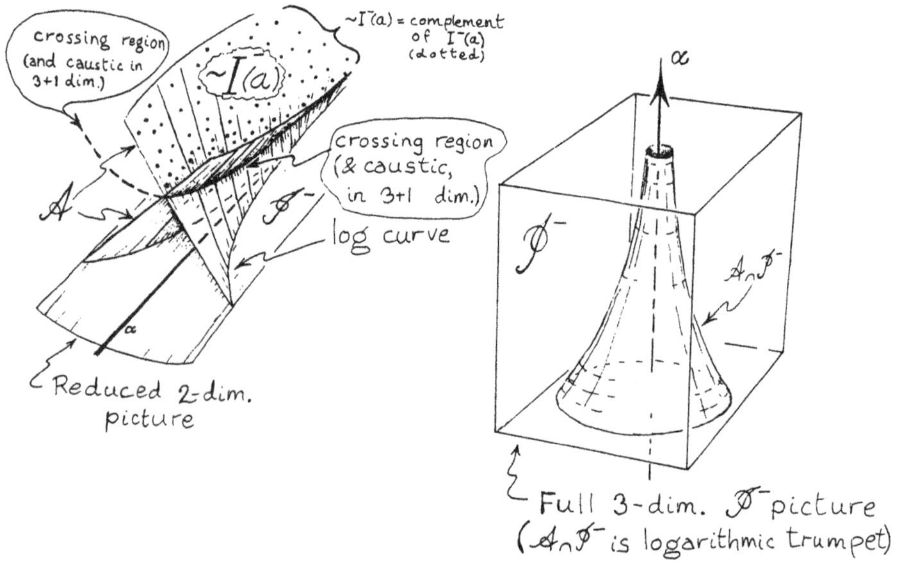

Now consider *negative* mass. The picture is like the one above but the other way up:

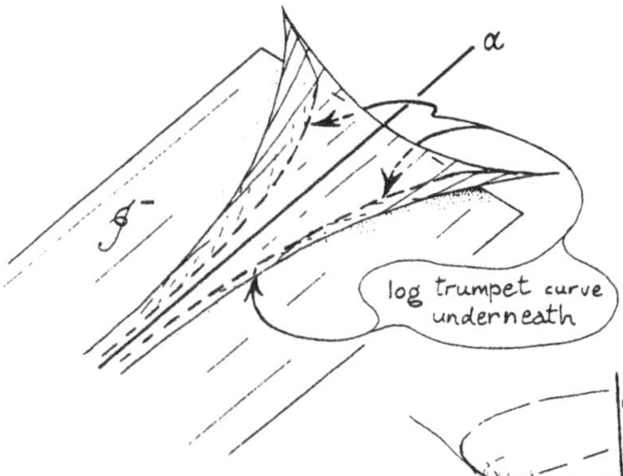

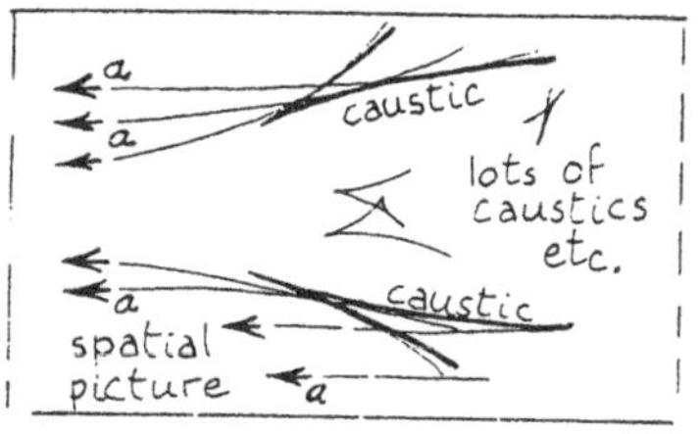

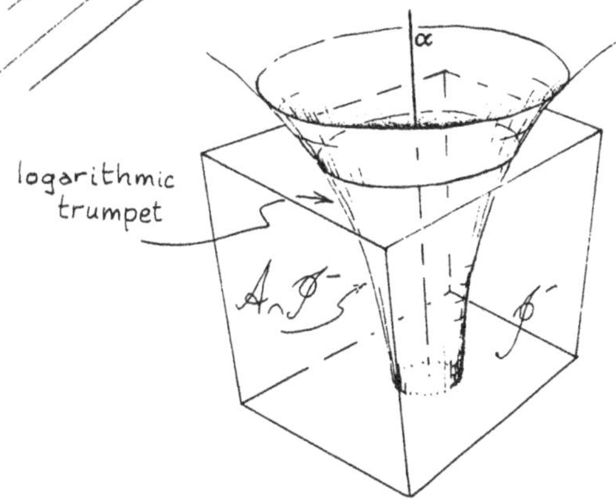

What's not so clear from this picture is the fact that $\mathscr{I}^- \subset \overline{I^-(a)}$ $(= I^-(a) \cup \partial I^-(a))$, as in the positive and zero mass cases $\partial I^-(a)$ does not intersect $\mathscr{I}^-$ transversally and we have no immediate contradiction with SNCC. To see what *does* happen, consider the following diagram.

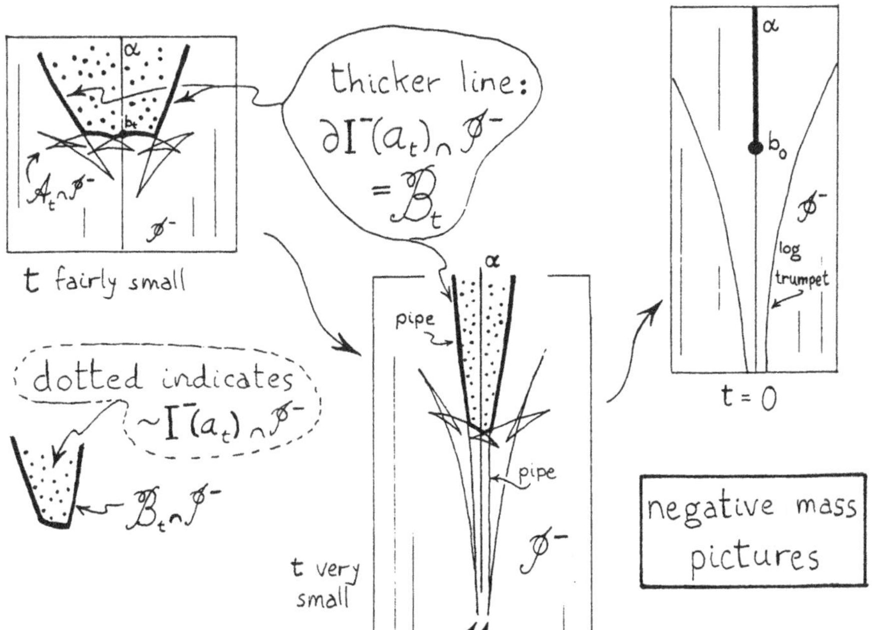

In this diagram we have a smooth family of points $a_t (t \geqslant 0)$ lying on (say) a null ray terminating at a, where $a = a_0$. For each $t > 0$, $\partial I^-(a)$ intersects $\mathscr{I}^-$ transversally in a (not necessarily smooth) cut of $\mathscr{I}^-$. The point $b_t = \alpha \cap \partial I^-(a_t)$ lies on a null ray on $\partial I^-(a_t)$ free of conjugate points (except possibly at end-points). As $t \to 0$ the intersection $\mathcal{A}_t \cap \mathscr{I}^-$ (where $\mathcal{A}_t$ is past light cone of a_t) looks like the above diagram. The region $\mathcal{B}_t = \partial I^-(a_t) \cap \mathscr{I}^-$ moves continuously with t, finally converging on a future-endless segment of α with past end-point $b_0 = b$.

One must show that the part of $\mathcal{A} \cap \mathscr{I}^-$ that lies away from the logarithmic trumpet and the narrower pipe within it actually stabilizes as $t \to 0$ so that b_t attains a limit. The pipe and trumpet come from the resemblance of $\mathcal{M}$'s i^0 to that of a negative mass Schwarzschild solution as can be seen from examining the spatial picture. Having b_0, we find the null ray on $\partial I^-(a_0)$ from b_0 to a_0 which must be free of conjugate points: the desired contradiction with SNCC.

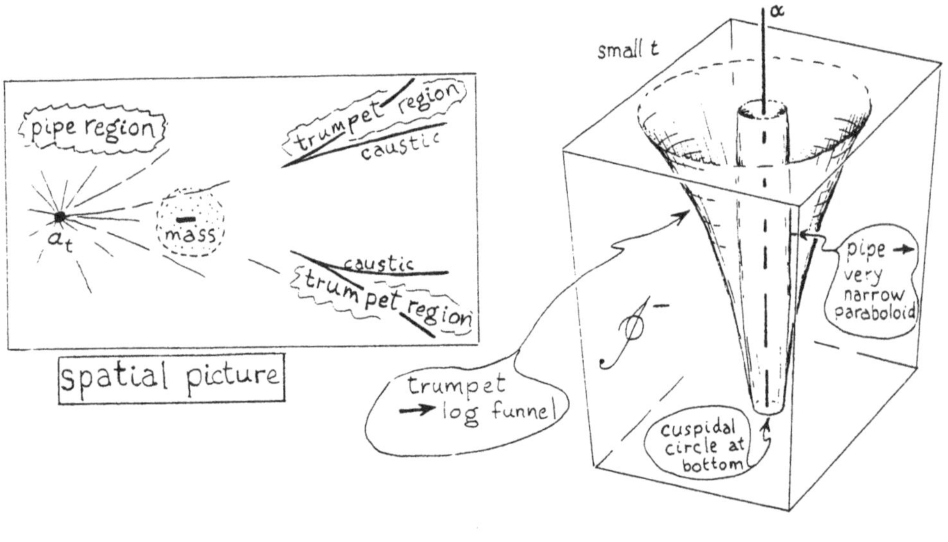

No doubt the argument can be strengthened in various ways (e.g. allowing conjugate points at $\mathscr{I}^{\pm}$ or the mass to be zero). Further details will appear elsewhere.

References

Ashtekar, A.& Hansen (1978) J. Math. Phys. **19**, 1542.

Hawking, S.W. & Penrose, R. (1970) Proc. Roy. Soc. Lond. **A314** 529.

Penrose, R. (1980) in the Taub Festschift volume,(F.Tipler ed.).

Penrose, R. (1986) *Spinors and space-time*, Vol.2, CUP.

Schoen & Yau, (1979) Phys. Rev. Lett. **43** (1979) 1457, and Commun. Math. Phys. **65**, 45.

Witten, (1981) Commun. Math. Phys. **80**, 381.

§II.3.11 Mass positivity from focussing and the structure of space-like infinity

by A.Ashtekar & R.Penrose (TN 31, October 1990)

In §II.3.10, one of us (RP) outlines an argument to show that in asymptotically simple space-times satisfying a certain assumption, the mass is non-negative. The assumption—called the *null conjugate-point condition* requires that every endless null geodesic contain a pair of conjugate points. This condition is not overly restrictive. For example, it is implied by completeness of null geodesics, a weak energy condition ($T_{ab}l^a l^b \geq 0$ if $l_a l^a = 0$) and genericity. The purpose of this contribution is to examine more closely what it is about the structure of spatial infinity, i^o, that the argument establishes. In particular, we will be able to establish that the mass in question is indeed the *ADM mass at* i^o. More precisely, our main result is that the ADM 4-momentum P_a of an asymptotically simple space-time satisfying the null conjugate point condition is necessarily a (future directed) causal 4-vector at i^o. (In this article, the assumption of asymptotic simplicity will include, in addition to asymptotic flatness at null infinity, that at spatial infinity. Thus, we assume that the space-time is AEFANSI in the sense of Ashtekar (1980) and that the parity condition of Ashtekar (1985) is satisfied at i^o.)

Let us begin by reviewing the result presented in §II.3.10. Let $\mathcal{M}$ be an asymptotically simple space-time. Fix a point a^+ on the future null infinity, $\mathcal{I}^+$ on $\mathcal{M}$. Let it lie on the generator α of $\mathcal{I}$. The past-directed light rays from a^+ will be said to *focus negatively* if they meet $\mathcal{I}^-$ in a family of points which recede indefinitely into the past (i.e., towards i^-). Let us suppose that this occurs. Then, if we examine the light rays neighbouring a given ray λ through a^+, as λ approaches α, the total shear of these rays along λ has the form of a convergence in the radial direction and a divergence in the transverse direction (see figure 1.⊖).

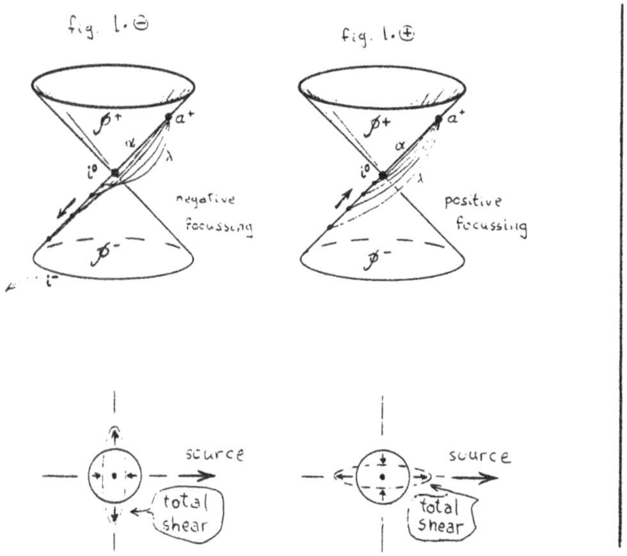

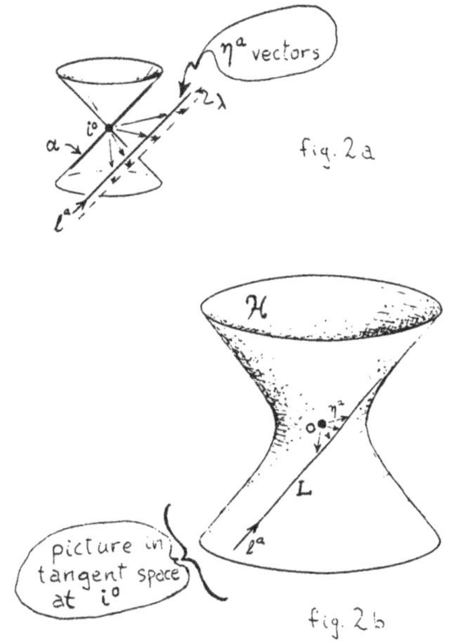

This is the behavior encountered in the negative mass Schwarzschild solution. In the positive mass Schwarzschild space-time, the situation is just the opposite. Now, we encounter *positive focussing.* In this case, light rays from a^+ meet $\mathcal{I}^-$ in a family of points that approaches i^o, there is a divergence of rays in the radial direction and convergence in the transverse direction. Thus, if we move in radially, the rays spray away from each other as i^o is approached, whereas transversally they pinch towards one another. (See fig 1.⊕.) Now, as the null geodesic λ gets closer to α, various non-linear terms die off and the shear is given by the integral, $\int_\lambda ds\, C_{abcd} l^a l^c \eta^b \eta^d$, evaluated along λ, where l^a is the parallelly propagated tangent to λ, s is the affine parameter, and where η^a is the *radial* connecting vector. Thus, if negative focussing occurs, the integral would be positive and if positive focussing occurred, the integral would be negative. It was shown in §II.3.10 that, *if $\mathcal{M}$ satisfies the null conjugate point condition, negative focussing cannot occur.* Hence, in this case, the integral is necessarily non-positive for all geodesics λ originating at some point a^+ on $\mathcal{I}^+$ which are sufficiently close to the generator α of $\mathcal{I}$ on which a^+ lies. In this article, we show that this result in turn implies that the ADM 4-momentum is a (future pointing) causal vector. Thus, the intuition derived from the Schwarzschild solution is indeed valid more generally.

Let us recall the structure available at i^o. The asymptotic conditions of Ashtekar (1980) imply that the tangent space at i^o is well-defined and carries a (universal) Minkowskian metric of signature $(-+++)$. Further, along any C^1 curve with tangent vector η^a at i^o, $\Omega^{\frac{1}{2}} C_{abcd}$ admits a limit at i^o, where C_{abcd} is the Weyl tensor of the unphysical metric g_{ab}. We can decompose the limit into its electric and magnetic parts using the unit space-like vectors η^a at i^o, and thus acquire two *symmetric, traceless* tensor fields $E_{ab}(\eta)$ and $B_{ab}(\eta)$ on the hyperboloid $\mathcal{H}$ of unit space-like vectors at i^o. (Note that, by their definition, the two fields are tangential to $\mathcal{H}$). Let us focus on the asymptotic electric field, E_{ab}. It satisfies the field equation $D_{[a}E_{b]c} = 0$ on $\mathcal{H}$, where D is the derivative operator compatible with the natural metric $h_{ab} = g^o{}_{ab} - \eta_a \eta_b$ on $\mathcal{H}$, where $g^o{}_{ab}$ is the Minkowski metric in the tangent space at i^o. The information about the ADM 4-momentum P_a is contained entirely in E_{ab}:

$$P_a V^a := -\tfrac{1}{8\pi} \oint_S dS^a\, E_{ab} V^b, \tag{1}$$

where V^a is any vector in the tangent space at i^o and S is any 2-sphere cross-section of $\mathcal{H}$. The field equation and the trace-free property of E_{ab} imply that it is divergence-free while the projection of V^a into the hyperboloid (forced by its contraction with E_{ab} in the integrand) is a conformal Killing field on $\mathcal{H}$, whence the surface independence of (1). The field equation on E_{ab} also implies that it admits a scalar potential, E:

$$E_{ab} = D_a D_b E + E h_{ab} \tag{2.a}$$

In any given conformal completion, the potential E can in fact be constructed explicitly from the asymptotic (unphysical) Ricci tensor. The fact that E_{ab} is trace-free implies that E must satisfy the (tachyonic) massive scalar field equation:

$$D^a D_a E + 3E = 0 \tag{2.b}$$

on $\mathcal{H}$. In the Schwarzschild space-time with 4-momentum $P_a = mt_a$, with $t_a t^a = -1$, E is given by:

$$\frac{E = m\, 1 + 2(t.\eta)^2}{\sqrt{(1 + t.\eta)^2}} \tag{3}$$

More generally, in physically interesting situations, the asymptotic magnetic field B_{ab} vanishes on $\mathcal{H}$ and the leading terms in the *physical* metric are dictated entirely by E. There exists a coordinate system in terms of which the physical metric has the asymptotic form:

$$d\hat{s}^2 = (1 + \frac{E}{\hat{\rho}})^2 d\hat{\rho}^2 + \hat{\rho}^2 \left(h^o{}_{\mathbf{ab}} + \frac{h^1{}_{\mathbf{ab}}}{\hat{\rho}} + \frac{h^2{}_{\mathbf{ab}}}{\hat{\rho}^2} + \dots \right) d\hat{\phi}^{\mathbf{a}} d\hat{\phi}^{\mathbf{b}}, \tag{4}$$

with $h^1{}_{\mathbf{ab}} = Eh^o{}_{\mathbf{ab}}$. Here $h^o{}_{\mathbf{ab}}$ is the metric on the unit time-like hyperboloid in Minkowski space. Thus, $(\hat{\rho}, \hat{\phi}^{\mathbf{a}})$ should be thought of as the "asymptotic hyperboloid" coordinates.

It is easy to check that the space of solutions to the equation $D_a D_b \bar{E} + \bar{E} h_{ab} = 0$ is precisely 4-dimensional. The solutions are of the form $\bar{E} = K_a \eta^a$, where K_a is a fixed vector in the tangent space of i^o. Thus, there is some gauge freedom in the choice of the potential; we can add to the natural potential E of E_{ab} any $\bar{E}$ without changing the value of the field E_{ab}. This freedom is intertwined with the fact that if one uses only the asymptotic conditions of Ashtekar (1980), there is some ambiguity in the conformal completion at i^o. Given a completion, one can obtain a four parameter family of inequivalent ones by logarithmic translations. In the physical space language, there are the transformations of the type:

$$\hat{x}^{\mathbf{a}} \rightarrow \hat{x^{\mathbf{a}}} + K^{\mathbf{a}} log\hat{\rho}, \tag{5}$$

where $\hat{x}^{\mathbf{a}}$ are the asymptotically Cartesian coordinates, $\hat{\rho}^2 = x^{\mathbf{a}} x_{\mathbf{a}}$ and where $K_{\mathbf{a}}$ are constants. The new completions are C^1 related at i^o. Therefore, i^o and the tangent spaces in the two completions are naturally identified. Under this identification, the field E_{ab} of one completion is mapped to that of the second. The potentials E, however, are not preserved. On $\mathcal{H}$ we have:

$$E \rightarrow E + K_a \eta^a \qquad \text{and} \qquad E_{ab} \rightarrow E_{ab} \tag{6}$$

Since the field is unaffected, so is the ADM 4-momentum (and, to next order, also angular momentum). Thus, the logarithmic translations may be thought of as "gauge" in this framework. In a large class of space-times, this gauge freedom can be eliminated. Suppose, as in [3], that the electric field E_{ab} is reflection symmetric on $\mathcal{H}$. Then, we can demand that the potential should also be reflection symmetric. (Note that E of the Schwarzschild solution (eq 3) automatically satisfies the condition.

In Minkowski space the requirement singles out the potential $E = 0$.) This requirement selects a unique potential and hence removes the logarithmic ambiguity in the completion. As a part of the boundary conditions at i^o we assume that E_{ab} is reflection symmetric and work in a completion in which the potential E is also even under reflection.

With this formalism at hand, let us return to the implication of §II.3.10. As the null geodesic λ approaches the generator α in the completed space-time (fig 2.1), the connecting vectors η^a become, in the limit, the position vectors of points on a null geodesic (straight line) L in $\mathcal{H}$, and the tangent vector l^a is now parallelly propagated with respect to h_{ab} (fig 2.b). Since the argument of §II.3.10 tells us that integral of $C_{abcd}\eta^a\eta^c l^b l^d$ along λ is non-positive, in the limit we conclude that $\int_L ds\, E_{ab}l^a l^b \leq 0$. Using the expression (2) of E_{ab}, we have $E_{ab}l^a l^b = (l^a D_a)^2 E$, so that the last condition becomes:

$$\dot{E}^+ - \dot{E}^- \leq 0, \tag{7.a}$$

where $\dot{E} = l^a D_a E$ and $\pm$ denote, respectively, the values at the future and past (ideal) end points of L. Now, since E is reflection symmetric, the two terms on the left side of (6) add and we have:

$$\dot{E}^+ \leq 0. \tag{7.b}$$

To see the implication of this condition, let us examine the asymptotic form of E. Let us foliate the tangent space of i^o by a family of planes t=const and consider the corresponding foliation of $\mathcal{H}$. Assuming that E admits a power series expansion of the type $\sum E^{(n)}(\theta,\phi)\, t^{-n}$, where n runs from some *finite* negative value to $+\infty$, the field equation implies that E must admit an asymptotic expansion of the following type:

$$E(t,\theta,\phi) = (a_0 + a_m Y_{1m})(\theta,\phi))t + \frac{a_m Y_{1m}(\theta,\phi)}{2t} + \frac{E^{(3)}(\theta,\phi)}{t^3} + \ldots \tag{8}$$

where $Y_{1m}(\theta,\phi)$ are the three $\ell = 1$ spherical harmonics. (We believe the required assumption is always satisfied in the reflection symmetric case.) The condition $\dot{E}^+ \leq 0$ implies that the coefficient of the first term is non-negative and hence the 4-vector $\bar{P}_a = a_0 t^a + V^a$ at i^o–with t^a the unit time-like vector orthogonal to the slices and the spatial vector V^a in the $t = 0$ slice, given by $V^a\eta_a = a_m Y_{1m}(\theta,\phi)$—is future-directed and causal. At first, we were misled into thinking that the coefficient of the first term is the mass aspect at the future end of $\mathcal{H}$ and therefore the argument would show that the mass aspect should be positive. This is incorrect. In fact, the electric field E_{ab} constructed from the leading order term (via equation 2.a) vanishes identically whence the term makes no contribution what so ever to the ADM 4-momentum integral. Rather, the mass-aspect is the third term, $E^{(3)}$ in the expansion. However, because we have restricted ourselves to even potentials E, it *does* follow that the ADM 4-momentum P_a constructed from the correct mass aspect is precisely given by the vector $\bar{P}_a$, which resides in the leading order term. Therefore, although we cannot conclude that the mass aspect should be positive, (7.b) does indeed imply that the ADM 4-momentum is a causal, future-directed vector.

We conclude with two remarks. First, one can show that the vector space obtained by superposing the asymptotic mass aspects of Schwarzschild solutions (3) (whose the 4-momentum is not restricted to be time-like) is dense in the space of all asymptotic mass aspects $E^{(3)}(\theta,\phi)$ arising from smooth solutions to (2.b). We believe, furthermore, that the same is true of the entire solutions E everywhere on $\mathcal{H}$. Thus, in the reflection-symmetric case, one can work with superpositions of (3) without loss of generality. The second remark has to do with the leading term in the asymptotic expansion (8). The fact that this term diverges is an indication that there is a mismatch at i^o between the following two limits: sliding down a generator of $\mathcal{I}^+$, and, approaching i^o along a space-like direction and then making an infinite boost. The mis-match is a measure of the ADM 4-momentum. There is a similar mis-match in the way $\mathcal{I}^-$ is attached to i^o. This mis-match should show up in the metric coefficients rather than the curvature. If one allows E to acquire non-symmetric part under reflection, one can, by a logarithmic translation, remove a mis-match between i^o and $\mathcal{I}^+$. However, then the mis-match with $\mathcal{I}^-$ is twice as big. There is something "cohomological" here. Can twistor considerations shed further light on it? A clearer treatment of this issue is needed to make further progress in a proper understanding of asymptotically flat space-times, e.g. along the lines initiated by Friedrich.

We have presented here only the overall picture. Some details are yet to be worked out fully. Also, the results can probably be generalized in a number of ways. A more complete account will appear elsewhere.

References

Ashtekar, A. (1980) In: *General Relativity and Gravitation*, Vol. 2, (A.Held ed.), Plenum, New York, 1980.

Ashtekar, A. (1985) Found. Phys. **15**, 419.

§II.3.12 **The Initial Value Problem in General Relativity by Power Series** *by V. Thomas* (TN 33, November 1991)

We work throughout in a curved vacuum space-time. At some point O, we take the light cone $\mathscr{C}$. Suppose we know the structure of $\mathscr{C}$, in the sense that we can express the component of the Weyl curvature spinor Ψ along any null ray. Assuming an analytic space-time we use a Taylor expansion, which when we take the component along the null ray gives, for any point P on $\mathscr{C}$,

$$\Psi_{ABCD}(P)\xi^A\xi^B\xi^C\xi^D = \Psi_{ABCD}(O)\xi^A\xi^B\xi^C\xi^D + r\nabla_{AA'}\Psi_{BCDE}(O)\xi^A\bar{\xi}^{A'}\xi^B\xi^C\xi^D\xi^E$$
$$+ \frac{r^2}{2!}\nabla_{AA'}\nabla_{BB'}\Psi_{CDEF}(O)\xi^A\bar{\xi}^{A'}\xi^B\bar{\xi}^{B'}\xi^C\xi^D\xi^E\xi^F + \ldots$$

where P is displaced by parameter value r from O along a null geodesic defined by the null vector $\xi^A\bar{\xi}^{A'}$ at O.

This form only requires the symmetric part of the derivatives of Ψ to find the component of Ψ at any point P on $\mathscr{C}$.

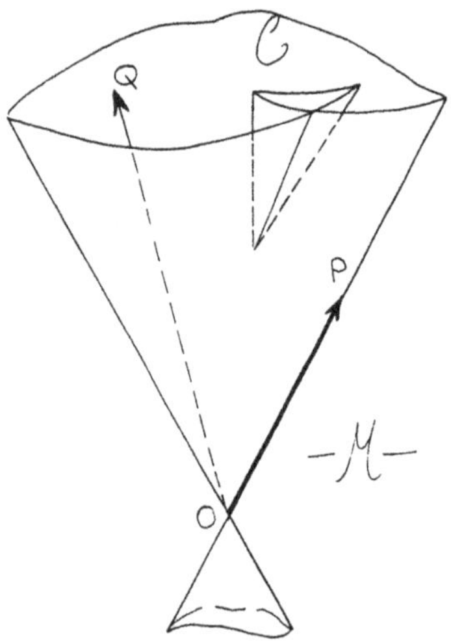

So we have the geometry of $\mathscr{C}$ determined by the set of all symmetric derivatives of Ψ. Now, by similarly doing a Taylor expansion for the whole space-time $\mathcal{M}$, we can express the whole Ψ at any point Q in terms of derivatives (unsymmetrised) of Ψ at O. But we know that the symmetric derivatives of Ψ form an exact set, an algebraic basis (Penrose & Rindler, 1984), so that knowing the elements of this set (i.e. the structure of the light cone)—

$$\{\Psi_{ABCD}, \Psi^{E'}_{ABCDE}, \Psi^{E'F'}_{ABCDEF}, \ldots\}$$

where

$$\Psi^{E'F'\ldots H'}_{ABCDEF\ldots H} = \nabla^{(E'}_{(E}\nabla^{F'}_{F}\ldots\nabla^{H')}_{H}\Psi_{ABCD)}$$

allows us to calculate the unsymmetrized derivatives of Ψ and thus determine the structure inside the light cone. This was first suggested by Penrose in 1963.

We want therefore to find expressions which give the unsymmetrised derivatives of Ψ, completely in terms of those symmetrised derivatives of Ψ; i.e.

$$\textit{unsymmetrised } n^{th} \textit{ derivative of } \Psi = \Psi_n + \textit{function in lower order } \Psi_n\textit{'s}$$

where Ψ_n represents the nth symmetrized derivative of Ψ.

The process of expressing the derivative of Ψ in this way will involve substituting for Ψ_n's of lower order which have already been calculated. The d'Alembertian operator on each of the Ψ_n will also be needed.

The calculation is iterative in form. Starting with the last Ψ_n calculated, a ∇ is applied and the indices symmetrised. The symmetrisation is done so that the resulting expression has all the indices still in alphabetical order. This is the canonical form which is set for spinor expressions to allow them to be simplified to their lowest form.

It is possible to simplify general spinor expressions by reducing them to canonical form, using the ε-identity to uncross indices.

This does produce a large number of terms. Also we need to eliminate those parts of an expression of the form $\alpha_A \alpha_B \varepsilon^{AB}$, which equal zero. This is done by symmetrising over α's indices.

$$\begin{aligned} \alpha_A \alpha_B \varepsilon^{AB} &\rightarrow \tfrac{1}{2}(\alpha_A \alpha_B \varepsilon^{AB} + \alpha_B \alpha_A \varepsilon^{BA}) \\ &\rightarrow \tfrac{1}{2}(\alpha_A \alpha_B \varepsilon^{AB} - \alpha_A \alpha_B \varepsilon^{AB}) \ \rightarrow 0 \end{aligned}$$

In practice however we are symmetrising over huge numbers of indices, giving large numbers of terms in the expression. In order to avoid having to do too much of these simplifications, all indices are kept in canonical form as the calculation proceeds.

Continuing from the symmetrised derivative of Ψ_n, we substitute for Ψ_n. This gives an expression containing terms of at least two derivatives of Ψ_{n-1}. Terms in this expression can be simplified using the Ricci identities, where the sum of two appropriate terms of two derivatives is replaced with a product of Ψ's, i.e.

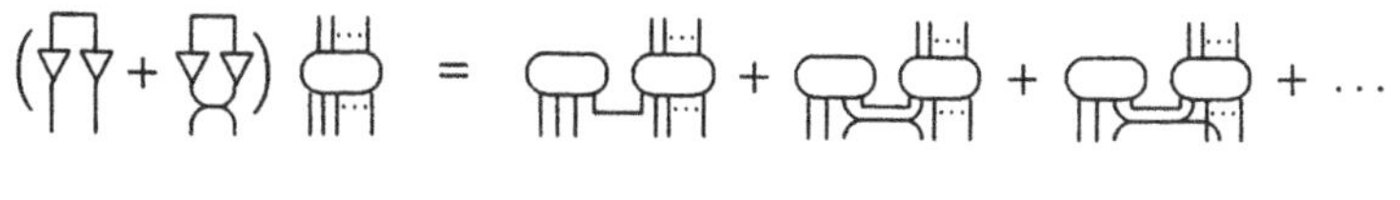

where 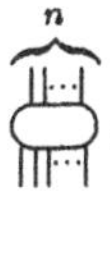 $= \Psi_n$.

There will also be terms of the form 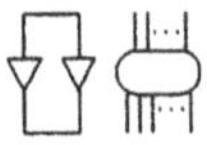 to be substituted for. This will use an expression for $\square\Psi_{n-1}$ which will have been calculated at the same time as Ψ_{n-1}, where

$$\square\Psi_{n-1} = \sum \text{ terms in } \Psi_{n-1} \text{ and lower order } \Psi\text{'s.}$$

The expression can thus be reduced to one containing only Ψ_n's and one term, the unsymmetrised n^{th} derivative of Ψ_{ABCD}. These methods are also applied in calculating $\square\Psi_{n-1}$ as part of the iterative process.

Starting with Ψ_{ABCD} and knowing that $\varepsilon^{AB}\nabla_{AA'}\Psi_{BCDE} = 0$, we can easily calculate $\Psi_1 = \nabla_{AA'}\Psi_{BCDE}$ and calculate $\square\Psi_{ABCD}$ in terms of Ψ_{ABCD}. This is then taken to calculate Ψ_2 in terms of Ψ_1, Ψ_0 plus $\nabla_{AA'}\nabla_{BB'}\Psi_{CDEF}$, and $\square\Psi_1$, in terms of Ψ_1 and Ψ_0, and so on.

This process has been implemented in **Mathematica**. The expresions obtained are extremely long and **Mathematica** encounters certain difficulties in managing expressions of this length effectively. Explorations are being made to transfer this work to another system.

A more specialised approach to the problem would be to set the light cone to be converging, by giving specific values for the symmetric derivatives which set the structure of $\mathscr{C}$. This can be achieved by

$$\Psi_{ABCD} = a\,\iota^A\iota^B\iota^C\iota^D \qquad \text{and} \qquad \Psi^{E'}_{ABCDE} = b\,o^{E'}o_E o_A o_B o_C o_D$$

and all others are zero.

This reconverging light cone structure would mean that there must be a singularity in the space-time. The structure of the space-time as the singularity is approached could be investigated using this method.

References

Penrose, R. (1963), Null hypersurface initial data for classical fields of arbitrary spin and for general relativity, in Aerospace Research Laboratories Report 63-56, (P.G.Bergmann ed.) reprinted in Gen. Rel. Grav. **12**, pp. 225-264 (1980).

Penrose, R. & Rindler, W. (1984) *Spinors and Space-Time, vol. I: Two-Spinor Calculus and Relativistic Fields*, Cambridge University Press.

Penrose, R. (1985) Numerical Relativity by Power Series, in *Galaxies, axisymmetric systems and relativity: essays presented to W.B.Bonnor on his 65th birthday*, (M.A.H.MacCallum ed.), CUP.

Chapter 4

Quasi-local mass

§II.4.1 **Two-surface twistors and quasi-local momentum and angular momentum** *by K.P. Tod.*

This section gathers together material on two-surface twistors and Penrose's quasi-local momentum and angular momentum construction. The articles give an incomplete view of work on the construction in that, by and large, the many successes of the construction are not well represented here. It should be remembered that where it worked, which was chiefly with 'non-contorted' 2-surfaces (defined below), the construction worked well. What is on display here is the struggle to find a way to deal with 'contorted' 2-surfaces: different avenues were explored and different modifications were suggested and it is this ferment of activity that is best represented. It is also the case that, for some of the results that follow, the original Twistor Newsletter article has until now been the only complete reference.

The construction as originally given by Penrose (§II.4.2; see also Penrose 1982 and Penrose & Rindler 1986) was intended to associated a kinematic twistor $A_{\alpha\beta}$ with an arbitrary (space-like, topologically spherical) 2-surface $\mathcal{S}$ in an arbitrary curved space-time $\mathcal{M}$ so that $A_{\alpha\beta}$ measured the total momentum and angular momentum, gravitational as well as material, threading through S. To construct $A_{\alpha\beta}$ one has first to define the 2-surface twistor space $\mathbb{T}^{\alpha}(\mathcal{S})$, the space of 2-surface twistors (briefly known as superficial twistors): a 2-surface twistor Z^{α} is a spinor-field ω^{A} on $\mathcal{S}$ satisfying the 2-surface twistor equations, which in turn are the parts of the usual twistor equation involving derivatives tangential to $\mathcal{S}$. Then $A_{\alpha\beta}$ is a bilinear form on $\mathbb{T}^{\alpha}(\mathcal{S})$, or equivalently a particular element of the dual of the symmetric tensor product $\mathbb{T}^{\alpha}(\mathcal{S})\odot\mathbb{T}^{\alpha}(\mathcal{S})$, defined by an integral over $\mathcal{S}$ of the curvature spinors of M contracted with a pair of elements of $\mathbb{T}^{\alpha}(\mathcal{S})$.

To extract a definition of momentum from $A_{\alpha\beta}$ one would need something equivalent to an infinity twistor; to extract a scalar representing total mass from $A_{\alpha\beta}$ one would need a norm, presumably obtained from an inner-product on $\mathbb{T}^{\alpha}(\mathcal{S})$. Recall that in flat space a twistor Z^{α} is represented by a field ω^A which varies on space-time and from which another field $\pi_{A'}$ is obtained by differentiation. Then the usual definition of inner product is

$$\Sigma = \Sigma(Z^\alpha, \bar{Z}_\alpha) = \omega^A \bar{\pi}_A + \pi_{A'} \bar{\omega}^{A'} \tag{1}$$

which turns out to be constant on space-time, so that it does define an inner-product on twistor space. This same expression can be tried with ω^A representing a 2-surface twistor and $\pi_{A'}$ obtained from it by differentiation just as in flat space. However, for generic $\mathcal{S}$, the quantity defined by (1) from a 2-surface twistor is not constant on $\mathcal{S}$ and so does not define an inner-product on $\mathbb{T}^\alpha(\mathcal{S})$. When it is constant, and does define an inner-product, the 2-surface $\mathcal{S}$ is said to be *non-contorted*; otherwise $\mathcal{S}$ is *contorted.* On a non-contorted $\mathcal{S}$, then, a quasi-local local mass can be calculated. Many examples have been worked out with non-contorted 2-surfaces and the results are quite satisfactory (for a review and bibliography see Tod 1990). On a contorted surface, on the other hand, it is generally believed that the original construction needs modification. The articles which follow can be roughly grouped according to theme as follows:

Mathematical developments. The original theory is introduced by Penrose (§II.4.2). The kinematic sequence from §II.4.2, which is related to the issue of identifying momentum from $A_{\alpha\beta}$, recurs in Hughston & Hurd (§II.4.3). The 2-surface twistor equations as an elliptic system are studied by Eastwood (§II.4.7) and Baston (§II.4.9); Tod (§II.4.10) looks at 2-surface twistors on tori.

Small spheres and large spheres. It is possible to calculate the quasi-local mass as a function of r for a sequence of 2-surfaces $\mathcal{S}(r)$ which collapse to a point when $r = 0$, and are approximately spheres for small r. This is done by Tod (§II.4.14), and the analogous calculation for small ellipsoids is carried out by Woodhouse (§II.4.16). At the other extreme, one can calculate the 2-surface twistors and the kinematic twistor on a sequence of surfaces $\mathcal{S}(r)$ tending to infinity. This may mean space-like infinity (Shaw §II.4.4, §II.4.13; the problem of defining angular-momentum fluxes through infinity is also studied here) or towards the infinity in an asymptotically anti-de-Sitter space-time (Kelly, §II.4.23: this work, which includes a positivity result for the kinematic twistor at infinity in an asymptotically anti-de-Sitter space-time, does not appear elsewhere in the literature, although it is described briefly in Tod 1990).

These calculations typically connect the quasi-local mass or kinematical quantities with other definitions or notions of these quantities, They also illuminate the problem of dealing with contorted 2-surfaces, since in both cases $\mathcal{S}(r)$ is usually contorted at some order in r. In this connection, Jeffryes (§II.4.19) shows that the expression for the volume-form on $\mathbb{T}^\alpha(\mathcal{S}(r))$ for $\mathcal{S}(r)$ near infinity, defined as an appropriate determinant contructed from 4 linearly independent 2-surface twistors, will typically give a complex function on $\mathcal{S}$ and not a complex constant: this fact closed one possible avenue for dealing with contortedness.

Embedding. It was realised early on that the 2-surface twistor space$\mathbb{T}^\alpha(\mathcal{S})$ gives a way of embedding $\mathcal{S}$ in complexified Minkowski space (since the Grassmanian of lines in $\mathbb{PT}^\alpha(\mathcal{S})$ is just complexified Minkowski space $\mathbb{CM}$, and a point p of $\mathcal{S}$ can be identified with the elements of $\mathbb{T}^\alpha(\mathcal{S})$ vanishing a p). This embedding is studied by Penrose (§II.4.22, see also §II.4.15). Then a non-contorted $\mathcal{S}$ is one for which the embedding goes into real Minkowski space (Jeffryes §II.4.8, §II.4.12; Tod §II.4.14).

Suggested modifications. Shaw (§II.4.7) suggested defining a sort of 3-surface-twistors from a variational principle and using them in a modified quasi-local $A_{\alpha\beta}$. Penrose (§II.4.15, 20) suggested modifications in the integrand defining $A_{\alpha\beta}$ motivated by the study of various contorted 2-surfaces. The idea of dealing with contortedness by complex conformal transformations (since $\mathbb{T}^\alpha(\mathcal{S})$, as noted above, embeds $\mathcal{S}$ in $\mathbb{CM}$), or by introducing torsion, appears in Jeffryes §II.4.8, 12, Tod §II.4.14 and Penrose §II.4.15.

Dougan & Mason §II.4.26 (see also Dougan & Mason 1991) propose a different construction aimed at defining a quasi-local momentum directly, motivated by the idea of momentum as generator of translations in Hamiltonian theory (Mason §II.4.11) and using a different definition of 2-surface twistors. The great strength of their construction is that they can use something like the Witten proof of positive energy (Witten 1981) to prove that their momentum is time-like and future-pointing.

Finally, there are articles that do not fit the above classification: Penrose (§II.4.4) sought a role for the kinematic twistor in an active deformation of some twistorial structure. Mason (§II.4.11a, b) sought an origin for the quasi-local mass construction as quasi-local Hamiltonians in the Hamiltonian formalism for general relativity (see also Mason 1989 and Mason & Frauendiener 1990); this feeds into §II.4.26. Jeffryes (§II.4.17) considered how $\mathbb{T}^\alpha(\mathcal{S})$ was influenced by the existence of isometries in the ambient space-time; in particular he showed that an axisymmetric

2-surface in a space-time with a hyper-surface-orthogonal axial symmetry is always non-contorted, an observation which has proved useful in numerial work. Penrose (§II.4.21) studied the counterpart of $\mathbb{T}^\alpha(\mathcal{S})$ for $\mathcal{S}$ of dimension 4 and codimension 2, and for $\mathcal{S}$ of dimension 2 and codimension 4. Penrose (§II.4.25) looked at the relation between 2-surface twistors on $\mathcal{S}$ and hypersurface twistors on a 3-surface containing S.

References

Dougan, A.J. & Mason, L.J. (1991) Phys. Rev. Lett., **67**, 16, 2119-2122.

Mason, L.J. (1989) Class. & Quant. Gravity, **6**, L7-13.

Mason, L.J. & Frauendiener, J. (1990) in *Twistors in Mathematics and Physics*, eds. T.N.Bailey & R.J.Baston, LMS Lecture Note Series 156, Cambridge, CUP.

Penrose, R. (1982) Proc. Roy. Soc. Lond., **A386**, 53-63.

Penrose, R. & Rindler, W. (1986) *Spinors and Space-time vol. 2*, Cambridge, CUP.

Tod, K.P. (1990) in *Twistors in Mathematics and Physics*, eds. T.N.Bailey & R.J.Baston, LMS Lecture Note Series 156, Cambridge, CUP.

Witten, E. (1981) Comm. Math. Phys. **80**, 381-402.

§II.4.2 **A theory of 2-surface ('superficial') twistors** *by R.Penrose* (TN 13, December 1981)

As had been noted some years back (Penrose 1968), the spin-lowering property of a solution in $\mathbb{M}^I$, of the symmetric 2-index twistor equation

$$\nabla_{A'}^{(A}\gamma^{BC)} = 0, \qquad \gamma^{AB} = \gamma^{BA} \tag{1}$$

can be used to reduce the expressions for the 10 conservation laws of linear gravity theory (field ϕ_{ABCD} subject to $\nabla^{AA'}\phi_{ABCD} = 0$ in the source-free region) to the charge conservation law of Maxwell theory (field $i\phi_{AB}$ subject to $\nabla^{AA'}\phi_{AB} = 0$ in a source-free space) where we set

$$\frac{G}{2}\,\phi_{AB} = \phi_{ABCD}\gamma^{CD} \tag{2}$$

Putting (in source-free region only)

$$K_{abcd} = \phi_{ABCD}\varepsilon_{A'B'}\varepsilon_{C'D'} + \bar{\phi}_{A'B'C'D'}\varepsilon_{AB}\varepsilon_{CD}$$

and

$$F_{ab} = i\phi_{AB}\varepsilon_{A'B'} - i\bar{\phi}_{A'B'}\varepsilon_{AB}\,,$$

so that

$$F^{*}_{ab} = \tfrac{1}{2}\,e_{abcd}F^{cd} = \phi_{AB}\varepsilon_{A'B'} + \bar{\phi}_{A'B'}\varepsilon_{AB} \tag{3}$$

we can express the conservation laws for K_{abcd}, for a region $\mathfrak{R}$ surrounding a source distribution, as the fact that

$$\eta = \oint_{\mathcal{S}} F^{*}_{ab}\; dx^a \wedge dx^b \qquad (\mathcal{S} \subset \mathfrak{R}) \tag{4}$$

measures the total source for F_{ab} intercepted by any compact 3-volume whose boundary is the 2-surface $\mathcal{S}$. As γ^{AB} varies, subject to (1), η will range over the various components of the source for K_{abcd}.

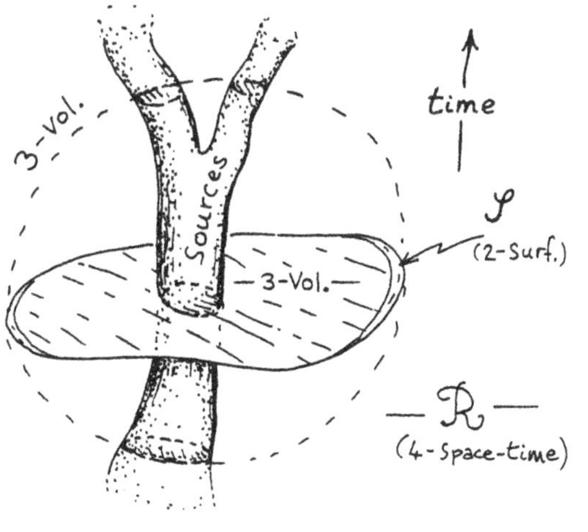

Let $G^{\alpha\beta}$ be the twistor ($\in \mathbb{T}^{(\alpha\beta)}$) whose primary part is γ^{AB}. $G^{\alpha\beta}$ has 10 independent complex components, corresponding to the fact that $\mathbb{T}^{(\alpha\beta)}$ is $\frac{\boxed{4}\boxed{3}}{\boxed{2}\boxed{1}} = \frac{4\times 5}{2} = 10$-dimensional, so there are 10 independent (complex) solutions of (1). The η of (4) is *real*, so we get 20 independent real 'charges' for K_{abcd} (10 of which will be zero whenever K_{abcd} arises in the normal way from a potential (= linearized metric)). The sources themselves are described by the various conserved currents

$$J^a = T^{ab}k_b \qquad (\nabla_a J^a = 0;\ \nabla_a T^{ab} = 0) \tag{5}$$

where k^a is a Killing vector, related to γ^{AB} by

$$k^a = \tfrac{4}{3}(i\nabla^A_{B'}\bar{\gamma}^{A'B'} - i\nabla^{A'}_{B}\gamma^{AB}). \tag{6}$$

T^{ab} is the usual energy-momentum tensor *source* for K_{abcd} $\left(K_{abcd} = K_{[cd][ab]},\ K_{[abc]d} = 0\right)$:

$$K_{acb}{}^{c} - \tfrac{1}{2}g_{ab}K_{cd}{}^{cd} = -8\pi G T_{ab} \tag{7}$$

$(G =$ gravitational constant$)$. In tensor terms, putting

$$S^{ab} = \gamma^{AB}\varepsilon^{A'B'} + \varepsilon^{AB}\bar{\gamma}^{A'B'} \tag{8}$$

we have

$$\nabla^{(a}S^{b)c} - \nabla^{(a}S^{c)b} + g^{a[b}\nabla_d S^{c]d} = 0 \tag{9}$$

as the translation of (1) (so equation (9) has 20 independent real solutions), and

$$e_{abcd}\nabla^b S^{cd} = \tfrac{3}{2}k_a \tag{10}$$

as the translation of (6). Equations (7), (9) and (10) imply

$$\nabla_{[a}(K_{bc]qd}S^{qd}) = \frac{4\pi G}{3}e_{abcd}T^{df}k_f \tag{11}$$

which becomes

$$\nabla_{[a}F^*_{bc]} = \frac{4\pi}{3}e_{abcd}J^d \tag{12}$$

(i.e. the Maxwell equation $\nabla_a F^{ab} = 4\pi J^b$) when

$$F^*_{ab} = \frac{1}{G}K_{abcd}S^{cd} \tag{13}$$

which reduces to (2) in source-free space.

In twistor terms, the relation between γ^{AB} and the Killing vector $k^{AB'}$(taken to be the primary part of the Hermitian twistor $K^{\alpha}{}_{\beta} \in \mathbb{T}^{\alpha}_{\beta}$) is given by

$$K^{\alpha}{}_{\beta} = 4\overline{G}_{\beta\gamma}I^{\alpha\gamma} + 4G^{\alpha\gamma}I_{\beta\gamma}. \tag{14}$$

(This is equation (6) in twistor form.) Note that the $K^{\alpha}{}_{\beta}$ of (14) is trace-free, and with this restriction $K^{\alpha}{}_{\beta}$ is determined by its primary part $k^{AB'}$. Equation (14) is part of the 'kinematic sequence' or 'moment sequence', which is (in the simplified form due to L.P.Hughston)

$$0 \to \mathbf{R} \to \mathbf{Q}^{[\alpha\beta]} \to \mathbf{H}^{\alpha}_{\beta} \to \mathbb{T}^{(\alpha\beta)} \to \mathbf{H}^{\alpha}_{\beta} \to \mathbf{Q}^{[\alpha\beta]} \to \mathbf{R} \to 0 \tag{15}$$

where these are considered as *real* vector spaces, of respective dimensions 0, 1, 6, 15, 20, 15, 6, 1, 0, $\mathbf{H}^{\alpha}_{\beta}$ consisting of trace-free Hermitian twistors H^{α}_{β} and $\mathbf{Q}^{[\alpha\beta]}$ consisting of twistor-real (i.e. $\overline{Q}_{\alpha\beta} = \frac{1}{2}\varepsilon_{\alpha\beta\gamma\delta}Q^{\gamma\delta}$) elements $Q^{\alpha\beta} \in \mathbb{T}^{[\alpha\beta]}$. Equation (14) is the $\mathbb{T}^{(\alpha\beta)} \to \mathbf{H}^{\alpha}_{\beta}$ part of (15), so twistors whose primary parts are Killing vectors constitute the image of this map. The kernel provides those γ^{AB}'s which give zero in (6) (i.e. in (10)). This kernel is the image of $\mathbf{H}^{\alpha}_{\beta} \to \mathbb{T}^{(\alpha\beta)}$, which is the set of twistors of the form

$$iH^{(\alpha}_{\gamma}I^{\beta)\gamma} \tag{16}$$

The η's of (4), with (13) substituted, arise as real parts of expressions like

$$\zeta = \frac{1}{16\pi G}\oint_{\mathscr{S}} K_{AA'BB'cd}\gamma^{AB}\varepsilon^{A'B'}dx^c \wedge dx^d \tag{17}$$

which we can write

$$\zeta = A_{\alpha\beta}G^{\alpha\beta} \tag{18}$$

where $A_{\alpha\beta}$ is the total *kinematic* or *angular momentum twistor* of the sources surrounded by $\mathscr{S}$ and satisfies

$$A_{\alpha\beta}I^{\gamma\beta} = \overline{A}^{\gamma\beta}I_{\alpha\beta}. \tag{19}$$

$A_{\alpha\beta}$ is dual (over the reals) to $G^{\alpha\beta}$, so it is located in the dual sequence to (15), which is another sequence of just the same form as (15), $A_{\alpha\beta}$ lying in the image of the corresponding map $\mathbb{H}^{\alpha}_{\beta} \to \mathbb{T}_{(\alpha\beta)}$, which is what (19) entails.

All this refers just to $\mathbb{M}^I$; we wish to do a corresponding thing in a general curved space $\mathcal{M}$. We cannot expect to solve (1) (or (9)) in general, but we need its appropriate analogue *only on the surface $\mathcal{S}$ itself.* However, if we try to restrict (1) to $\mathcal{S}$, we find only *two* equations that are entirely tangential to $\mathcal{S}$, which is insufficient to determine the *three* complex components of γ^{AB} on $\mathcal{S}$. The key idea is not to try to define the γ's directly but to consider them as tensor products of 1-index objects. Any solution of (1) is, indeed, of the form

$$\gamma^{AB} = \sum \underset{1}{\omega}^{(A} \underset{2}{\omega}^{B)} \tag{20}$$

in $\mathbb{M}^I$, where each $\underset{i}{\omega}^A$ satisfies the 1-index twistor equation

$$\nabla^{(A}_{A'} \omega^{B)} = 0 \tag{21}$$

and it is (21) rather than (1) that we restrict to $\mathcal{S}$. We take $\mathcal{S}$ to be *spacelike* and (normally) to have the *topology of a sphere* S^2. Using the GHP formalism (Geroch, Held & Penrose 1973) we find, for the two components tangential to $\mathcal{S}$ (in a spin-frame family whose flagpoles are orthogonal to $\mathcal{S}$), that

$$\eth' \omega^0 = \sigma' \omega^1, \quad \eth \omega^1 = \sigma \omega^0. \tag{22}$$

It follows from the Atiyah-Singer index theorem that equations (22) always have at least 4 (complex-) independent solutions and that *generically* equations (22) have *precisely* 4 independent solutions. (In the 'canonical' case, $\sigma = \sigma' = 0$, and the adjoint equations have no solutions.) The solution space $\mathbb{T}^{\alpha}(\mathcal{S})$ is the *2-surface twistor space* of $\mathcal{S}$, which will be a complex 4-dimensional vector space, assuming that we are in this normal situation. Then the symmetric tensor product $\mathbb{T}^{(\alpha\beta)}(\mathcal{S})$ will be 10-complex-dimensional. Each element of $\mathbb{T}^{(\alpha\beta)}(\mathcal{S})$ can be interpreted as a spinor field γ^{AB} over $\mathcal{S}$ (the tensor product of the ω^A's being taken pointwise on $\mathcal{S}$.) The quantities (17) can now be defined in a *general* $\mathcal{M}$, as before but with Riemann's R_{abcd} replacing K_{abcd}.

For a 1-term sum (20) we get

$$\zeta = \frac{-i}{4\pi G}\oint\{(\Psi_1-\Phi_{10})\underset{1}{\omega}{}^0\underset{2}{\omega}{}^0+(\Psi_2-\Phi_{11}-\Lambda)(\underset{1}{\omega}{}^0\underset{2}{\omega}{}^1+\underset{1}{\omega}{}^1\underset{2}{\omega}{}^0)+(\Psi_3-\Phi_{21})\underset{1}{\omega}{}^1\underset{2}{\omega}{}^1\}\underset{\sim}{\mathscr{S}} \tag{23}$$

($\underset{\sim}{\mathscr{S}}$ being the surface-area 2-form on $\mathscr{S}$). This defines $A_{\alpha\beta}\in\mathbb{T}_{(\alpha\beta)}(\mathscr{S})$ via (18), which has 10 *complex* components. To define mass-momentum and angular momentum *quasi-locally* we need an analogue of (19) so as to have only 10 *real* components. So far this is lacking; we do not have a convincing $I_{\alpha\beta}\in\mathbb{T}_{[\alpha\beta]}(\mathscr{S})$, nor a convincing $Z^\alpha\mapsto\overline{Z}_\alpha$ (through tentative definitions exist for the latter—not conformally invariant whereas $\mathbb{T}^\alpha(\mathscr{S})$ is). Things are rosier at $\mathscr{I}^+$, where $\sigma'=0$. $I_{\alpha\beta}$ exists via

$$I_{\alpha\beta}\underset{1}{Z}{}^\alpha\underset{2}{Z}{}^\beta = i\underset{1}{\omega}{}^0\underset{2}{\pi}{}_{1'}-i\underset{2}{\omega}{}^0\underset{1}{\pi}{}_{1'}=\underset{2}{\omega}{}^0\eth\underset{1}{\omega}{}^0-\underset{1}{\omega}{}^0\eth\underset{2}{\omega}{}^0 \tag{24}$$

being constant over $\mathscr{S}\subset\mathscr{I}^+$, where

$$\pi_{0'}=i\eth'\omega^1-i\rho\omega^0,\quad \pi_{1'}=i\eth\omega^0-i\rho'\omega^1. \tag{25}$$

Still $Z^\alpha\mapsto\overline{Z}_\alpha$ is not totally convincing, but the *Bondi-Sachs* 4-momentum is reproduced exactly and a *new improved angular momentum at* $\mathscr{I}^+$ is obtained.

Note K.P.Tod's tantalizing formula for (23), using (25), $\zeta=\frac{-i}{4\pi G}\oint_{\mathscr{S}}(\underset{1}{\pi}{}_{0'}\underset{2}{\pi}{}_{1'}+\underset{1}{\pi}{}_{1'}\underset{2}{\pi}{}_{0'})\underset{\sim}{\mathscr{S}}$.

More work in progress; see Penrose (1982). Thanks especially to I.M.Singer and K.P.Tod.

References

Geroch, Held & Penrose (1973) J. Math Phys, **14**, 874.

Penrose, R. (1968) Int. J. Theor. Phys. **1**, 61.

Penrose, R. (1982) Proc. Roy. Soc. Lond. **A381**, 53.

§II.4.3 **The kinematic sequence (revisited)** *by L.P.Hughston & T.R.Hurd* (TN 13, December 1981)

We begin with the observation that

$$0 \longrightarrow \mathbf{C} \xrightarrow{A^i} \mathbf{C}^i \xrightarrow{A^j} \mathbf{C}^{[ij]} \xrightarrow{A^k} \mathbf{C}^{[ijk]} \xrightarrow{A^l} \mathbf{C}^{[ijkl]} \xrightarrow{A^m} \mathbf{C}^{[ijklm]} \xrightarrow{A^n} \mathbf{C}^{[ijklmn]} \longrightarrow 0 \qquad (1)$$

with A^i $(i = 0, 1, \ldots, 5)$ a fixed point of $\mathbf{C}^6$, is an exact sequence of vector space maps, the vector spaces $\mathbf{C}$, $\mathbf{C}^i$ etc. being irreducible representations of $GL(6, \mathbf{C})$. If we restrict attention to the subgroup $SO(6, \mathbf{C})$ preserving a non-degenerate quadratic form Ω_{ij}, then some of these vector spaces decompose in an interesting fashion. To see this, rewrite the sequence in 'skew bitwistor' indices. Then, for instance,

$$\begin{array}{lll} \mathbf{C}^{[ij]} \longleftrightarrow & \mathbf{C}^{[\alpha\beta][\gamma\delta]} & \xrightarrow{\simeq} \mathbf{C}_{\rho}^{\alpha\,TF} \quad \text{(trace-free)} \\ & b^{\alpha\beta\gamma\delta} & \longmapsto b_{\rho}^{\alpha} = b^{\alpha\beta\gamma\delta}\Omega_{\rho\beta\gamma\delta} \end{array}$$

and

$$\begin{array}{lll} \mathbf{C}^{[ijk]} \longleftrightarrow & \mathbf{C}^{[\alpha\beta][\gamma\delta][\rho\sigma]} & \xrightarrow{\simeq} \mathbf{C}^{(\alpha\sigma)} \oplus \mathbf{C}_{(\mu\nu)} \\ & b^{\alpha\beta\gamma\delta\rho\sigma} & \longmapsto \left(b^{\alpha\beta\gamma\delta\rho\sigma}\Omega_{\beta\gamma\delta\rho},\ b^{\alpha\beta\gamma\delta\rho\sigma}\Omega_{\mu\alpha\beta\gamma}\Omega_{\nu\delta\rho\sigma}\right) \end{array}$$

where $\Omega_{\alpha\beta\gamma\delta} = \Omega_{[\alpha\beta\gamma\delta]} \longleftrightarrow \Omega_{ij}$. The exact sequence (1) now takes the form

$$0 \longrightarrow \mathbf{C} \longrightarrow \mathbf{C}^{[\alpha\beta]} \longrightarrow \mathbf{C}_{\gamma}^{\alpha\,TF} \longrightarrow \mathbf{C}^{(\alpha\beta)} \oplus \mathbf{C}_{(\gamma\beta)} \longrightarrow \mathbf{C}_{\gamma}^{\alpha\,TF} \longrightarrow \mathbf{C}_{[\alpha\beta]} \longrightarrow \mathbf{C} \longrightarrow 0 \qquad (2)$$

where the six non-trivial maps here are given by

$$b \longrightarrow bA^{\alpha\beta}$$

$$b^{\alpha\beta} \longrightarrow b^{\alpha\beta}A_{\gamma\beta} - \tfrac{1}{4}(b^{\rho\beta}A_{\rho\beta})\delta_{\gamma}^{\alpha}$$

$$b_{\gamma}^{\alpha} \longrightarrow \left(b_{\gamma}^{(\alpha}A^{\beta)\gamma},\ b_{(\gamma}^{\alpha}A_{\beta)\alpha}\right)$$

$$(b^{\alpha\beta}, d_{\gamma\beta}) \longrightarrow b^{\alpha\beta}A_{\gamma\beta} - d_{\gamma\beta}A^{\alpha\beta}$$

$$b_{\gamma}^{\alpha} \longrightarrow b_{[\gamma}^{\alpha}A_{\beta]\alpha}$$

$$b_{\gamma\beta} \longrightarrow b_{\gamma\beta}A^{\gamma\beta}$$

This we shall call the *kinematic sequence.* Note, however, that Penrose's well-known cyclic sequence looks rather similar:

$$\cdots \longrightarrow \mathbb{C}^{\alpha}_{\gamma} \xrightarrow{I} \mathbb{C}^{[\alpha\beta]} \xrightarrow{I} \mathbb{C}^{\alpha}_{\gamma} \xrightarrow{I} \mathbb{C}^{(\alpha\beta)} \oplus C_{(\gamma\beta)} \xrightarrow{I} \mathbb{C}^{\alpha}_{\gamma} \xrightarrow{I} \mathbb{C}^{[\alpha\beta]} \longrightarrow \cdots \qquad (3)$$

where $\mathbb{C}^{\alpha}_{\gamma} \simeq \mathbb{C}^{\alpha}_{\gamma}{}^{TF} \oplus \mathbb{C}$ and where $I_{\alpha\beta}$ is the vertex of $\mathfrak{I}$. In fact, this can be cut into an infinite number of finite chunks, each of which is like (2): the sequences (3) and (2) contain the same information! In particular, the generators of the conformal group are parametrized by $\mathbb{C}^{\alpha}_{\gamma}{}^{TF}$ and the Poincaré generators are given by the kernel of $I: \mathbb{C}^{\alpha}_{\gamma}{}^{TF} \longrightarrow \mathbb{C}_{[\gamma\beta]}$.

Nothing we have said so far has depended on whether or not $A^{\alpha\beta}$ is simple ($A^{\alpha\beta}A^{\gamma\delta}\Omega_{\alpha\beta\gamma\delta} = 0$). However, in the case when $A^{\alpha\beta}$ is *not* simple, the sequence (2) splits into two pieces:

$$0 \longrightarrow \mathbb{C} \xrightarrow{A} \mathbb{C}^{[\alpha\beta]} \xrightarrow{A} \mathbb{C}^{\alpha}_{\gamma}{}^{TF} \xrightarrow{A} \mathbb{C}^{[\alpha\beta]} \longrightarrow 0$$

and its dual

$$0 \longrightarrow \mathbb{C}_{[\alpha\beta]} \xrightarrow{A} \mathbb{C}^{\alpha}_{\gamma}{}^{TF} \xrightarrow{A} \mathbb{C}_{[\alpha\beta]} \xrightarrow{A} \mathbb{C} \longrightarrow 0.$$

Note incidentally that the Koszul complex is an exact sheaf sequence on $\mathbb{C}^6$ which is analogous to (2):

$$0 \longrightarrow \mathcal{O} \xrightarrow{X} \mathcal{O}^{[\alpha\beta]} \xrightarrow{X} \mathcal{O}^{\alpha}_{\gamma}{}^{TF} \xrightarrow{X} \mathcal{O}^{(\alpha\beta)} \oplus \mathcal{O}_{(\alpha\beta)} \xrightarrow{X} \mathcal{O}^{\alpha}_{\gamma}{}^{TF} \xrightarrow{X} \mathcal{O}_{[\alpha\beta]} \xrightarrow{X} \mathcal{O} \longrightarrow 0,$$

where the $X^{\alpha\beta}$ are $\mathbb{C}^6$-coordinates. This sequence and a similar one on $\mathbb{CP}^5$ are useful in the study of conformally invariant space-time fields.

Comment (1994). For further information on the kinematic (or moment) sequence see Penrose & Rindler (1986) §6.5, Hurd (1982) and Hughston & Hurd (1983).

References

Hughston, L.P. & Hurd, T.R. (1983) A $\mathbb{CP}^5$ calculus for space-time fields, Phys. Rep. **100**, 273-326.

Hurd, T.R. (1982) Conformal geometry and its applications to relativistic quantum theory, D.Phil. thesis, Oxford.

Penrose, R. & Rindler, W. (1986) *Spinors and space-time*, Vol. 2, C.U.P.

§II.4.4 **Two-surface twistors, angular momentum flux and multipoles of the Einstein-Maxwell field at $\mathscr{I}^+$** *by W.T.Shaw* (TN 14, July 1982)

The twistor approach to the momentum and angular momentum of linearised gravity on Minkowski space Penrose & MacCallum (1973) is: given the Weyl spinor Ψ_{ABCD} and a solution W^{AB} of the twistor equation $\nabla^{(A}_{A'}W^{BC)}=0$, construct the 2-form F_{ab} whose spinor equivalent is

$$\phi_{AB}=\Psi_{ABCD}W^{CD},$$

and evaluate the flux linking a given 2-surface S as

$$L=\oint_S F^*_{ab}\,dx^a\wedge dx^b.$$

To use this at $\mathscr{I}^+$ and in the full non-linear theory as in Penrose (1982) one needs to project the twistor equation onto an arbitrary cut. Write

$$W^{AB}=W^0o^Ao^B+W^1o^{(A}\iota^{B)}+W^2\iota^A\iota^B.$$

Then ϕ_{AB} has components

$$\begin{aligned}\phi_0&=W^0\Psi_0+W^1\Psi_1+W^2\Psi_2\\ \phi_1&=W^0\Psi_1+W^1\Psi_2+W^2\Psi_3\\ \phi_2&=W^0\Psi_2+W^1\Psi_3+W^2\Psi_4.\end{aligned}$$

With the correct normalization to reproduce weak field results and conformally mapping all the spinors to $\mathscr{I}^+$, the integral becomes:

$$L=Re\frac{-i}{4\pi G}\oint_{\Sigma\in\mathscr{I}^+}d\Sigma\left[W^0\Psi_1+W^1\Psi_2+W^2\Psi_3\right]$$

Immediate projection of the twistor equation onto directions tangential to a cut gives

$$\bar{\eth}W^0=0,\qquad\qquad \eth W^2=\sigma W^1.$$

Use of the Ricci identity and the twistor equation (or, as in Penrose 1982 writing W^{AB} as a symmetric product of two valence 1 twistors) yields a third equation:

$$\eth^2 W^1 = \sigma\eth W^0 + 2\eth(\sigma W^0)$$

These 3 equations are conformally invariant and have a 10 complex dimensional solution space: the 2-surface twistor space $\mathbb{T}^{(\alpha\beta)}$ associated with the cut. With the twistor space one associates a complex Minkowski space $\mathbb{CM}(\Sigma)$ and this space provides a 4-dimensional space of origins with respect to which the angular momentum and other multipoles may be defined.

Note that writing $\Psi_3 = \eth N$ and integrating by parts, one obtains

$$L_\Sigma = Re \frac{-i}{4\pi G}\int d\Sigma\left[W^0\Psi_1 + W^1(\Psi_2 - \sigma N)\right]$$

Remarks:

a) if $W^0 = 0$, $\eth^2 W^1 = 0$ and the correct expression for the Bondi mass is obtained.

b) if $W^0 = 0$ and if there is a solution α of $\sigma = \eth^2\alpha$, then the general solution for W^1 is

$$W^1 = \eta + 2W^0\eth\alpha - \alpha\eth W^0$$

where $\eth^2\eta = 0$, and inserting the values of W^0, W^1 yields the complex components of angular momentum with respect to an origin in $\mathbb{CM}$ labelled by η.

The angular momentum is defined with respect to an origin in $\mathbb{CM}(\Sigma)$. Under a change of origin represented by $X(\zeta,\tilde{\zeta})$ at $\mathscr{I}^+$ then

$$W^0 \to W^0; \qquad W^1 \to W^1 + X\eth W^0 - 2W^0\eth X$$

and one can evaluate the change in L. For a stationary spacetime the result simplifies, for then, if the direction of stationarity is given by $T(\zeta,\tilde{\zeta})$ then $\Psi_2 \propto T^{-3}$ and

$$L_\Sigma(X) - L_\Sigma(0) = -3\int_\Sigma d\Sigma\, W^0\Psi_2 T\eth\left[T^{-1}X\right].$$

c.f. Bramson (1975). As noted in Penrose 1982 the origin dependence is decoupled from the cut.

Evolution and Angular Momentum Flux. Suppose one evaluates L on two different cuts Σ_1 and Σ_2. By the divergence theorem applied to ϕ_{AB},

$$\mathrm{L}_{\Sigma_2} - \mathrm{L}_{\Sigma_1} = \int_{\Sigma_1}^{\Sigma_2} d^3s\left[\text{Þ}'\phi_1 - (\eth - \tau)\phi_2\right]$$

and making use of the Bianchi identities at $\mathscr{I}^+$, but not having imposed any part of the twistor equation gives

$$\int_{\Sigma_1}^{\Sigma_2} d^3S[\Psi_1 \cdot \text{Þ}'W^0 + \Psi_2\cdot[\text{Þ}'W^1 - (\eth + 2\tau)W^0] + \Psi_3\cdot[\text{Þ}'W^2 - (\eth+\tau)W^1 + 2\sigma W^0] + \Psi_4\cdot[\sigma W^1 - \eth W^2]]$$

Now, the twistor equation, projected up the null generators of $\mathscr{I}^+$gives

$$\left.\begin{aligned} \text{Þ}'W^0 &= 0 \\ \text{Þ}'W^1 &= (\eth + 2\tau)W^0 \\ \text{Þ}'W^2 &= (\eth + \tau)W^1 - 2\sigma W^0 \end{aligned}\right\} \text{ evolution equations}$$

so if all the twistor equations are satisfied $L_{\Sigma_1} = L_{\Sigma_2}$. However, it turns out that when $\Phi_{02} \neq 0$, i.e. when there is radiation, the evolution equations do not preserve the 2-surface equations. Explicitly, using the G.H.P. commutators:

$$\begin{aligned} &\text{Þ}'(\bar{\eth}W^0) = 0 \\ &\text{Þ}'(\eth W^2 - \sigma W^1) = W^1\Phi_{02} \\ &\text{Þ}'(\eth^2 W^1 - \sigma\eth W^0 - 2\eth(\sigma W^0)) = 3\eth W^0\Phi_{02} + 2W^0(\eth - \tau)\Phi_{02}. \end{aligned}$$

But to meaningfully compare the angular momenta at Σ_1 and Σ_2 one needs to remain in 2-surface twistor space as one moves from Σ_1 to Σ_2 along an intermediate slicing. To effect this, consider adding f, g to the evolution equations for W^1 and W^2 respectively and choosing f and g so as to preserve the 2-surface equations. Then the change in L is

$$L_{\Sigma_2} - L_{\Sigma_1} = \int_{\Sigma_1}^{\Sigma_2} d^3S\,[f\Psi_2 + g\Psi_3]$$

(first form of flux law).

The expression $f\Psi_2 + g\Psi_3$ is interpreted as an effective momentum-angular momentum current flowing 'out' of $\mathcal{I}^+$. To preserve the 2-surface equations f, g satisfy a pair of differential equations on the intermediate slicing, viz.

$$(\eth - \tau)g = \sigma f - W^1 \Phi_{02}$$
$$\eth^2 f - 2\tau\eth f - f(\eth\tau - \tau^2) = -3\Phi_{02}\eth W^0 - 2W^0(\eth - \tau)\Phi_{02}.$$

Integrating by parts using $\Psi_3 = \eth N$ and the first equation gives

$$L_{\Sigma_2} - L_{\Sigma_1} = \int_{\Sigma_1}^{\Sigma_2} d^3S\left[f(\Psi_2 - \sigma N) + W^1 N\Phi_{02}\right].$$

When $W^0 = 0$, can take $f = 0$, so $\text{Þ}' W^0 = 0$ and $\text{Þ}' W^1 = 0$ and the usual Bondi-Sachs momentum loss formula is obtained: $(\overline{N} = \Phi_{02})$

$$\int_{\Sigma_2} W^1(\Psi_2 - \sigma N) = \int_{\Sigma_1} W^1(\Psi_2 - \sigma N) + \int_{\Sigma_1}^{\Sigma_2} d^3S\, W^1 N\overline{N}.$$

When $W^0 \neq 0$, the equation for f may be solved by noting that it is possible to write $\Phi_{02} = \text{Þ}'\eth^2 q$, where q has conformal weight $+1$ and spin weight 0. Then the solution for f is

$$f = -\text{Þ}'[2W^0\eth q - q\eth W^0].$$

Thus finally

$$L_{\Sigma_2} - L_{\Sigma_1} = \int_{\Sigma_1}^{\Sigma_2} d^3S\left[W^1 N\overline{N} + (\Psi_2 - \sigma N)\ \text{Þ}'(q\eth W^0 - 2W^0\eth q)\right].$$

Unfortunately the value of W^1 on Σ_2 and also on the intermediate slicing is not unique. This is because there is no unique choice of q to generate the radiation Φ_{02}. Indeed, in a given conformal scaling of $\mathcal{I}^+$, if one expands q in terms of spherical harmonics the components of harmonics with $l = 1, 0$ are arbitrary. Recalling the formula for the change in W^1 under a change of origin in $\mathbf{CM}(\Sigma)$ $(\eth^2 X = 0)$

$$W^1 \to W^1 + X\eth W^0 - 2W^0\eth X$$

it is easily seen that the indeterminacy in q corresponds to losing all information about the position of the origin in $\mathbb{CM}$ on the intermediate slicing and on Σ_2. This at least is mild compared to supermomentum ambiguities, and we have the following picture:

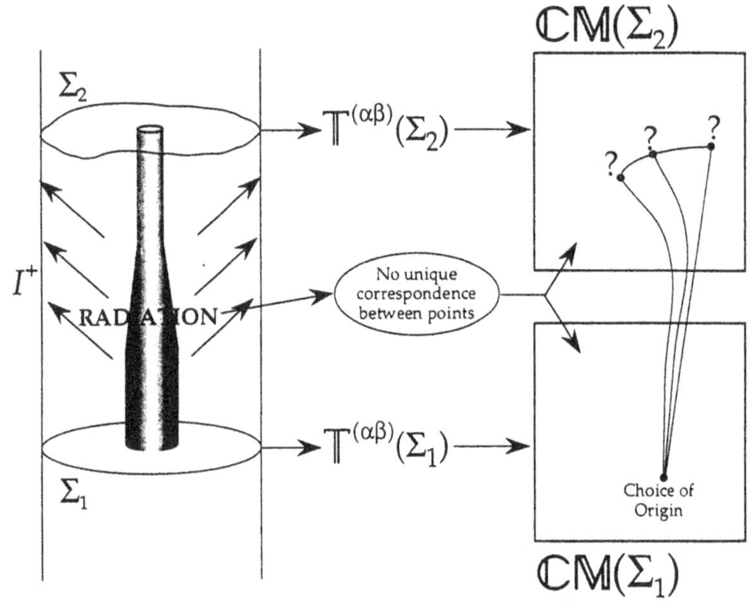

The above is clearly unsatisfactory as it stands. Two approaches to resolving the difficulty:

(a) Pick a conformal scaling on $\mathcal{I}^+$. This means abandoning conformal (Lorentz) invariance and treating angular momentum as defined with respect to a timelike direction c.f. other multipoles (see below). If we do this all redundant spherical harmonics in q can be removed, a unique value of W^1 on Σ_2 obtained and thus a unique answer for the flux derived. The angular momentum is then a tensor under the rotation group only. It is not clear how to do this consistently as there has to be some way of choosing the scaling—e.g. one might align the timelike direction with the Bondi momentum, but this would not work for a system radiating momentum.

(b) evaluate the Pauli-Lubanski spin vector. Once a choice of real origins has been made this is constant over $\mathbb{CM}$ (at least for stationary regions of $\mathcal{I}^+$). This should ignore the ambiguous components. This is a similar approach to Bramson (1978) but with a different concept of origin.

It is hoped that a more concrete picture will emerge through perhaps one of these approaches. The attitude in (a) is clearly applicable to multipole theory in general, and following Curtis (1978), it seems possible to give supertranslation independent definitions of the multipole moments of *stationary* Einstein-Maxwell fields at $\mathcal{I}^+$.

The number of definable multipole moments depends on the smoothness of ϕ_0 (Maxwell) and Ψ_0 (gravitation). The electromagnetic dipole and gravitational quadrupole moments exist provided ϕ_0 and Ψ_0 exist and here we restrict attention to just these two.

For the definition let T denote the direction of stationarity at $\mathcal{I}^+$. Following Curtis (1978) for the electromagnetic dipole, define the $\tilde{\Psi}_1$ and $\tilde{\Psi}_2$ components of a 4-index spinor $\tilde{\Psi}_{ABCD}$ by

$$\tilde{\Psi}_p = \left(\frac{4-p}{2T}\right)\phi_{p-1} \qquad\qquad p = 1, 2.$$

and let $\tilde{\phi}_{AB} = W^{CD}\tilde{\Psi}_{ABCD}$. The corresponding flux integral is then

$$D_\Sigma = \mathrm{Re}\ -\mathrm{i}\int_\Sigma \mathrm{d}\Sigma \left[\frac{3W^0\phi_0 + 2W^1\phi_1}{2T}\right]$$

(un-normalized as yet). This has the following properties:

(a) If at a cut Σ a choice of origin is made, and then Σ undergoes a supertranslation to a new cut Σ', then

$$D_{\Sigma'} = \mathrm{D}_\Sigma\,.$$

(b) at a fixed cut Σ under a change of origin on $\mathbb{CM}$ then

$$D_\Sigma(\mathrm{X}) - \mathrm{D}_\Sigma(0) = -3\int_\Sigma d\Sigma\, W^0\phi_1\eth[T^{-1}X]$$

(note the angular momentum origin dependence).

The gravitational quadrupole is similar: one forms a 6 index spinor field $\tilde{\Psi}_{ABCDEF}$ from Ψ_{ABCD} and turns it into a 2-form using a 4 index twistor Ω^{CDEF}. The components of Ω^{ABCD} satisfy various differential equations on a given cut (up to fourth order).

Given the solution for the Ω^p the quadrupole is then given by (un-normalized)

$$Q_\Sigma = Re\ -i\int d\Sigma \left[\frac{5\Omega_0\Psi_0 + 4\Omega_1\Psi_1 + 3\Omega_2\Psi_2}{2T}\right]$$

and is again supertranslation independent. Under a change of origin in $\mathbb{CM}$ (neglecting $O(X^2)$)

$$Q_\Sigma(X) - Q_\Sigma(0) = \int_\Sigma d\Sigma \left\{-10\Omega_0\Psi_1 - 6\Omega_1\Psi_2\right\} \eth[T^{-1}X].$$

The higher moments may be obtained by running down the coefficients in expansions of ϕ_0 and Ψ_0 in powers of r^{-1} (in the physical space), assuming such an expansion exists. Thus, for example the electromagnetic quadrupole is obtained from ϕ_0^1. The behaviour of these higher moments is being checked and the relationship to other definitions considered. As usual the answers for the Kerr solution as in Curtis (1978) agree with other definitions.

Many thanks to R. Penrose and K.P. Tod for continued ideas and assistance.

References:

Penrose, R. and MacCallum, M.A.H. (1973) Phys. Rep. **6C**, 241-315.

Penrose, R. §II.4.2 and (1982) Proc. Roy. Soc. **A381**, 53.

Bramson, B.D. (1975) Relativistic Angular Momentum, Proc. Roy. Soc. **A341**, 463-490.

Bramson, B.D. (1978) The Invariance of Spin, Proc. Roy. Soc. **A364**, 383-392.

Curtis, G.E. (1978) Twistors and Multipole Moments, Proc. Roy. Soc. **A359**, 133-149.

§II.4.5 **General-relativistic kinematics??** *by R.Penrose* (TN 15, January 1983)

Recall that the standard flat-space *kinematics* of a relativistic object (in the sense of its energy-momentum-angular-momentum structure, the term 'dynamics' applying only when a specific Hamiltonian—or equivalent—has been provided governing the equations of motion of the structure of the object) can be described by the symmetric *kinematic twistor* $A_{\alpha\beta}$ $(= A_{\beta\alpha})$ subject to

$$A_{\alpha\beta}I^{\beta\gamma} = \overline{A_{\beta\gamma}I^{\alpha\gamma}}. \qquad (1)$$

The Hermiticity condition (1) is needed to reduce the number of independent components of $A_{\alpha\beta}$ from 10 complex ones (i.e. 20 real ones) to 10 real ones. The obtaining of an appropriate analogue of (1) has proved to be a major stumbling block in the interpretation of the '$A_{\alpha\beta}(\mathcal{S})$' that arises in the general-relativistic theory of 2-surface ('superficial') twistors, this $A_{\alpha\beta}(\mathcal{S})$ being associated with a spacelike topological 2-sphere $\mathcal{S}$ in (curved) space-time and being intended to describe the 'kinematics' (in the above sense) of the (total) object surrounded by $\mathcal{S}$ (cf. Penrose in §II.4.2 and Penrose (1982)).

I have always regarded (1) (or its equivalent) as a rather unpleasant looking relation, from the point of view of twistor theory, for such a basic system of physical quantities as that described by $A_{\alpha\beta}$. Perhaps there is some deep-seated connection between this awkwardness and the difficulties in obtaining an appropriate analogue of (1) in general relativity, for a finite 2-surface $\mathcal{S}$. What I have in mind is the possibility that the 2-surface twistor space $\mathbb{T}^{\alpha}(\mathcal{S})$ may have a structure slightly different from that of the standard $\mathbb{T}^{\alpha}$ associated with Minkowski space $\mathbb{M}$, and that the $A_{\alpha\beta}$ in $\mathbb{M}$ may be describing an *infinitesimal change* away from this standard Minkowskian twistor space structure. The idea is that the $A_{\alpha\beta}(\mathcal{S})$, for a finite $\mathcal{S}$ in a curved $\mathcal{M}$ would have a somewhat different structure from the $A_{\alpha\beta}$ in $\mathbb{M}$ and would perhaps describe a *finite* change away from the standard Minkowskian twistor space structure.

How can we associate the Minkowskian $A_{\alpha\beta}$ with an infinitesimal change in twistor space structure? A clue may be found in the 'kinematic sequence' (cf. Penrose & Rindler, 1986, p. 198, and Hughston & Hurd in §II.4.3) where it is found that the space of $A_{\beta\alpha}$'s plays a dual role, either as a subspace of $\mathbb{T}_{(\alpha\beta)}$ (subject to 1), or as a factor space of the trace-free Hermitian part $\hat{\mathbb{H}}^{\alpha}_{\beta}$ of $\mathbb{T}^{\alpha}_{\beta}$. We can regard $\hat{\mathbb{H}}^{\alpha}_{\beta}$ as the space of real conformal Killing vectors k^a in $\mathbb{M}$, these providing infinitesimal motions of $\mathbb{T}^{\alpha}$ preserving the twistor norm $Z^{\alpha}\overline{Z}_{\alpha}$. The relation between k^a and the corresponding angular momentum structure, given by M^{ab} regarded as a field (a function of the origin point), may be expressed as

$$M^{ab} = \nabla^{[a}k^{b]} \tag{2}$$

(cf. Penrose & MacCallum 1973 and Penrose & Rindler 1986). The required dependence on position of M^{ab} is a consequence of (2) (where $\nabla^{(a}k^{b)} = \frac{1}{4}g^{ab}\nabla_c k^c$) and all M^{ab}'s arise in this way. The translations and dilations (and their compositions) are the k^a's which are killed by (2) so we may regard the space of M^{ab}-fields (i.e. the space of $A_{\alpha\beta}$'s) as the space of conformal Killing vectors for $\mathbb{M}$ *factored out* by the translations and dilations. Thus to describe this space (or its non-linear 'exponentiation') we seek some structure in $\mathbb{M}$ which is invariant precisely under these translations and dilations and look for the family of conformal transforms (infinitesimal or finite) of this structure.

One such structure is the 'projective π-projection' of $\mathbb{T}$:

$$\mathbb{T}^{\alpha} \rightarrow \{\mathbb{S}_{A'}\} \tag{3}$$

where $\{\mathbb{S}_{A'}\}$ stands for the *projective* space of $\pi_{A'}$'s, taken as a *given* space. Thus a proposal for a 'non-linear $A_{\alpha\beta}$' would be a map (3) which is a general $SU(2,2)$-transform of the standard one. Geometrically, (2) is described by a line $\hat{I}$ in $\mathbb{PN}$, where the family of planes through $\hat{I}$ is given a parameterization as a *standard* Riemann sphere $\mathbb{CP}^1$ (namely $\{\mathbb{S}_{A'}\}$), the fibres of the projection being the individual planes. It is perhaps a little easier to think in dual terms, where the injection

$$\{\mathbb{S}^{A}\} \rightarrow \{\mathbb{T}^{\alpha}\} \tag{4}$$

plants a standard Riemann sphere $\mathbb{CP}^1$ as a parameterized projective line $\hat{I}$ in $\mathbb{PN}$, and the different ways of doing this give the required space. Note that there is a 10-parameter (real) freedom in doing this (4 for the line $\hat{I}$; 6 for the different choices of parameterization on $\hat{I}$). When $\hat{I}$ lies infinitesimally separated from I, with an infinitesimal shift in parameterization, we recover the original space of 'linear' $A_{\alpha\beta}$'s.

I am trying to put forward a minor modification of this tentative suggestion. In place of (3) and (4), consider the possible exact sequences

$$0 \rightarrow \mathbb{S}^{A} \rightarrow \mathbb{T}^{\alpha} \rightarrow \mathbb{S}_{A'} \rightarrow 0 \tag{5}$$

where $\mathbb{S}^{A}$ is a *given* 2-dimensional complex vector space and $\mathbb{S}_{A'}$ is its complex conjugate dual. $\mathbb{T}^{\alpha}$ is also taken to be a *given* space, namely a particular 4-dimensional complex vector space with standard $(+ + - -)$ twistor norm $Z^{\alpha}\overline{Z}_{\alpha}$, but the maps can vary. (5) is to be compatible with this structure of $\mathbb{T}^{\alpha}$ in the (standard) sense that the dual sequence

$$0 \leftarrow \mathbb{S}_{A} \leftarrow \mathbb{T}_{\alpha} \leftarrow \mathbb{S}^{A'} \leftarrow 0 \tag{6}$$

to (5) is also the complex conjugate of (5) (in the reverse order). In terms of indexed quantities, the sequences (5), (6) can be expressed by use of objects

$$e_{A'\alpha}\,, \quad e_A{}^{\alpha} \tag{7}$$

where

$$\overline{e_{A'\alpha}} = e_A{}^{\alpha} \tag{8}$$

and

$$e_{A'\alpha}\, e_B{}^{\alpha} = 0, \tag{9}$$

$\mathbb{S}^A \to \mathbb{T}^\alpha$ being achieved by $\omega^A e_A{}^\alpha = Z^\alpha$ and $\mathbb{T}^\alpha \to \mathbb{S}_{A'}$ by $Z^\alpha e_{A'\alpha} = \pi_{A'}$, etc.

Suppose we have two such pairs of objects, namely (7), subject to (8), (9) and

$$f_{A'\alpha}\,, \quad \text{with } f_A{}^\alpha = \overline{f_{A'\alpha}}\,, \quad f_{A'\alpha} f_B{}^\alpha = 0. \tag{10}$$

Let us define

$$A_{\alpha\beta} = -iC e_{A'(\alpha} f^{A'}{}_{\beta)} \tag{11}$$

for some positive constant C, and

$$B_{\alpha\beta} = e_{A'[\alpha} f^{A'}{}_{\beta]} \tag{12}$$

(index raised with the 'standard' $\varepsilon^{A'B'}$ of $\mathbb{S}_{A'}$). When $f_{A'\alpha} = e_{A'\alpha}$ = the *canonical* $\pi_{A'}$-projection for $\mathbb{M}$, then $A_{\alpha\beta} = 0$ and $B_{\alpha\beta} = I_{\alpha\beta}$. When $f_{A'\alpha}$ differs from the canonical $e_{A'\alpha}$ by a small quantity of order ε, then $A_{\alpha\beta}$ is of order ε and is, to 1st order in ε, a quantity having the structure of a kinematic twistor (satisfying (1) to first order in ε), where $I_{\alpha\beta} = B_{\alpha\beta} + 0(\varepsilon)$. For a general $A_{\alpha\beta}$, $B_{\alpha\beta}$ given by (11), (12) we have the Hermiticity property

$$A_{\alpha\gamma}\bar{B}^{\beta\gamma} = \bar{A}^{\beta\gamma} B_{\alpha\gamma}, \tag{13}$$

like (1), but $B_{\alpha\beta}$ is not normally 'simple'. (Certain restrictions may be put on $e_{A'\alpha}$, $f_{A'\alpha}$ if $\varepsilon_{\alpha\beta\gamma\delta}$ is regarded as part of the structure of $\mathbb{T}^\alpha$ e.g.

$$e_{A'[\alpha}\, e^{A'}{}_{\beta]} = \tfrac{1}{2}\varepsilon_{\alpha\beta\gamma\delta}\, e_A{}^\gamma e^{A\delta}, \tag{14}$$

and the corresponding relation for the fs, but the status of these is unclear.)

Recall now, the Tod form of the expression for $A_{\alpha\beta}(\mathscr{S})$:

$$A_{\alpha\beta}(\mathcal{S}) = -\frac{i}{4\pi G}\oint (\underset{\alpha}{\pi}_{0'}\underset{\beta}{\pi}_{1'} + \underset{\alpha}{\pi}_{1'}\underset{\beta}{\pi}_{0'})\mathcal{S} \tag{15}$$

($\mathcal{S}$ = surface-area 2-form on $\mathcal{S}$) (cf. Penrose in §II.4.2 and Penrose 1982), and also the local expression on $\mathcal{S}$

$$\underset{\alpha}{\pi}_{0'}\underset{\beta}{\pi}_{1'} - \underset{\alpha}{\pi}_{1'}\underset{\beta}{\pi}_{0'} \tag{16}$$

for $I_{\alpha\beta}$. This tantalizingly suggests a relation between (15) and (11), and between (16) and (12), with something like an integral of $\underset{\alpha}{\pi}_{0'}\underset{\beta}{\pi}_{1'}$ over $\mathcal{S}$ producing the quantity

$$Se_{A'\alpha}f^{A'}{}_{\beta} = SB_{\alpha\beta} + \frac{iS}{C}A_{\alpha\beta} \tag{17}$$

(S being the total surface area of $\mathcal{S}$, and $C = \frac{S}{4\pi G}$). For this to make sense, the integral of $\underset{\alpha}{\pi}_{0'}\underset{\beta}{\pi}_{1'}$ would have to have matrix rank 2 (as an element of $\mathbb{T}_{\alpha\beta}$). I have not been able to prove (or disprove) this, and the whole scheme remains highly conjectural.

References

Penrose, R. (1982) Quasi-local mass and angular momentum in general relativity, Proc. Roy. Soc. **A381**, 53.

Penrose, R. & MacCallum, M.A.H. (1973) Twistor theory: an approach to quantization of fields and space-time, Phys. Rep. **6C**, 241-315

Penrose, R. & Rindler W. (1986) *Spinors and space-time*, Vol. 2 , C.U.P.

§II.4.6 **Spinors, ZRM fields and twistors at spacelike infinity** *by W.T.Shaw* (TN 15, January 1983)

To describe spinor fields at spacelike infinity (ι^0), it is necessary to construct an $SU(1,1)$ spin structure on K, the hyperboloid of spacelike directions at ι^0. Then, given spinor differential equations on the space-time, one can compute the asymptotic spinor differential equations induced on the hyperbolic structure. One way of doing this is to reduce $SL(2,\mathbb{C})$ spinors to $SU(1,1)$ spinors on a foliation of timelike hypersurfaces contracting to a point at ι^0.

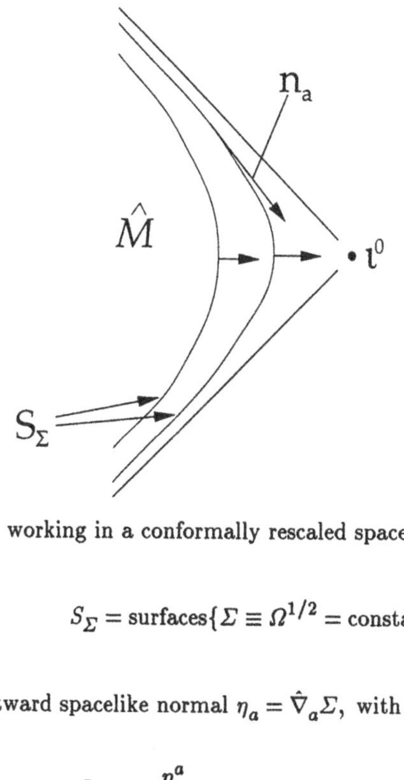

$M \to \hat{M}$

$g_{ab} \to \Omega^2 g_{ab}$

If one is working in a conformally rescaled space-time a natural foliation is the family:

$$S_\Sigma = \text{surfaces}\{\Sigma \equiv \Omega^{1/2} = \text{constant}\}$$

with outward spacelike normal $\eta_a = \hat{\nabla}_a \Sigma$, with normalised form

$$n^a = \frac{\eta^a}{||\eta^b \eta_b||^{1/2}}.$$

$SL(2,\mathbb{C})$ may be reduced to $SU(1,1)$ by the standard procedure: let $\lambda^{\dagger A} = \sqrt{2}\ n^{AB'}\overline{\lambda}_{B'}$ then $(\lambda^\dagger)^{\dagger A} = +\lambda^A$ and an inner product defined by $<\alpha,\beta> = \alpha^{\dagger A}\beta_A$. Define (Witten operator)

$$\mathbb{D}_{AB} = \sqrt{2}\ n^{B'}_{(A}\hat{\nabla}_{B)B'}.$$

$\mathbb{D}_{AB}$ is related to the intrinsic spinor covariant derivative D_{AB} preserving

$$h_{ab} = \hat{g}_{ab} + n_a n_b = h_{(AC)(BD)} = -\tfrac{1}{2}(\varepsilon_{AB}\varepsilon_{CD} + \varepsilon_{CB}\varepsilon_{AD})$$

by

$$\mathbb{D}_{AB}\lambda_C = D_{AB}\lambda_C + \frac{1}{\sqrt{2}}\pi_{ABCD}\lambda^D \qquad (*)$$

where $\pi_{ab} = \pi_{(AC)(BD)}$ is the second fundamental form, which can be decomposed into shear and expansion parts as

$$\pi_{ABCD} = \sigma_{(ABCD)} - \frac{\theta}{6}(\varepsilon_{AC}\varepsilon_{BD} + \varepsilon_{BC}\varepsilon_{AD}) \,.$$

One then computes the spinor differential equations induced on S_Σ, imposes a regularity condition on the spinors, and defines

$$\overset{K}{D}_{AB} = \lim_{\to \iota^0} \Sigma D_{AB}$$

Then the curvature of D^K is that of a unit timelike hyperboloid, and the limit $\Sigma \to 0$ can be taken. Firstly this is done for zero rest mass fields, and secondly for twistors.

1. Zero Rest Mass Fields. Let $\phi_{AB\cdots CD}$ be a field with $2n$ indices and helicity $s = -n$ satisfying

$$\hat{\nabla}^A_{A'}\phi_{AB\cdots CD} = 0 \qquad \text{in } \hat{M} - (\text{image of sources in } M).$$

and suppose that

$$\tilde{\phi}_{AB\cdots LM} = \Sigma^{2r}\phi_{AB\cdots LM}$$

is finite but direction dependent at ι^0. Then applying the above and computing π_{ab} to $O(\Sigma^{-1}) + O(1)$, one finds that the limit $\overset{\circ}{\phi}$ of $\tilde{\phi}$ satisfies

$$\overset{K}{D}_{DA}\overset{\circ}{\phi}{}^{A}{}_{B\cdots LM} = \frac{1}{\sqrt{2}}[n+1-2r]\overset{\circ}{\phi}_{DB\cdots LM}$$

For positivity helicity fields $\Psi_{A'B'\cdots C'D'}$, let

$$\overline{\Psi}_{AB\cdots CD} = 2^n\, n_A{}^{A'} n_B{}^{B'} \cdots n_D{}^{D'}\Psi_{A'B'\cdots C'D'}$$

and a similar computation yields

$$\overset{K}{D}_{DA}\,\overset{\circ}{\bar{\Psi}}{}^{A}{}_{B\cdots LM} = -\frac{1}{\sqrt{2}}[n+1-2r]\,\overset{\circ}{\bar{\Psi}}{}_{DB\cdots LM}$$

Reconstituting these (for $n \in \mathbb{Z}$) into vector equations and then recombining as electric and magnetic parts the leading order equations of Geroch (1976), Sommers (1979), and Ashtekar & Hansen (1978) can be recovered.

2. Twistors. The same procedure can be applied to the equation:

$$\nabla_{A'}^{(A}\omega^{B)} = 0$$

provided that the leading order magnetic part of the Weyl curvature vanishes, i.e.

$$0 = B_{ab} = \lim_{\to \iota^0}(\Sigma \hat{C}_{acbd}n^c n^d)$$

(This will be demonstrated shortly). The most straightforward way of doing the calculation is to firstly work it out in $\mathbb{M}$ and then ask when the resulting equations generalise to a conformally curved space. Writing in $\mathbb{M}$,

$$\hat{\omega}^A = \overset{0}{\omega}{}^A + \Sigma\overset{1}{\omega}{}^A$$

then one finds a pair of differential equations on K, viz.

$$\overset{K}{D}_{CA}\overset{0}{\omega}_B = \frac{1}{\sqrt{2}}\overset{0}{\omega}_{(A}\varepsilon_{C)B} \tag{1}$$

$$\overset{K}{D}_{CA}\overset{1}{\omega}_B = -\frac{1}{\sqrt{2}}\overset{1}{\omega}_{(A}\varepsilon_{C)D} - \frac{\mu}{\sqrt{2}}K_{CABD}\overset{0}{\omega}{}^D \tag{2}$$

[μ is coefficient $O(1)$] where K_{ab} is the $O(1)$ part of π_{ab}. In a *conformally curved* space K_{ab} acts as a tensor potential for B_{ab} via

$$B_{ab} \propto \varepsilon_a{}^{mn}\overset{K}{D}_m K_{bn}\,.$$

(1) and (2) are the asymptotic twistor equations for $\mathbb{M}$. When do they generalise to a curved space? Now, the Ricci identity for K is

$$2\overset{K}{D}_{E(F}\overset{K}{D}_{A)}{}^{E}\lambda_C = \lambda_{(F}\varepsilon_{A)C}$$

and applying this to (2) gives immediately $\overset{K}{D}_{(A}{}^{B}K_{C)BEF} = 0$ so $B_{ab} = 0$. These equations have many interesting properties. For example, asymptotic twistors act as helicity-raising operators on asymptotic ZRM fields. One can do this twice, in which case acting on a ZRM field, one is essentially contracting the field with a vector

$$W^a = W^{AB} = \omega^{(A}\tilde{\omega}^{B)}$$

for a pair ω^A, $\tilde{\omega}^A$ of solutions to (1) and (2). These vectors are essentially the conformal Killing vectors on K. To see this, one can first perform a SPI supertranslation (Ashtekar & Hansen 1978),

$$\Sigma \to \Sigma(1+\alpha\Sigma)$$

and set $K_{ab} = 0$. Let $W^{AB} = \omega^{(A}\tilde{\omega}^{B)}$ where $\lambda, \tilde{\lambda} = \pm 1$. We have

$$\overset{K}{D}_{CA}\omega_B = \frac{\lambda}{\sqrt{2}}\omega_{(A}\varepsilon_{C)B} \quad \text{and} \quad \overset{K}{D}_{CA}\tilde{\omega}_B = \frac{\tilde{\lambda}}{\sqrt{2}}\tilde{\omega}_{(A}\varepsilon_{C)B}\,.$$

There are two possibilities, $\lambda = -\tilde{\lambda}$ or $\lambda = +\tilde{\lambda}$. If (a) $\lambda = -\lambda$, then W^a satisfies

$$D_a W_b + \left(\frac{\lambda}{\sqrt{2}}\omega_C\tilde{\omega}^C\right)h_{ab} = 0$$

If (b) $\lambda = \bar{\lambda}$ then

$$D_a W_b = \lambda i\varepsilon_{abc}W^C.$$

It is well known that the ten *real* conformal Killing fields on K are precisely (a) 4 curl-free conformal Killing fields and (b) 6 Killing fields with rotation. So, in fact, when $K_{ab} = 0$.

$$\mathbb{T}^{\alpha}(\iota^0) \otimes_s \mathbb{T}^{\alpha}(\iota^0) \simeq \mathbb{C}\{\text{conformal Killing fields}\}$$

When $K_{ab} \neq 0$ the translational (type a) fields acquire additional terms unless one of $\overset{o}{\omega}^A$ or $\overset{o}{\tilde{\omega}}^A$ vanish.

These ideas have applications to the theory of conserved quantities. Certainly in linearised theory one can recover the ADM mass and a supertranslation invariant notion of angular momentum. For the curved asymptotically flat space the ADM mass is again recovered and a new notion of angular momentum obtained. Thanks to R. Penrose and K.P. Tod.

Comment (1994). Fuller details of the treatment of spinor fields at space-like infinity are given in Shaw (1983a), for the treatment of ι^0 by conformal methods. In some respects, the treatment is more straightforward in a non conformal picture, as given in Beig & Schmidt (1982) and Shaw (1983b). A generalization to include space-times with non-vanishing stress tensor at ι^0 was given in Shaw (1986).

References

Ashtekar, A. and Hansen, R.O. (1978) J. Math. Phys. **19**(7) 1542.

Beig, R. and Schmidt, B.G. (1982) Comm. Math. Phys. **87**, 65.

Geroch, R. (1976) in *Asymptotic Structure of Spacetime*, (Esposito & Witten eds.), Plenum.

Penrose, R. §II.4.2

Shaw, W.T. (1983a) Gen Rel. Grav. **15**, 12, 1163.

Shaw, W.T. (1983b) Proc. Roy. Soc. **A390**, 191.

Shaw, W.T. (1986) Class. Quant. Grav. **3**, L77.

Sommers, P. (1978) J. Math. Phys. **19**(3), 549.

§II.4.7 **The 'normal situation' for superficial twistors** *by M.G.Eastwood* (TN 15, January 1983)

In §II.4.2 and Penrose (1982), reference is made to the 'normal situation' for superficial twistors as that in which the solution space of the superficial twistor equations, $\eth'\omega^0 = \sigma'\omega^1$ and $\eth\omega^1 = \sigma\omega^0$, has $\mathbb{C}$-dimension four. Since

$$D \equiv \begin{bmatrix} \eth' & -\sigma' \\ -\sigma & \eth \end{bmatrix}$$

is an elliptic operator on the sphere with adjoint

$$D^* = \begin{bmatrix} \eth & \overline{\sigma}' \\ \overline{\sigma} & \eth' \end{bmatrix},$$

its index, $\dim \ker D - \dim \ker D^*$, is independent of σ, σ'. In this case the spin weights involved ($-\frac{1}{2}$ and $\frac{1}{2}$ for ω^0 and ω^1, $-\frac{3}{2}$ and $\frac{3}{2}$ for the adjoint) imply that, in case $\sigma = 0 = \sigma'$, $\dim \ker D = 4$, $\dim \ker D^* = 0$. Thus in general (as observed in Penrose 1982) $\dim \ker D \geq 4$ with equality generic and occurring exactly when $\ker D^* = 0$. The aim of this note is to give an explicit argument with explicit bounds to show that the 'normal' situation occurs for small shears σ and σ'.

If the sphere under consideration is regarded as the usual Riemann sphere then the adjoint equations become

$$\begin{pmatrix} \overline{\partial} f \\ \partial g \end{pmatrix} = x \begin{pmatrix} f \\ g \end{pmatrix} \quad \text{for } f \in \Gamma(H^3),\ g \in \Gamma(\overline{H}^3),$$

and

$$x \in \operatorname{Hom} \left(\begin{matrix} H^3 \\ \oplus \\ \overline{H}^3 \end{matrix}, \begin{matrix} \Lambda^{0,1}(H^3) \\ \oplus \\ \Lambda^{1,0}(\overline{H}^3) \end{matrix} \right)$$

where H is the Hopf bundle and $\Lambda^{p,q}$ denotes forms of type (p, q). Here, x is the matrix $\begin{bmatrix} 0 & -\overline{\sigma} \\ -\overline{\sigma}' & 0 \end{bmatrix}$ although we may as well consider the seemingly more general case of x arbitrary. [That this greater generality is illusory may be seen by using integrating factors (change of gauge) $f \mapsto e^u f$, $g \mapsto e^v g$ for arbitrary smooth functions u and v whence X is changed according to

$$\begin{bmatrix} \alpha & \beta \\ \gamma & \delta \end{bmatrix} \mapsto \begin{bmatrix} \alpha + \overline{\partial} u & e^{u-v}\beta \\ e^{v-u}\gamma & \delta + \partial v \end{bmatrix}$$

which, since $H^1(\text{Sphere}, \mathcal{O}) = 0$, may be used to eliminate the diagonal entries. This may be rephrased as the complex structure on H^3 being unique—note that the sphere is special in this respect, an added difficulty in defining superficial twistors with respect to other 2-surfaces.] If we choose an affine coordinate patch on the sphere then these equations reduce to equations on genuine functions on the C-plane:

$$\left.\begin{aligned}\frac{\partial F}{\partial \bar z} &= AF + BG\\ \frac{\partial G}{\partial z} &= CF + DG\end{aligned}\right\}$$

where the twisted nature at infinity is precisely reflected in the existence of the limits of z^3F, $\bar z^3 G$, z^2A, $z^5\bar z^{-3}B$, $\bar z^5 z^{-3}C$, and $\bar z^2 D$ as $z \to \infty$. Note that, in particular, F and G are $O(1/|z|^3)$ and $X = \begin{bmatrix} A & B \\ C & D\end{bmatrix}$ is $O(1/|z|^2)$ as $z \to \infty$. Recall the Cauchy integral formula

$$u(\zeta) = \frac{1}{2\pi i}\left[\int_{\partial\Omega} \frac{u\,dz}{z-\zeta} + \int_{\Omega} \frac{\partial u}{\partial \bar z}\frac{d\bar z \wedge dz}{z-\zeta}\right]$$

for any u smooth on $\bar\Omega$. If $u(z) \to 0$ as $z \to \infty$ then we may take $\Omega = \mathbb{C}$ and obtain

$$u(\zeta) = \frac{1}{\pi}\int_{\mathbb{C}} \frac{\frac{\partial u}{\partial \bar z}}{z-\zeta}\,.$$

Applying this to F for (F, G) a solution of the adjoint equations, we obtain

$$F(\zeta) = \frac{1}{\pi}\int_{\mathbb{C}} \frac{\mathrm{AF}+\mathrm{BG}}{z-\zeta}.$$

Thus if we set

$$a = \sup_{\zeta\in\mathbb{C}} \frac{1}{\pi}\int_{z\in\mathbb{C}} \left|\frac{A(z)}{z-\zeta}\right|$$

etc., assuming for the moment that these exist, we conclude that $\sup|F| \le a\sup|F| + b\sup|G|$. A similar bound holds for $\sup|G|$ and so

$$\begin{pmatrix}\sup|F|\\ \sup|G|\end{pmatrix} \le \begin{bmatrix} a & b\\ c & d\end{bmatrix}\begin{pmatrix}\sup|F|\\ \sup|G|\end{pmatrix}$$

which admits no solutions if $\begin{bmatrix} a\, b\\ c\, d\end{bmatrix}$ is small (for example distance decreasing) except $F = 0 = G$.

Note that in this argument we have used only that $|F|$ and $|G|$ vanish at infinity whereas we really know more rapid decay. Perhaps this can be used to improve the bounds. Indeed, it is quite conceivable that D^* never has non-trivial kernel, i.e. the normal situation always occurs. It remains to shew that

$$\int_{z \in \mathbb{C}} \left| \frac{A(z)}{z - \zeta} \right|$$

is bounded uniformly in ζ. Without loss of generality we may assume that A is circularly symmetric and a monotone decreasing function of distance from the origin. Dividing the integration into two pieces according to region of integration as indicated in the following diagram:

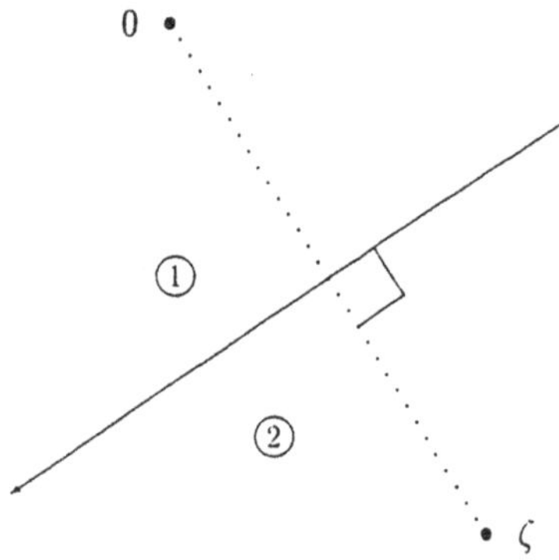

it is easy to see that each is bounded by $\int_{\mathbb{C}} \left|\frac{A}{z}\right|$, which exists because the integrand decays like $\frac{\text{constant}}{|z|^3}$ at infinity.

References

Penrose, R. (1982) Quasi-local mass and angular momentum in general relativity, Proc. Roy. Soc. **A381**, 53.

§II.4.8 **'Maximal' twistors & local and quasi-local quantities** *by W.T. Shaw* (TN 16, August 1983)

The basis of Penrose's 2-surface twistor construction, (Penrose 1982 cf. §II.4.2), is the generalisation to curved space-time of the right hand side of the identity

$$L[K] \equiv \int_{\Sigma} T_{ab}K^b d\Sigma^a = \frac{1}{4\pi G}\int R_{abcd}\Omega^{cd}\, dS^{ab} \tag{1}$$

where $\Omega^{ab} = \omega^{AB}\varepsilon^{AB'}$, $\nabla^{BB'}\omega^{CD} = -i\varepsilon^{B(C}K^{D)B'}$. K^a is a Killing vector of $\mathbb{M}$, and T^{ab} and R_{abcd} are the energy-momentum and curvature tensors of a weak gravitational field on $\mathbb{M}$. Here Σ is a spacelike hypersurface with boundary $\partial\Sigma$.

It is not known in general, in full general relativity, how the right hand side of (1) is related to local expressions involving the energy-momentum tensor. When Σ admits a four dimensional family of 3-surface twistors, one recovers a form of (1). (See Tod 1983, where this is done explicitly for hypersurfaces in conformally flat space-times.) However, for 3-surface twistors, one requires $B_{ab} = 0 = \dot{B}_{ab}$, where B_{ab} is the magnetic part of the Weyl curvature with respect to Σ and $\dot{B}_{ab}$ its normal derivative. Is there a construction which generalises the *whole* of the identity (1) to curved space-time?

One answer appears to arise from simply doing 'the best one can' with the twistors. Let t^a be the unit normal to Σ, $(t^a t_a = 1)$, and let $h_{ab} = g_{ab} - t_a t_b$ be the induced 3-metric, $D_a = h_a{}^b\nabla_b$ be the projection of the full connection onto Σ. Indices may be freely converted from primed to unprimed type using $t^{AA'}$. Thus, if $V_a \in T^*(\Sigma)$

$$V_{AB} = \sqrt{2}\, t^{A'}_{(B}V_{A)A'}, \quad V_{AA'} = -\sqrt{2}\, t^{B}_{A'}V_{AB}$$

Let $\Delta = D^a D_a$, G_{ab} be the Einstein tensor, and K_{ab} the extrinsic curvature of Σ. Adjoints are defined by $\alpha^{\dagger A} = \sqrt{2}\, t^{AA'}\bar{\alpha}_{A'}$, so that $\alpha_A\alpha^{\dagger A} = \sqrt{2}\, t^{AA'}\alpha_A\bar{\alpha}_{A'} \geq 0$.

The basic identity required is that for all spinors ω^A, $\alpha_{A'}$

$$-D^{AA'}[-3t^{CC'}\alpha_{C'}t^{B}_{A'}D_{(AB}\omega_{C)}] = \tfrac{3}{2}[D^{(AB}\bar{\alpha}^{C)}]^{\dagger}D_{(AB}\omega_{C)} -$$

$$\sqrt{2}t^{CC'}\alpha_{C'}\Big[\Delta\omega^C + \tfrac{1}{2}t^{D'}_{C}t^{A'E}G_{A'ED'F}\omega^F + \sqrt{2}K_{AA'CD}D^{AA'}\omega^D\Big] \tag{2}$$

'Doing the best one can' means considering the following variational problem: Let $S(\partial\Sigma)$ be the set of all spinor fields ω^A with ω^A specified to have some form on $\partial\Sigma$. Minimise, over $\omega^A \in S(\partial\Sigma)$

$$\mathcal{L}[\omega] = \int_\Sigma [D^{(AB}\omega^{C)}]^\dagger D_{(AB}\omega_{C)}\sqrt{-g}\, d^3x \tag{3}$$

Using (2), the Euler-Lagrange equations for ω^A are

$$\Delta\omega^C + \sqrt{2}K_{AA'D}{}^C D^{AA'}\omega^D = 4\pi G\, t^{D'}_C T_{D'BEE'} t^{EE'}\omega^B \tag{4}$$

where $G_{ab} = -8\pi G T_{ab}$. With ω^A specified on $\partial\Sigma$, since (3) is finite (if $\partial\Sigma$ is finite) and positive definite, one expects (4) to have solutions in Σ. Attention will be restricted to *2-surface twistor* boundary conditions on $\partial\Sigma$. Then the solutions to (4) are maximal twistor fields on Σ, and by design, reduce to ordinary twistor fields wherever possible.

The $\pi_{A'}$ part of the MTF is defined by

$$\pi_{A'} = \frac{2i}{3} D_{AA'}\omega^A \tag{5}$$

Then a short calculation using (4) shows that

$$D_{AA'}\pi^{A'} = -i4\pi G\, t^{A'}_A T_{A'BCC'} t^{CC'}\omega^B \tag{6}$$

Note that in vacuum $\pi_{A'}$ satisfies Witten's equation

$$D_{AA'}\pi^{A'} = 0 \tag{7}$$

By introducing a potential

$$V(\omega) = -\frac{1}{\sqrt{2}} G_{AA'BB'} t^{AA'}\omega^B \bar{\omega}^{B'} \tag{8}$$

into $\mathcal{L}$ one can arrange that (7) is always satisfied. The coupled system (5) and (7) is of some interest in its own right and has also been considered by G.T. Horowitz. Here attention will be restricted to the system (4), (5), (6).

Now, if $\omega^{AB} = \omega^{(A}\tilde{\omega}^{B)}$, the vector determined by ω^{AB} is

$$K^{AA'} = \tilde{\omega}^A \pi^{A'} + \omega^A \tilde{\pi}^{A'} \tag{9}$$

The charge associated with K^a is

$$Q[K] = \int_{\Sigma} T_{ab} K^b \mathrm{d}\Sigma^a \tag{10}$$

Using (6) and (9) this is just

$$\frac{-i}{2\pi G} \int_{\partial\Sigma} dS^{AA'} t_A{}^{B'} \pi_{A'} \tilde{\pi}_{B'} \tag{11}$$

The system (4), (5), (9), (10) and (11) could be said to constitute an alternative curved space-time generalisation of (1). (11) is strikingly similar to Tod's expression for the 2-surface kinematic twistor but now $\pi_{A'}$ is not defined by the 2-surface twistor equation, but by the 3-surface contraction (5). The two types of $\pi_{A'}$ agree if and only if

$$n^{AB} D_{(AB}\omega_{C)} = 0 \quad \text{on } \partial\Sigma \tag{12}$$

where n^a is normal to both t^a and $\partial\Sigma$. Putting $\alpha_{C'} = \overline{\omega}_{C'}$ in (2) shows that if ω^A satisfies (4), if (12) holds then

$$D_{(AB}\omega_{C)} \equiv 0 \quad \text{on } \Sigma \tag{13}$$

Thus (11) defines an expression which one knows will give good answers in several of the examples computed by Tod, but which may be useful in more general situations. Applications of this approach are discussed in Shaw (1985).

Many thanks to R. Penrose, K.P.Tod and G.T.Horowitz.

References

Penrose, R.(1982) Proc. Roy. Soc. **A381**, 53 and §II.4.2.

Tod, K.P. (1983) Proc. Roy. Soc. **A388**, 457.

Shaw, W.T. (1985) Class. Quant. Grav. **2**, 189.

§II.4.9 **The index of the 2-twistor equations** *by R.J.Baston* (TN 17, January 1984)

To obtain information about the dimension of a two-surface twistor space on a spacelike two-surface not necessarily homeomorphic to a sphere, one may compute the *index* of the two-surface twistor operator T. This is the difference in dimension between ker T and ker T^* where T^* is a suitably defined adjoint operator. To compute this index, one makes use of the Riemann-Roch formula.

Given such a two-surface, S, with an outward sense in a spacetime M (and hence an orientation of S), the metric in M induces a positive definite (smooth) metric on S and hence S has the structure of a compact Riemann surface, of genus g, say. There is a C^*-principal bundle over S consisting of orthonormal spinor dyads (o^A, ι^A) (i.e. $o^A \iota_A = 1$) at points of S such that $o^A o^{A'}$and $\iota^A \iota^{A'}$ are respectively outgoing and incoming null directions orthogonal to S. Let $E(s, w)$ be the complex line bundles on S induced from this bundle by the representation

$$\rho(s, W): \ C^* \to C^*; \qquad \lambda \to (\lambda\bar{\lambda})^w (\lambda/\bar{\lambda})^s$$

These are the spin s and conformal weight w functions on the sphere.

The connection on the spin bundles $\mathcal{O}^{A'}$ etc. on M induces differential operators $\eth(s, w)$ and $\eth'(s, w)$ on $E(s, w)$—edth and its complex conjugate, respectively. $\eth'(s, w)$ induces a holomorphic structure on $E(s, w)$ by requiring the exactness of

$$0 \longrightarrow \mathcal{O}(s, w) \longrightarrow E(s, w) \xrightarrow{\eth'(s, w)} E(s-1, w) \longrightarrow 0$$

(so that $\eth'(s, w)$ is a $\bar{\partial}$ operator—see Eastwood & Tod 1982). $\mathcal{O}(s, w)$ is the sheaf of holomorphic sections of $E(s, w)$ (which is itself confused with its sheaf of germs of smooth sections). This short exact sequence is a resolution of $\mathcal{O}(s, w)$ and can be used to compute its cohomology. In particular, we have, by the Riemann-Roch formula, that

$$\text{index } \eth'(s, w) = \dim H^0(S, \mathcal{O}(s, w)) - \dim H^1(S, \mathcal{O}(s, w)) = C_1(\mathcal{O}(s, w)) + 1 - g$$

where $c_1(\mathcal{O}(s, w))$ is the first (integral) Chern class of $\mathcal{O}(s, w)$. This is readily computed: the result is independent of w (since $\mathcal{O}(0, w)$ is trivial), additive in s (via the tensor property) and $\mathcal{O}(-1, 0) = \kappa$ is the canonical bundle whose first Chern class is $2(g-1)$. Thus

$c_1(\mathcal{O}(s,w)) = 2s(1-g)$ and

$$\text{index } \eth'(s,w) = (2s+1)(1-g)$$

Similarly, taking conjugates (when $\eth'(s,w) \to \eth(-s,w)$):

$$\text{index } \eth(s,w) = (-2s+1)(1-g)$$

Now the two-surface twistor operator T is

$$\begin{bmatrix} -\sigma & \eth' \\ \eth & -\sigma' \end{bmatrix} : \quad \underset{\omega^0}{E(-\tfrac{1}{2}, -\tfrac{1}{2})} \oplus \underset{\omega^1}{E(\tfrac{1}{2}, \tfrac{1}{2})} \longrightarrow E(\tfrac{3}{2}, \tfrac{1}{2}) \oplus E(-\tfrac{3}{2}, -\tfrac{1}{2})$$

(where $\eth' = \eth'(\frac{1}{2}, \frac{1}{2})$ and $\eth = \eth(-\frac{1}{2}, -\frac{1}{2})$) and therefore

$$\text{index } T = \text{index } \eth' + \text{index } \eth = 4(1-g)$$

(index T depends only on the top order symbol of T).

Since the dimension of the space of two-surface twistors is at least four in Minkowski space, one can expect that the adjoint two-surface twistor equations in Minowski space have a multiplicity of solutions when $g > 0$. In general, however, for $g > 0$, there need not even be *any* non-trivial two-surface twistors!

Thanks to K.P.Tod and M.G.Eastwood.

References

Eastwood & Tod (1982) Math. Proc. Camb. Phil. Soc. **92**, 317.

§II.4.10 **An occurrence of Pell's equation in twistor theory** *by K.P.Tod* (TN 17, January 1984)

The index of the two-surface twistor equations on a two-surface has been shown by a number of people to be $4(1-g)$ where g is the genus of the two-surface. On a sphere, $g=0$ and there is always at least a four-parameter family of solutions. However, on a torus $g=1$ and there may not be any solutions or the number of solutions may behave in a strange way. For a torus in flat space there will *always* be at least four solutions defined by restriction. I want to describe some simple tori *not* in flat space, where the number of solutions does indeed behave in a strange way.

I shall work in toroidal coordinates (ξ,η,ϕ) defined from cylindrical polars (r,z,ϕ) by: set $w=r+iz$, $\zeta=\xi+i\eta$ and $\frac{w+a}{w-a}=e^{-i\zeta}$ for some real a. The metric becomes

$$dr^2+dz^2+r^2d\phi^2=\frac{a^2}{X^2}(d\eta^2+d\xi^2+\sinh^2\eta\, d\phi^2)$$

where $X=\cosh\eta-\cos\xi$ (see e.g. Margenau & Murphy). The surfaces of constant η are tori obtained from circles centre $r=a\coth\eta$, $z=0$ radius $a\operatorname{cosech}\eta$, and the surfaces of constant ξ are spheres centre $r=0$, $z=a\cot\xi$, radius $a\operatorname{cosec}\xi$. The coordinate ranges are $0\le\eta<\infty$, $0\le\xi<2\pi$, $0\le\phi<2\pi$.

Focus attention on T_0, the torus $\eta=\eta_0$. This has unit normal $N=\frac{X}{a}\partial_\eta$ and null tangent $\delta=\frac{X}{a\sqrt{2}}(\partial_\xi+i\operatorname{cosech}\eta_0\partial_\phi)$. Form a null tetrad with δ, $\overline{\delta}$, $D=\frac{1}{\sqrt{2}}(\partial_t+N)$ and $\Delta=\frac{1}{\sqrt{2}}(\partial_t-N)$.

It is a simple matter to write down the two-surface twistor equations on T_0 and separate the ϕ-dependence. Let

$$\omega^0=X^{-1/2}\exp(im\phi)F(\xi)\qquad\text{and}\qquad\omega^1=X^{-1/2}\exp(im\phi)G(\xi)$$

then

$$\left(\frac{d}{d\xi}+\frac{m}{\sinh\eta_0}\right)F=-\tfrac{1}{2}\coth\eta_0 G;\qquad\left(\frac{d}{d\xi}-\frac{m}{\sinh\eta_0}\right)G=\tfrac{1}{2}\coth\eta_0 F\tag{1}$$

whence

$$\frac{d^2F}{d\xi^2}=-\omega_m^2F;\qquad\omega_m^2=\frac{\cosh^2\eta_0-4m^2}{4\sinh^2\eta_0}\tag{2}$$

To get well-defined spinor fields on T_0 we require ω^0 and ω^1 to change sign when $\phi\to\phi+2\pi$ or $\xi\to\xi+2\pi$. Thus $m=\frac{1}{2}k$ and $\omega_m=\frac{1}{2}n$ for odd integers k, n and (2) becomes

$$(n^2 - 1)\sinh^2\eta = 1 - k^2 \tag{3}$$

The only solution of (3) is $k^2 = 1$, $n^2 = 1$ and all η. This gives four solutions to (1) on each torus of constant η (and no more).

Now we wish to perturb the space-time containing T_0 and see what happens to the solutions. The simplest way to do this is to decree that ϕ no longer have period 2π but, say, $2\pi(1+\delta)$. In other words, we introduce a conical singularity in the tz-plane. This changes the condition on m and (3) becomes

$$(n^2 - 1)\sinh^2\eta = 1 - k^2(1+\delta)^{-2} \tag{4}$$

At once we see that, if $\delta < 0$, there are *no* solutions to (4) and thus no 2-surface twistors. This small addition of curvature eliminates all solutions!

If instead $\delta > 0$, then for each odd $n > 1$ with $k^2 = 1$ there is one choice of η satisfying (4) and so one paticular torus with four solutions. Most tori have none but there is a sequence with 4, with increasing eigenvalue n.

The next case of interest is when δ reaches 2. (Note that this is a very large periodicity in ϕ) Solutions with $n = 1$, $k = \pm 3$ exist on all tori but special ones exist with $k^2 = 1$ and n and η related by (4). These special tori have 8 solutions.

When δ reaches 4, solutions with $n = 1$, $k = \pm 5$ exist on all tori and extra ones exist with $k = \pm 1$, $k = \pm 3$ respectively at solutions of

$$({n_1}^2 - 1)\sinh^2\eta = \frac{24}{25}; \qquad ({n_2}^2 - 1)\sinh^2\eta = \frac{16}{25}. \tag{5}$$

This gives two different families of tori each with 8 solutions. We ask: can these two families intersect to give a torus with 12 solutions? For this we need a simultaneous solution of (5) i.e.

$$3{n_2}^2 - 2{n_1}^2 = 1 \tag{6}$$

(6) has the obvious solution $(n_2, n_1) = (1, 1)$. This is not acceptable as a solution of (5), but its existence is important. The way to proceed is to regard (6) as the condition for (n_2, n_1) to be a unit time-like vector in an integral Lorentzian metric and to seek an integral Lorentz transformation L. Since $(1, 1)$ is known to be a unit vector, the result of applying L to it any

number of times will also be a unit vector and this will give infinitely many solutions to (6).

In general, with the metric $At^2 - Bz^2$, the required Lorentz transformation is

$$\begin{pmatrix} a & bB \\ bA & a \end{pmatrix} \quad \text{where} \quad a^2 - ABb^2 = 1 \tag{7}$$

which is Pell's equation!

Solutions to (7) are guaranteed to exist provided AB is not a perfect square. They are related to the continued fraction of $\sqrt{AB}$. In this example $AB = 6$ so $(a,\ b) = (5,\ 2)$ will do. Now $L = \begin{pmatrix} 5 & 4 \\ 6 & 5 \end{pmatrix}$ and we generate infinitely many solutions to (6) and so infinitely many tori with 12 solutions, by repeatedly applying this to $(1,1)$.

At the next stage ($\delta = 6$) there are three equations like (5) and so three infinite families of tori with 8 solutions. Each pair of families gives an equation like (6) so that there are three infinite families with 12 solutions. One might hope that these families would intersect giving infinitely many tori with 16 but in fact this doesn't happen. To see why, consider some of the equations which arise with $\delta = 8$:

$$10{n_2}^2 - 7{n_1}^2 = 3\,; \qquad 5{n_3}^2 - 2{n_1}^2 = 3\,. \tag{8}$$

We know that each of these has infinitely many solutions in integers, each of which gives a torus with 12 solutions to (1). If we write them as

$$\frac{p^2}{q^2} - \frac{7}{10} = \frac{3}{10q^2}; \qquad \frac{r^2}{q^2} - \frac{2}{5} = \frac{3}{5q^2}$$

then we see that the solutions provide rational approximations to the irrationals $\left(\frac{7}{10}\right)^{1/2}$ and $\left(\frac{2}{5}\right)^{1/2}$ respectively with accuracy $O(q^{-2})$. Simultaneous solutions to (8) would provide simultaneous rational approximations to the same order, but this is forbidden by a theorem of Schmidt (1990). The simultaneous rational approximation of n irrationals is allowed precisely to order $q^{-1-1/n}$! Thus (8) and similar sets of simultaneous equations can have at most finitely many simultaneous solutions.

In conclusion, from this particular way of choosing tori in curved space we *cannot* get infinite families with $16, 20, \ldots$ solutions. However, there may well be odd ones with more. The largest number I have found corresponds to the solution $(n_1,\ n_2,\ n_3) = (49,\ 41,\ 31)$ of (8) which gives a torus with 16 independent 2-surface twistors.

I am grateful to Tony Scholl for telling me about the work of Schmidt and for discussions on number theory.

Reference

Schmidt, W.M. (1970) in Proceedings 1970 Int. Cong. of Math. Nice, vol. 1.

§II.4.11 **The Sparling 3-form, the Hamiltonian of general relativity and quasi-local mass**

by L.J.Mason (TN 17, January 1984 and TN 18, July 1984)

The purpose of this article is twofold. The first is to show how the Sparling 3-form $\Gamma = iD\lambda_{A'} \wedge D\bar{\lambda}_A \wedge dx^{AA'}$ can be interpreted as the Hamiltonian density of general relativity. Here D is the covariant exterior derivative. Notation is otherwise as in Penrose & Rindler (1986). The second is to use this fact to show that the components of the angular-momentum twistor of Penrose's quasi-local mass construction can be interpreted as the values of Hamiltonians that generate certain motions of some spanning surface.

The Sparling 3-form as Hamiltonian density of general relativity. In the canonical formalism of general relativity we take the space of positive definite 3-metrics g_{ij} on a hypersurface Σ to be the configuration space and the corresponding canonically conjugate momenta as given by symmetric tensor densities $\pi_{ij} = \sqrt{g}(g_{ij}\, tr\, K - K_{ij})$ where K_{ij} is the extrinsic curvature $(i, j = 1, 2, 3)$.

The Hamiltonian density is usually given as:

$$\mathcal{H}_{total} = \mathcal{H}N + \mathcal{H}_i N^i$$

where N is the lapse function, N_i the shift vector field, $\mathcal{H}$ is the superhamiltonian and $\mathcal{H}_i$ is the super momentum. These are most easily defined as $\mathcal{H} = G_{\perp\perp}\sqrt{g}$ and $\mathcal{H}_i = G_{\perp i}\sqrt{g}$ where $G_{\alpha\beta}$ is the Einstein tensor and $\perp$ denotes the direction orthogonal to the given hypersurface and i

indexes directions tangent to the hypersurface (cf. Misner, Thorne & Wheeler 1973, pp. 515, 521).

If Σ is asymptotically flat we cannot just use

$$\int_{\Sigma} \mathcal{H}_{total}\, d^3x$$

as the Hamiltonian since the variation of H should not contain (non-vanishing) surface integrals (these would lead to a pathological and incorrect evolution equation on the surface, see below). We must therefore add a surface integral to $\int_{\Sigma} \mathcal{H}_{total}\, d^3x$. When the field equations are satisfied and N and N_i are asymptotically constant, the value of the Hamiltonian is that of the surface integral (as $\mathcal{H}$ and $\mathcal{H}_i$ are zero when the field equations are satisfied) and yields $P_0 N + P_i N_i$ where (P_0, P_i) is the ADM momentum (Regge & Teitelboim 1974).

If we take null lapse and shift (i.e. $N^2 = N^i N_i$) then we can write $(N, N_i) = \bar{\lambda}^A \lambda^{A'}$ for some spinor field $\lambda^{A'}$, and the surface integral can be written as follows (Witten 1981):

$$\oint_{\partial\Sigma} i\lambda_{A'} \nabla_b \bar{\lambda}_A\, dx^b \wedge dx^{AA'}.$$

Thus our expression for the Hamiltonian H can be written:

$$H = \int_{\Sigma} \bar{\lambda}^A \lambda^{A'} G_{AA'\beta}\, d\Sigma^{\beta} + \oint_{\partial\Sigma} i\lambda_{A'} \nabla_b \lambda_A\, dx^b \wedge dx^{AA'}.$$

However, in Sparling (1981) we are given the equation:

$$\Gamma = iD\lambda_{A'} \wedge D\bar{\lambda}_A \wedge dx^{AA'} = \lambda^{A'} \bar{\lambda}^A G_{AA'b}\, d\Sigma^b + d(i\lambda_{A'} D\bar{\lambda}_A \wedge dx^{AA'})$$

where

$$d\Sigma^a = \varepsilon^a{}_{bcd}\, dx^b \wedge dx^c \wedge dx^d.$$

So we see that if we integrate Γ over the section of the spin bundle given by $\lambda_{A'}$ over Σ, we will obtain the correct Hamiltonian with the surface integral automatically incorporated.

Note that H is still regarded as a functional of g_{ij}, π_{ij}, N_i and N. However when dealing with spinors it is convenient to extend the phase space by using the set of orthonormal tetrads on Σ as configuration space. In this formalism we must use 6 extra constraints and add an extension to the Hamiltonian (which is constrained to vanish). The interesting question is whether we can introduce twistorial coordinates on the gravitational phase space using the metric, and treat Hamiltons equations for general relativity twistorially.

Related descriptions of the gravitational Hamiltonian are put forward in Ashtekar & Horowitz (1982) and Nestor (1983). The idea is implicit in Witten (1981).

A Hamiltonian interpretation of the Angular Momentum twistor. The angular momentum twistor of a 2-surface S in a space-time M is usually taken to describe the momentum and angular momentum of the sources and gravitational field on a 3-surface spanning S.

In symplectic mechanics there is a direct correspondence between observables and canonical transformations of the phase space in question. For an observable H, this is given by Lie dragging along a vector field X_H given by:

$$X_H \lrcorner \omega + dH = 0$$

where ω is the symplectic form on the phase space and H is the observable in question.

Observables such as angular momentum and momentum generate canonical transformations that correspond to Lie dragging the field along a killing vector. The purpose here is to show how an analogous viewpoint can be taken for the angular momentum twistor in general relativity.

The phase space we are interested in is that of the gravitational field and source fields, this is usually given by initial data on some 3-surface. In curved space, though, we do not in general have any killing vectors. However a 2-surface twistor provides us with a notion of a killing vector at the 2-surface:

$$Z^\alpha = (\omega^A, \pi_{A'}) \quad \rightarrow \quad K^{AA'} = \omega^A \pi^{A'}.$$

It turns out that this is enough to define the corresponding component of momentum/angular momentum of the gravitational field and source linking the 2-surface. This is because the total Hamiltonian density, when the field equations are satisfied, is an exact 3-form so that the value of the Hamiltonian only depends on the values of the 'quasi killing vector' on the bounding 2-surface.

In the usual Hamiltonian treatment of general relativity one casts off boundary terms at will. Clearly we must keep track of them for the above approach to make sense. Regge & Teitelboim (1974) established criteria for finding the correct surface integral.

If H has the wrong surface integral attached to it then

$$\delta H = \int_{\Sigma} (\dot{g}_{ij}\delta\pi_1^{ij} - \dot{\pi}^{ij}\delta g_{ij})\sqrt{g}\, d^3x + \text{ a surface integral}$$

Clearly Hamilton's equations will only hold if the surface integral vanishes. As explained above, the Sparling 3-form provides a Hamiltonian density satisfying this criteria, at least in the asymptotic context. (The Sparling 3-form Γ is defined on the spin bundle. To evaluate the Hamiltonian we restrict Γ to a section of the spin bundle $\pi_{A'} = \lambda_{A'}(x)$ where $\lambda_{A'}\bar{\lambda}_A$ defines the vector field corresponding to the desired lapse and shift, and then we integrate Γ over this section, restricted to a hypersurface Σ of the spacetime).

The main equation satisfied by the Sparling 3-form is:

$$\Gamma = id\lambda_{A'} \wedge d\bar{\lambda}_A \wedge dx^{AA'} = d(i\lambda_{A'} d\bar{\lambda}_A \wedge dx^{AA'}) + \lambda_{A'}\bar{\lambda}_A G_b^{AA'} d^3x^b$$

Now add on the matter Hamiltonian density $\lambda_{A'}\bar{\lambda}_A T_b^{AA'} d^3x^b$ to obtain the total Hamiltonian density, $\mathcal{H}_{total}$. If we impose the Einstein field equations the stress energy tensor cancels against the Einstein tensor to give

$$\mathcal{H}_{total} = d(i\lambda_{A'} d\bar{\lambda}_A \wedge dx^{AA'}).$$

Thus the Hamiltonian is

$$H_{total} = \int_{\Sigma} \mathcal{H}_{total} = \oint_{\substack{\mathcal{S} = \partial\Sigma \\ \pi = \lambda}} i\lambda_{A'} d\bar{\lambda}_A \wedge dx^{AA'}$$

which is just a boundary integral as promised.

We now write down the value of the Hamiltonian that generates a motion of the 3-surface Σ along a complex vector field whose value on $\mathcal{S} = \partial\Sigma$ is the 'quasi-Killing vector' $\pi^{A'}\omega^A$. We complexify so that $\bar{\lambda}_A \to \tilde{\lambda}_A$, and set $\lambda_{A'} = \pi_{A'}$ (abuse of notation above) and $\tilde{\lambda}_A = \omega_A$.

Now

$$d\tilde{\lambda}^A \Big|_{\mathcal{S},\, \tilde{\lambda}^A = \omega^A} = \nabla_{BB'}\omega^A dx^{BB'} = \varepsilon_B{}^A \pi_{B'} dx^{BB'}$$

by the 2-surface twistor equation. Thus

$$\mathcal{H}_{total} = \oint_{S} i\pi_{A'}\pi_{B'} dx^{A'}{}_A \wedge dx^{AB'}$$

which is K.P.Tod's formula for $A_{\alpha\beta}Z^{\alpha}Z^{\beta}$. Thus the component A_{00} of the angular momentum twistor is the value of the Hamiltonian that generates motions along a complex vector field on Σ whose boundary values on $\mathcal{S}$ is $\underset{0}{\pi}^{A'}\underset{0}{\omega}^{A}$ and similarly for the other components.

Comment. This article originally appeared as two twistor newsletter articles in TN17 and TN 18 respectively. Expanded and improved versions appear in Mason (1989) and Mason & Frauendiener (1990).

References

Ashtekar, A. & Horowitz, G. (1982) On the canonical approach to quantum gravity, Phys. Rev. **D26**, 3342-53.

Mason, L.J. (1989) Class. & Quant. Gravity, **6**, L7-13

Mason, L.J. & Frauendiener, J. (1990) in *Twistors in Mathematics and Physics*, (T.N.Bailey and R.J.Baston, eds), LMS Lecture Note Series 156, CUP.

Misner, C.W., Thorne, K.S. & Wheeler, J.A. (1973) Gravitation, Pages 515 & 521, Freeman, San Francisco.

Nestor, J. (1983) The Gravitational Hamiltonian, p. 155-63 in *Asymptotic behaviour of Mass and Space-Time Geometry*, ed. F.J. Flaherty, Springer Lecture Notes in Physics 202, Springer, Berlin.

Regge, T. & Teitelboim, C. (1974) The role of surface integrals in the Hamiltonian of general relativity, Annals of Phys. **88**, 286-318.

Sparling, G.A.J. (1981) The Einstein Vacuum Equations, Pittsburgh preprint.

Witten, E. (1981) A New Proof of the Positive Energy Theorem, Comm. Maths. Phys., 381-402.

§II.4.12 **Dual two-surface twistor space** *by B.P.Jeffryes* (TN 17, January 1984)

This paper is a short treatment of the space dual to two-surface twistor space, and how this relates to 'norms' and conformal embeddings of the two-surface in flat space-time. We assume throughout that only 4 solutions exist to the two-surface twistor equations.

Introduction. The two-surface twistor $Z^\alpha = (\omega^A, \pi_{A'})$ is defined as solutions to the following equations:

$$\eth'\omega^0 = \sigma'\omega^1 \qquad \eth\omega^1 = \sigma\omega^0 \tag{1}$$

$$-i\pi_{1'} = \eth\omega^0 - \rho'\omega^1 \qquad -i\pi_{0'} = \eth'\omega^1 - \rho\omega^0 \tag{2}$$

Two $\pi_{A'}$ derivatives may be obtained by applying the commutator $[\eth, \eth']$ to ω^0, ω^1

$$\begin{aligned} -i(\eth\pi_{0'} + \rho\pi_{1'}) &= X_2\omega^1 + X_1\omega^0 \\ -i(\eth'\pi_{1'} + \rho'\pi_{0'}) &= X_2\omega^0 + X_3\omega^1, \end{aligned} \tag{3}$$

where

$$X_1 = \Psi_1 - \phi_{01}, \qquad X_2 = \Psi_2 - \phi_{11} - \Lambda, \qquad X_3 = \Psi_3 - \phi_{21}.$$

Since there are 4 solutions the other $\pi_{A'}$ derivatives serve to define 8 functions s, B, C, D and their primed versions

$$\begin{aligned} -i\eth\pi_{1'} &= i(\overline{s}'\pi_{0'} - B'\pi_{1'}) + C'\omega^1 + D'\omega^0 \\ -i\eth'\pi_{0'} &= i(\overline{s}\pi_{1'} - B\pi_{0'}) + C\omega^0 + D\omega^1. \end{aligned} \tag{4}$$

Applying $[\eth, \eth']$ to $\pi_{0'}, \pi_{1'}$ gives equations satisfied by s, B, C, D

$$\eth B + (\overline{s}\overline{s}' - \overline{\sigma}\overline{\sigma}') = X_2 - \overline{X}_2 \tag{5}$$

$$C + \eth'\rho - \eth\overline{s} = B\rho + B'\overline{s} \tag{6}$$

$$\eth C + \sigma D - \overline{s}D' - \eth' X_1 = -BX_1 \tag{7}$$

$$\eth D - \overline{s}C' + \rho' C - \sigma' X_1 + \rho X_3 - \eth' X_2 = -BX_2 \tag{8}$$

and their primed versions.

The Dual. There are two approaches to the dual: the algebraic and the differential. Algebraically we can pick a basis for the twistor space and hence have a matrix of functions over the two-surface:

$$
{}_i\mathbb{T}^\alpha = \begin{bmatrix} {}_1\omega^0 & {}_1\pi_{1'} & {}_1\omega^1 & {}_1\pi_{0'} \\ {}_2\omega^0 & {}_2\pi_{1'} & {}_2\omega^1 & {}_2\pi_{0'} \\ {}_3\omega^0 & {}_3\pi_{1'} & {}_3\omega^1 & {}_3\pi_{0'} \\ {}_4\omega^0 & {}_4\pi_{1'} & {}_4\omega^1 & {}_4\pi_{0'} \end{bmatrix}. \tag{9}
$$

To obtain the dual space we simply invert the matrix, obtaining (to give agreement with the flat case)

$$
{}^i\tilde{\mathbb{T}}_\alpha = \begin{bmatrix} {}_2\tilde{\pi}_0 & {}_2\tilde{\omega}^{1'} & {}_2\tilde{\pi}_1 & {}_2\tilde{\omega}^{0'} \\ {}_1\tilde{\pi}_0 & {}_1\tilde{\omega}^{1'} & {}_1\tilde{\pi}_1 & {}_1\tilde{\omega}^{0'} \\ {}_4\tilde{\pi}_0 & {}_4\tilde{\omega}^{1'} & {}_4\tilde{\pi}_1 & {}_4\tilde{\omega}^{0'} \\ {}_3\tilde{\pi}_0 & {}_3\tilde{\omega}^{1'} & {}_3\tilde{\pi}_1 & {}_3\tilde{\omega}^{0'} \end{bmatrix} \tag{10}
$$

with a basis dual to our original choice. Differentially we note that

$$
{}^i\tilde{\pi}_0 = \frac{\varepsilon^{ijkl}\,{}_j\omega^1\,{}_k\pi_{0'}\,{}_l\pi_{1'}}{\varepsilon^{pqrs}\,{}_p\omega^0\,{}_q\omega^1\,{}_r\pi_{0'}\,{}_s\pi_{1'}} \tag{11}
$$

and so on, and we can differentiate, obtaining equations for *dual twistors*

$$
\eth\tilde{\omega}^{0'} = \overline{s}'\tilde{\omega}^{1'} \qquad\qquad \eth'\tilde{\omega}^{1'} = \overline{s}\,\tilde{\omega}^{0'} \tag{12}
$$

$$
i\tilde{\pi}_0 = \eth\tilde{\omega}^{1'} - \rho\tilde{\omega}^{0'} - B'\tilde{\omega}^{1'} \qquad\qquad i\tilde{\pi}_1 = \eth'\tilde{\omega}^{0'} - \rho'\tilde{\omega}^{1'} - B\tilde{\omega}^{0'}. \tag{13}
$$

Note also that

$$
B' = \eth\,\ln(\varepsilon^{pqrs}\,{}_p\omega^0\,{}_q\omega^1\,{}_r\pi_{0'}\,{}_s\pi_{1'}) \tag{14}
$$

and similarly for B. We now note that $\sigma' = 0$ implies that for 2 solutions $\omega^0 = 0$. If we take $\rho = \rho' = 0$ (we may so do by making a conformal transformation such that

$$
\Omega\big|_{2\text{-}surface} = 1, \qquad \frac{\text{Þ}\Omega}{\Omega} = \rho, \qquad \frac{\text{Þ}'\Omega}{\Omega} = \rho'\ ).
$$

this does not affect σ, σ', s, s' and we see that the upper left hand corner of ${}_i\mathbb{T}^\alpha$ vanishes, and thus the bottom right hand corner of $\tilde{\mathbb{T}}_\alpha$ vanishes, implying that $s' = 0$. The converse is obviously also true, as is the equivalence of σ vanishing and s vanishing.

Norms and embedding.

Lemma 1. The following are equivalent:

(a) $\Sigma_0 = \omega^0\overline{\pi}_0 + \omega^1\overline{\pi}_1 + \text{c.c.} = \text{constant}.$

(b) $\Psi_2 = \overline{\Psi}_2,\ s = \sigma,\ s' = \sigma'$

(c) The two-surface is embeddable in real conformally flat space.

Proof: (a) implies (b) by differentiation. (b) implies (c) by observing that $B = B' = 0$ since $\eth B = \eth\eth'(\ln \operatorname{Det}\mathbb{T}^\alpha) = 0$ and is a conformal Laplacian on the Riemann sphere, and that equations (6), (7), (8) are then precisely the Bianchi identities in a conformally flat space-time. This tells us that a conformal factor exists such that X_1, X_2, X_3 may be made to vanish, and that this two-surface may be placed in flat space-time. (c) implies (a) because (a) is true in flat space-time and is conformally invariant.□

By a similar argument we obtain:

Lemma 2.

(a) $\Sigma_1 = \omega^0\overline{\pi}_0 + \omega^1\overline{\pi}_1 + iB'\omega^0\overline{\omega}^{1'} + \text{c.c.} = \text{constant}.$

(b) $s = \sigma,\ s' = \sigma'$

(c) The two-surface is embeddable in real conformally flat space-time, with torsion introduced by a complex conformal factor:

$$g_{ab} \to g_{ab}, \qquad \varepsilon_{AB} \to e^{i\theta}\varepsilon_{AB}, \qquad \varepsilon_{A'B'} \to e^{-i\theta}\varepsilon_{A'B'}.$$

Lemma 3. All spacelike two-surfaces of spherical topology with $\sigma = \sigma' = 0$ are conformal to a metric two-sphere in flat space-time.

Proof: We have $s = \sigma,\ s' = \sigma'$, thus (b) in Lemma 2 is satisfied and our two-surface is conformal to one with $\sigma = \sigma' = 0$ in flat space-time. All such two-surfaces have scalar curvature $= \rho\rho'$ constant, and our two-surface is thus a sphere.□

The generic embedding of the two-surface in complex conformally flat space-time has been introduced before (Tod 1983). This can be made explicit by noting that in general equations

(6), (7), (8) are the Bianchi identities in a conformally flat complex space-time with conformal torsion generated by

$$\Omega = (\mathrm{Det}\,\mathbb{T}^{\alpha})^{1/2}, \qquad g_{ab} \to g_{ab}, \qquad \varepsilon_{AB} \to \Omega\varepsilon_{AB}, \qquad \varepsilon_{A'B'} \to \frac{1}{\Omega}\varepsilon_{A'B'}.$$

The curvature on unprimed spinors is as for the two-surface in its curved space time, but the shears of the primed basis spinors are given by $\overline{s}$, $\overline{s}'$ not $\overline{\sigma}$, $\overline{\sigma}'$.

Dual Angular Momentum. Starting from the dual twistor space one may seek to define a dual angular-momentum twistor $\tilde{A}^{\alpha\beta}$. Writing $\tilde{Z}_{\alpha} = (\tilde{\omega}^{0'}, \tilde{\pi}_0)$ one might write a similar expression to the Tod form of $A_{\alpha\beta}Z^{\alpha}Z^{\beta}$. Just as

$$A_{\alpha\beta}\underset{1}{Z}^{\alpha}\underset{2}{Z}^{\beta} \propto \int (\underset{1}{\pi}_{0'}\underset{2}{\pi}_{1'} + \underset{2}{\pi}_{0'}\underset{1}{\pi}_{1'})\,dS \tag{15}$$

we define

$$\tilde{A}^{\alpha\beta}\underset{1}{\tilde{Z}}_{\alpha}\underset{2}{\tilde{Z}}_{\beta} \propto \int (\underset{1}{\tilde{\pi}}_{0}\underset{2}{\tilde{\pi}}_{1} + \underset{2}{\tilde{\pi}}_{0}\underset{1}{\tilde{\pi}}_{1})\,dS \tag{16}$$

This may be rewritten

$$\tilde{A}^{\alpha\beta}\underset{1}{\tilde{Z}}_{\alpha}\underset{2}{\tilde{Z}}_{\beta} \propto \int (C\underset{1}{\tilde{\omega}}^{0'}\underset{2}{\tilde{\omega}}^{0'} + X_2(\underset{1}{\tilde{\omega}}^{0'}\underset{2}{\tilde{\omega}}^{1'} + \underset{1}{\tilde{\omega}}^{1'}\underset{2}{\tilde{\omega}}^{0'}) + C'\underset{1}{\tilde{\omega}}^{1'}\underset{2}{\tilde{\omega}}^{1'})\,dS \tag{17}$$

if one considers instead a surface for which equations (12), (13) are the complex conjugate angular-momentum twistor, with curvature terms defined as if $\overline{s}$, $\overline{s}'$ were the complex conjugate shears, instead of $\overline{\sigma}$, $\overline{\sigma}'$. Thus one might define

$$\begin{aligned}\tilde{A}^{\alpha\beta}\underset{1}{Z}_{\alpha}\underset{2}{Z}_{\beta}\int\big[(\eth\overline{s} - \eth'\rho)\underset{1}{\tilde{\omega}}^{0'}\underset{2}{\tilde{\omega}}^{0'} &+ (\overline{X}_2 - s\overline{s}' + \overline{\sigma}\overline{\sigma}')(\underset{1}{\tilde{\omega}}^{0'}\underset{2}{\tilde{\omega}}^{1'} + \underset{1}{\tilde{\omega}}^{1'}\underset{2}{\tilde{\omega}}^{0'}) \\ &+ (\eth'\overline{s}' - \eth\rho')\underset{1}{\tilde{\omega}}^{1'}\underset{2}{\tilde{\omega}}^{1'}\big]\,dS\end{aligned} \tag{18}$$

Integration by parts shows that expressions (17) and (18) are equal. One may now define a quantity of dimension mass by

$$\mu^2 = -\tfrac{1}{2}\tilde{A}^{\alpha\beta}A_{\alpha\beta} \tag{19}$$

though how useful this is remains to be seen. Thanks to K.P.Tod, W.T.Shaw and R. Penrose.

Comment (1994). A development of this work was published as part of Jeffryes (1987). The definition of dual angular momentum given here gives the wrong answer when applied to null infinity. Following this discovery no further use has been made of the definition.

References

Jeffryes, B. (1987) 2-Surface Twistors, Embeddings and Symmetries, Proc. Roy. Soc. Lond. **A411**, 59-83.

Tod, K.P. (1983) Proc. Roy. Soc. **A 388**, 457-472.

§II.4.13 **Symplectic geometry of $\mathscr{I}^+$ and 2-surface twistors** *by W.T. Shaw* (TN 17, January 1984)

In trying to understand what form of angular momentum flux law may be associated with Penrose's 2-surface twistors (Penrose 1982) defined on cross-sections of $\mathscr{I}^+$, one is led to make a comparison with the theory of fluxes given by Ashtekar & Streubel (1981) (denoted A-S from here on), who were able to associate fluxes with generators of the BMS group. The purpose of this note is to explain the relationship of the two approaches, and to show how the twistor approach supplies an element missing from the A-S analysis—the construction of *charge* integrals for the fluxes.

A-S first construct the phase space of radiative modes of the gravitational field on $\mathscr{I}^+$. They show that the induced action of the BMS group on this space preserves the symplectic structure thereon, and compute the Hamiltonians generating these canonical transformations. In $\mathbb{M}$, an analogous analysis of electromagnetic theory suggests that the corresponding Hamiltonians are integrals of a Hamiltonian density which is just the flux associated with the generators V of the transformations. Transferring the interpretation to full general relativity, one arrives at a Hamiltonian density ${}^{H}F_V$ regarded as a local flux. In this symplectic approach the ${}^{H}F_V$ arise as primary quantities and one has to integrate them to obtain the 2-sphere charge integrals ${}^{H}Q_V[S]$, with

$$^{H}Q_V[S_2] - {^{H}Q_V}[S_1] = \int_{S_1}^{S_2} {^{H}F_V}\, d^3S.$$

Such an integration was carried out by A-S for the case when V is a BMS super-translation. The general case, for V a rotation or boost also, was unresolved. It turns out that one can use the theory of 2-surface twistors to solve this problem.

To see what is involved, I need to describe the quantity $^{H}F_V$ in more detail. $^{H}F_V$ is constructed from quantities defined intrinsically on $\mathscr{I}^+$. Let q_{ab} denote the pull-back to $\mathscr{I}^+$ of the (rescaled) space-time metric g_{ab}, and let q^{ab} be any symmetric tensor field within $\mathscr{I}^+$ satisfying

$$q_{am}q^{mn}q_{bn} = q_{ab}.$$

Let N_{ab} denote the News tensor field on $\mathscr{I}^+$, that is, the pull-back to $\mathscr{I}^+$ of

$$-R_{ab} + \frac{R}{6} g_{ab}$$

in a conformal frame in which the metric of $\mathscr{I}^+$ is a unit sphere. Let D denote the torsion-free connection on $\mathscr{I}^+$ induced by ∇. For any BMS generator V, then

$$^{H}F_V = (16\pi G)^{-1} N_{ab}[(\pounds_V D_c - D_c \pounds_V) l_d + l_c D_d v]\, q^{ac} q^{bd},$$

where, if n^a is the null generator of $\mathscr{I}^+$, l_a is any covector satisfying $l_a n^a = 1$, and v is defined by

$$\pounds_V q_{ab} = 2\, v\, q_{ab}.$$

Firstly this must be translated into spin-coefficients. Complete a tetrad by introducing m^a, $\overline{m}^a$ with

$$m^a l_a = 0 = m^a n_a = m^a m_a; \quad m^a \overline{m}_a = -1$$

and let (as usual) $\sigma = m^b m^a D_a l_b$; $\tau = m^b n^a D_a l_b$. The News tensor is then N, where

$$\overline{N} = \eth\tau - \tau^2 - \text{Þ}'\sigma$$

employing GHP notation for simplicity, and

$$q^{ac}q^{bd}N_{ab} = 2Nm^c m^d + 2\overline{N}\overline{m}^c\overline{m}^d.$$

Now let k^a be any self-dual BMS generator, that is,

$$k^a = k_m m^a + k_n n^a, \quad \overline{\eth} k_m = 0 = \text{Þ}' k_m, \quad \text{Þ}' k_n = \frac{1}{2}(\eth + 2\tau)k_m.$$

A tedious calculation gives

$${}^H F_k = \frac{-1}{8\pi G}[N\{\eth^2 k_n + k_n\overline{N} - \frac{\sigma}{2}\eth k_m - \eth(\sigma k_m)\} + \overline{N}\{\overline{\eth}^2 k_n + k_n N - \frac{3}{2}k_m\eth\overline{\sigma} + \frac{1}{2}\eth(k_m\sigma)\}].$$

Recall that the kinematic twistor is defined as follows. Let S be a 2-surface cross-section of $\mathcal{I}^+$ with l^a chosen to be orthogonal to S. Let (o, ι) be the associated spin frame. Let $\omega^A = \overset{0}{\omega} o^A + \overset{1}{\omega}\iota^A =$ define a 2-surface twistor, so that

$$\overline{\eth}\overset{0}{\omega} = 0; \qquad \eth\overset{1}{\omega} = \sigma\overset{0}{\omega}$$

on S. The kinematic twistor $A^S_{\alpha\beta}$ of S is then given, for a pair of twistors $\underset{1}{Z}^\alpha$ and $\underset{2}{Z}^\beta$ labelled by solutions $\underset{1}{\omega}^A$ and $\underset{2}{\omega}^B$ of the above by

$$A^S_{\alpha\beta}\underset{1}{Z}^\alpha\underset{2}{Z}^\beta = \frac{-i}{4\pi G}\oint_S\left[\underset{1}{\overset{0}{\omega}}\underset{2}{\overset{0}{\omega}}\Psi^0_1 + (\underset{1}{\overset{0}{\omega}}\underset{2}{\overset{1}{\omega}} + \underset{1}{\overset{1}{\omega}}\underset{2}{\overset{0}{\omega}})(\Psi^0_2 - \sigma N)\right] dS$$

where the $\Psi_i^{\,0}$ are the usual components of the (appropriately rescaled) Weyl tensor. Now $\underset{1}{\omega}^A$ and $\underset{2}{\omega}^B$ define a symmetric 2-index twistor with principal part $\omega^{AB} = \underset{1}{\omega}^{(A}\underset{2}{\omega}^{B)}$. In flat space-time ω^{AB} would immediately generate a self-dual Killing vector of $\mathbb{M}$, given, in the unrescaled space-time (denoting this by carets) by the equation

$$\hat{\partial}_{CC'}\hat{\omega}_{AB} = -i\hat{\varepsilon}_{C(A}\hat{k}_{B)C'}.$$

The BMS vector associated with ω^{AB} is the vector k^a given by the smooth extension of $\mathcal{I}^+$ of $\hat{k}$, viz, $k^a = \hat{k}^a$. Applying the required conformal rescaling yields

$$k^a = 2i\underset{1}{\overset{0}{\omega}}\underset{2}{\overset{0}{\omega}}m^a + i(\underset{1}{\overset{0}{\omega}}\underset{2}{\overset{1}{\omega}} + \underset{1}{\overset{1}{\omega}}\underset{2}{\overset{0}{\omega}})n^a = k_m m^a + k_n m^a \text{ say.}$$

We may use this to associate, at S, self-dual BMS vectors with 2-surface twistors. The kinematic twistor, as a function $L^S[k]$ of these k, is just

$$L^S[k] = \frac{-1}{8\pi G}\oint_S \left[k_m \Psi_1^0 + 2k_n(\Psi_2^0 - \sigma N)\right] dS.$$

However, these vectors k are not general, but are constrained by the twistor equations, which imply

$$\bar{\eth} k_m = 0; \qquad \eth^2 k_n = \frac{\sigma}{2}\eth k_m + \eth(\sigma k_m)$$

at S. The latter equation constrains k to lie in a particular Poincaré subgroup of the BMS group. For this particular group choice there are some interesting consequences. Dray and Streubel [3] have shown that the above equations on k imply that, for all values of the constants a and b,

$$L^S[k] = \frac{-1}{8\pi G}\oint_S \left[k_m(\Psi_1^0 + a\eth(\sigma\bar{\sigma}) + 2a\sigma\eth\bar{\sigma}) + 2k_n(\Psi_2^0 - \sigma N + a\eth^2\bar{\sigma} + b\bar{\eth}^2\sigma)\right] dS$$

In particular, L^S is equivalent to L_0^S, where L_0^S is given by

$$L_0^S[k] = \frac{-1}{8\pi G}\oint_S \left[k_m(\Psi_1^0 + \tfrac{1}{2}\eth(\sigma\bar{\sigma}) + \sigma\eth\bar{\sigma}) + 2k_n(\Psi_2^0 - \sigma N + \tfrac{1}{2}\eth^2\bar{\sigma} - \tfrac{1}{2}\bar{\eth}^2\sigma)\right] dS.$$

We can extend this to a $\mathbb{C}$-linear function of any self-dual BMS vector k. Consider now $L_0^S[k]$ as a function of general BMS vectors k. That is, k satisfies the BMS evolution equations

$$\text{Þ}' k_m = 0, \qquad \text{Þ}' k_n = \tfrac{1}{2}(\eth + 2\tau)k_m$$

and is not constrained by the twistor equations at any particular 2-surface. Then, if S_1 and S_2 are two 2-sphere cross-sections of $\mathcal{I}^+$, a long and tedious calculation shows that

$$L_0^{S_2}[k] - L_0^{S_1}[k] = \int_{S_1}^{S_2} {}^H F_k d^3S.$$

It is straightforward to establish this for the easy case when S_2 is obtained from S_1 by a time translation. For the general case one should note that the formulae for the maps are given in spin frames adapted to the 2-surface, so that in evolving care is required in keeping track of which frames are employed. This resolves the problem of obtaining charge integrals for the A-S flux. The corresponding formula for real BMS vectors can be obtained by taking real parts.

The remaining issue is now clear. If one solves the twistor equations at S_2, the self-dual vectors so obtained are not the same as those obtained by BMS propagation from vectors obtained from 2-surface twistors at S_1, unless $\mathrm{N} = 0$ in the region between S_1 and S_2. This is the old problem—the natural Poincaré group for defining angular momentum shifts in the presence of gravitational radiation, relative to the BMS group. The twistor equations extend the notion of 'natural' from stationary space-times to general ones. One may say, interpreting the twistor approach in conventional terms, that the flux associated with the Penrose formula is a synthesis of the Hamiltonian flux with the effect of a continual supertranslation of the origins of rotations. However, the bug in this idea is the same as that noted in §II.4.4. The two sets of BMS vectors at S_1 and S_2 can be related, but not, apparently, in a unique conformally invariant way. This corresponds to a translational freedom in identifying the origins in the natural $\mathbb{CM}$ associated with $\mathbb{T}^\alpha(S_1)$ with the origins in the corresponding $\mathbb{CM}$ associated with $\mathbb{T}^\alpha(S_2)$. This freedom can be removed by picking conformal frames associated with the Bondi energy-momentum in which to fix various functions, but it would be satisfying to have an alternative (and conformally invariant) procedure.

Very many thanks to T.Dray, with whom parts of these calculations were done, and also to R.Penrose, K.P.Tod, and G.T.Horowitz for useful discussions. See Shaw (1984) and Dray (1984) for fuller details.

References

Ashtekar, A. & Streubel, M. (1981) Proc. Roy. Soc. **A 376**, 585.

Dray, T. (1984) Class. & Quant. Grav. **2**, L7.

Dray, T. & Streubel, M. (1984), Class. & Quant. Grav. **1**, 15.

Penrose, R. (1982) Proc. Roy. Soc. Lond. **A 381**, 53, see also §II.4.4.

Shaw, W.T. (1984) Class. & Quant. Grav. **1**, L33.

§II.4.14 **More on quasi-local mass** *by K.P. Tod* (TN 18, July 1984)

In Tod (1983), I calculated some examples of Penrose's quasi-local mass for a variety of specific two-surfaces in particular space-times. What made the calculations manageable was that each of the two-surfaces considered could be embedded in conformally flat space with the same first and second fundamental forms. Call such a two-surface conformally embeddable (c.e.). Then evidently on a c.e. two surface the 'usual twistor norm'

$$\Sigma = \omega^A \bar{\pi}_A + \bar{\omega}^{A'} \pi_{A'} \tag{1}$$

is actually constant when $(\omega^A, \pi_{A'})$ satisfy the two-surface twistor equations, and so Σ defines a norm on $\mathbb{T}(\mathcal{S})$, the two-surface twistor space of the two-surface $\mathcal{S}$. The converse of this statement is also true so that the norm defined by (1) is constant on $\mathcal{S}$ if and only if $\mathcal{S}$ is c.e. This has been shown by Ben Jeffryes and me. Thus c.e. surfaces can be regarded as understood at least as far as the quasi-local mass is concerned. (There is still e.g. the question of angular momentum). For non-c.e. surfaces, which Roger Penrose suggests we call 'contorted', things are more difficult.

Another question raised in Tod (1983) was the calculation of the quasi-local mass for a 'small sphere'. Recall that to define a small sphere you pick a time-like unit vector t^a at a point p in space-time. Now use t^a to normalise the null vectors at p. If r is the affine parameter on the null cone at p then a small sphere is a surface of constant (small) r. Write $\mathbb{T}(\mathcal{S}(r))$ for the twistor space of the small sphere $\mathcal{S}(r)$. Then one can seek to calculate $A_{\alpha\beta}(r)$ as a power series in r. This is a very messy calculation! In Tod (1983) I remarked that the first non-zero term was $O(r^5)$ and quadratic in the Weyl tensor, though I didn't know at that time the precise form of it. However, I eventually worked it out and the calculation has also been done by Ron Kelly (and we agree on the answer).

There is an obvious coordinatisation of $\mathbb{T}(\mathcal{S}(r))$ and in this coordinatisation $A_{\alpha\beta}$ has the form:

$$A_{\alpha\beta} = r^5 \begin{bmatrix} \lambda_{AB} & P_B{}^{A'} \\ P_A{}^{B'} & 0 \end{bmatrix} + O(r^6)$$

with

$$P_a = \frac{c^2}{90G} \Psi_{ABCD} X_{A'B'C'D'}\, t^{BB'} t^{CC'} t^{DD'} \tag{2}$$

where

$$X_{A'B'C'D'} = \Psi_{A'B'C'D'} - 4\Psi_{ABCD}\, t^A_{A'} t^B_{B'} t^C_{C'} t^D_{D'} = -4iH_{AA'BB'}\, t^A_{C'} t^B_{D'}. \tag{3}$$

It is tempting to call P_a the momentum, but this would not have invariant significance. To calculate the norm of $A_{\alpha\beta}$ one needs the norm on $\mathbb{T}(\mathscr{S}(r))$. Since (2) is the first non-zero term in $A_{\alpha\beta}$, we can use the first term in Σ given by (1) which *is* constant on $\mathscr{S}$. We find that m_p^2, the Penrose mass, is just P_aP^a so that λ_{AB} is not needed for this. (Note that λ_{AB} is *not* to be thought of as angular momentum since it is in the 'wrong' corner!)

Now, this is rather bad news. To see why, consider the case of small spheres in the Schwarzschild solution. Any three-surface of spherical symmetry (i.e. of the form $f(r, t) = 0$) can be shown to admit three-surface twistors. Consequently any two-surface in a spherically symmetric three-surface gives one of two answers for m_p^2: either m_s^2, the Schwarzschild mass, if it goes round the hole or zero if it doesn't. If the vector t^a used to define the small spheres lies in the (t, r)-plane then the small spheres will be non-contorted and P_a will have to be zero. This is a strong constraint on P_a but from (2) and (3) it can be seen to be satisfied. (In (3), $H_{ab} = H_{AA'BB'}$ is the magnetic part of the Weyl tensor at the point p.) However if t^a does *not* lie in the (t, r) plane then (2) turns out to define a space-like vector so that $m_p^2 < 0$.

What should we make of this? Given the earlier successes of the construction with non-contorted surfaces the obvious idea is to find some kind of modification for contorted surfaces. The small spheres are contorted in that the norm defined by (1) is constant at $O(1)$ but varies at $O(r)$. One expects that a complex conformal transformation might help matters! Recall that under the conformal transformations

$$\varepsilon_{AB} \to \Omega\varepsilon_{AB}, \qquad \varepsilon_{A'B'} \to \tilde{\Omega}\varepsilon_{A'B'}$$

where both Ω and $\tilde{\Omega}$ are complex we find

$$\Sigma \to \Sigma + i(\Upsilon_{AA'} - \overline{\Upsilon}_{AA'})\omega^A\overline{\omega}^{A'}$$

where $\Upsilon_a = \nabla_a \ln \Omega$ and $\varepsilon_{\alpha\beta\gamma\delta} \to \Omega\tilde{\Omega}^{-1}\varepsilon_{\alpha\beta\gamma\delta}$ for the twistor four-form. For the small spheres, Ron Kelly and I find that

$$\Sigma + i(\Upsilon_a - \overline{\Upsilon}_a)\omega^A\overline{\omega}^{A'} = \text{constant} + O(r^2)$$

if

$$\Omega = 1 + \frac{ir^2}{2}H_{ab}l^al^b \tag{4}$$

where H_{ab} is as before and l^a is the normalised null vector at p. Thus the small sphere $\mathcal{S}(r)$ is contorted by an amount related to the magnetic part of the Weyl tensor.

Next we consider the determinant of four solutions of the two-surface twistor equation, i.e. the determinant of the 4×4 matrix whose rows are the four sets of $(\omega^0, \omega^1, \pi_{0'}, \pi_{1'})$. If $\mathcal{S}$ is non-contorted, this is a constant on $\mathcal{S}$ and defines $\varepsilon_{\alpha\beta\gamma\delta}$. We find

$$\det = \text{constant} \times (1 - \frac{ir^2}{3} H_{ab} l^a l^b) \tag{5}$$

to this order. Thus

$$\Omega\tilde{\Omega}^{-1} = 1 + \frac{ir^2}{3} H_{ab} l^a l^b + \text{higher order}$$

and again the conformal factors are related to the magnetic part of the Weyl tensor. Now returning to the quasi-local mass, the suggested modification is to include in the integral a factor $\tilde{\Omega}\Omega^{-1}$ to undo the effect of the contortedness on the four-form, i.e. to define

$$A_{\alpha\beta} Z^\alpha Z^\beta = -\frac{ic^2}{4\pi G} \int \Psi_{ABCD} \omega^{AB} (\tilde{\Omega}\Omega^{-1}) d\sigma^{CD}.$$

Applied to the small sphere by Ron Kelly this precisely eliminates the $O(r^5)$ term in P_a! Thus the mass becomes zero at this order. Of course, these suggestions (which are under active investigation) raise almost as many questions as they answer but they do indicate a way to deal with contorted surfaces.

References

Tod, K.P. (1983) Some examples of Penrose's quasi-local mass construction, Proc. Roy. Soc. Lond. **A388** 457.

§II.4.15 **'New improved' quasi-local mass, and the Schwarzschild solution** *by R. Penrose* (TN 18, July 1984)

As has been pointed out by K.P.Tod in the previous article (§II.4.14), my original definition of quasi-local mass $m(\mathscr{S})$, for spacelike $\mathscr{S} \simeq S^2$ in a general space-time $\mathscr{M}$ (§II.4.2 and Penrose 1982) seems to run into difficulties in cases when the spacelike 2-surface $\mathscr{S}$ is *contorted* (i.e. not embeddable in a conformally flat space-time without changing its intrinsic or extrinsic curvatures). In view of the excellent results that K.P.Tod has earlier obtained with that definition (Tod 1983 and Shaw 1984) it would seem most unreasonable simply to abandon the quasi-local mass, but rather one should be seeking some simple modification of the original definition which gives agreement with these earlier results and which avoids the anomalies that can arise in contorted cases.

One of the more striking results that K.P.Tod had earlier obtained was that for any $\mathscr{S}$ drawn in any 3-surface of revolution $t = f(r)$ in the Schwarzschild solution $\mathscr{M}_m$, for mass m, the (original) quasi-local mass is always zero if $\mathscr{S}$ does not surround the source and is always m if $\mathscr{S}$ does surround it (just once). These are all *uncontorted* surfaces $\mathscr{S}$. The natural *conjecture* had been that the same result ought to hold if $\mathscr{S}$ does not lie in any $t = f(r)$ (and is therefore *contorted*). If one accepts the results for uncontorted $\mathscr{S}$'s in $\mathscr{M}_m$ then any other result in the contorted cases seems physically unreasonable. The result $(m(\mathscr{S}))^2 < 0$ for (certain) small contorted spheres, that K.P.Tod actually obtained (and was confirmed by R.Kelly) simply adds to this physical unreasonableness.

Just before the results of K.P.Tod's calculation were at hand, I had embarked on a seemingly promising-looking line of reasoning aimed at proving the above *conjecture*. But then K.P.Tod's results had appeared to invalidate this (then partly conjectural) line of argument. On re-examining my argument somewhat later I was surprised to find that it actually leads to a *modification* of my original definition for $A_{\alpha\beta}(\mathscr{S})$. R.Kelly and K.P.Tod then found that with such a modified definition, $m(\mathscr{S})$ indeed vanishes to 5th order for small contorted spheres in (e.g.) $\mathscr{M}_m$, as it presumably should.

My argument depended upon the existence of a *Killing spinor* (The sign '$\mathrel{\overline{\underline{\smash{\cdot}}}}$' means that the LHS and the RHS are related by a constant factor which I haven't been bothered to sort out.)

$$X_{AB} \mathrel{\overline{\underline{\smash{\cdot}}}} \psi^{-1/3}\alpha_{(A}\beta_{B)},$$

where

$$\Psi_{ABCD} = \psi\alpha_{(A}\alpha_B\beta_C\beta_{D)}, \qquad \alpha_A\beta^A = 1$$

in $\mathcal{M}_m$ (cf. Walker & Penrose 1970) which satisfies

$$\nabla_{A'(A}X_{BC)} = 0.$$

This much holds also in any {22} vacuum (e.g. Kerr, NUT, C-metric). But I also require:

Property K. The field X^{AB}, restricted to $\mathcal{S}$, belongs to the symmetric tensor product of the 2-surface twistor space $\mathbb{T}(\mathcal{S})$ with itself.

The argument consists of two parts:

(1) Show that *if* property K holds, then the modified (i.e. 'new improved') quasi-local mass expression vanishes if $\mathcal{S}$ does not surround the source and gives the Schwarzschild mass m if $\mathcal{S}$ does surround it (just once). [The latter still presents problems.]

(2) Show that property K holds for the Schwarzschild solution.

Part (2) originated as a tentative conjecture. It is now known that his conjecture is false: property K must indeed fail for a general (contorted) $\mathcal{S}$ in the Schwarzchild solution (as follows from a calculation by N.M.J.Woodhouse, Woodhouse (1987), cf. §II.4.14 and Kelly, Tod & Woodhouse (1986)Nevertheless, I beleive that there is some merit in giving my argument for (1) here and for outlining my attempted argument aimed in the direction of (2). Perhaps there are pointers here towards finding out what is *really* going on.

Part (1). We work in $\mathbb{T}^*(\mathcal{S})$, which is a flat (dual) twistor space, and consider the twistor contour integral

$$A_{\alpha\beta} \equiv \oint \frac{W_\alpha W_\beta\, d^4W}{(K^{\rho\sigma}W_\rho W_\sigma)^3}. \tag{A}$$

Here $K^{\alpha\beta}$ is a twistor in $\mathbb{T}(\mathcal{S}) \odot \mathbb{T}(\mathcal{S})$ whose primary part agrees with X^{AB} on $\mathcal{S}$. For this we need property K. Also we need $\mathcal{S}$ to be embedded in $\mathbb{M}(\mathcal{S})$—the complex compact Minkowski space associated with $\mathbb{T}^*(\mathcal{S})$ (i.e. with $\mathbb{T}(\mathcal{S})$). The embedding is easily achieved, where we identify each point P of $\mathcal{S}$ with the linear 2-space in $\mathbb{T}^*(\mathcal{S})$ whose elements take the local form $(\lambda_A, 0)$ at P (and so annihilate the 2-surface twistor fields ω^A which vanish at P). We also need a choice of twistor $\varepsilon^{\alpha\beta\gamma\delta}$ to define the d^4W:

$$d^4W = \frac{1}{24} dW_\alpha \wedge dW_\beta \wedge dW_\gamma \wedge dW_\delta \varepsilon^{\alpha\beta\gamma\delta}. \tag{B}$$

To perform the integral (A) we can introduce a basis $\delta_0^\alpha, \delta_1^\alpha, \delta_2^\alpha, \delta_3^\alpha$ for $\mathbb{T}(\mathcal{S})$ and take components with respect to it:

$$\lambda_0 = W_0 = W_\alpha \delta_0^\alpha, \quad \lambda_1 = W_1 = W_\alpha \delta_1^\alpha, \quad \mu^{0'} = W_2 = W_\alpha \delta_2^\alpha, \quad \mu^{1'} = W_3 = W_\alpha \delta_3^\alpha. \tag{C}$$

Note that each of $\delta_0^\alpha, \cdots, \delta_3^\alpha$ is a 2-surface twistor and so has a description in terms of '$(\omega^0, \omega^1, \pi_{0'}, \pi_{1'})$' which varies over $\mathcal{S}$ according to the standard 2-surface twistor equations on $\mathcal{S}$. If at one point Q on $\mathcal{S}$ we choose

$$\varepsilon^{\alpha\beta\rho\sigma} = 24\ \delta_0^{[\alpha}\delta_1^\beta\delta_2^\rho\delta_3^{\sigma]}, \qquad \text{so } \varepsilon^{0123} = 1 \tag{D}$$

then this equation will hold at all points of $\mathcal{S}$. But in terms of the standard local '$(\omega^0, \cdots, \pi_{1'})$' on $\mathcal{S}$ we shall find that the components of the spinor parts $(\cdots, \varepsilon^{AB}\varepsilon_{R'S'}, \cdots)$ may vary from point to point on $\mathcal{S}$. If $\mathcal{S}$ is uncontorted then these components are in fact constant over $\mathcal{S}$. But in the general case they vary. There is just one overall factor to be concerned with—a scalar field ν on $\mathcal{S}$. Then the components of these spinor parts are just $\pm\nu$ and 0 (This ν is just the $\tilde{\Omega}\Omega^{-1}$ of Tod's article). Tod conjectures that ν has constant modulus over $\mathcal{S}$. This would be important. Then fixing $\nu = 1$ at Q, say, we should have ν of *unit modulus* over $\mathcal{S}$.

Reverting, for the moment, to twistor components with respect to the 'constant' twistor frame of (C), we can write (with component indices underlined)

$$\mu^{\underline{A}'} = -ix^{\underline{A}\underline{A}'}\lambda_{\underline{A}} \tag{E}$$

where $x^{\underline{A}\underline{A}'}$ are Minkowski cooordinates for $\mathbb{M}(\mathcal{S})$. Then we can write

$$d^4W = d^2\lambda \wedge d^2\mu = -\lambda_{\underline{A}}\lambda_{\underline{B}} dx^{\underline{A}0'} \wedge dx^{\underline{B}1'} \wedge d\lambda_0 \wedge d\lambda_1 \tag{F}$$

and perform the λ-integral first, in (A), to obtain

$$A_{\alpha\beta}Z^{\alpha}Z^{\beta} \doteq \oint \frac{\omega^{\underline{A}}\omega^{\underline{B}}\lambda_A\lambda_B\lambda_C\lambda_D}{(X^{\underline{PQ}'}\lambda_{\underline{P}}\lambda_{\underline{Q}})^3}\, dx^{\underline{C}0'} \wedge dx^{\underline{D}1'} \wedge d\lambda_0 \wedge d\lambda_1$$

$$\doteq \oint \Psi_{\underline{ABCD}}\omega^{\underline{A}}\omega^{\underline{B}}\; dx^{\underline{C}0'} \wedge dx^{\underline{D}1'} \qquad \text{(G)}$$

since

$$\Psi_{ABCD} \doteq \oint \frac{\lambda_A\lambda_B\lambda_C\lambda_D}{(X^{PQ}\lambda_P\lambda_Q)^3}\, d\lambda_0 \wedge d\lambda_1 \qquad [\text{at each fixed point of } \mathcal{S}] \qquad \text{(H)}$$

(by direct calculation, if desired). Now the expression in the final right hand side of (G) is just the standard (original) quasi-local mass expression (in the case of vacuum) *except* for the fact that it is the differentials $dx^{\underline{C}0'} \wedge dx^{\underline{D}1'}$, for the embedding of $\mathcal{S}$ in $\mathbb{M}(\mathcal{S})$, which appear, rather than the local surface element for $\mathcal{S}$. In order to rewrite this integral in terms of this local surface element, we now return to descriptions in terms of the *local* (o^A, ι^A) spinor frame. This entails that the factor ν now appears in the integral (G).

Guessing that this applies generally for an arbitrary space-time which need not be vacuum, we are led to the '*new improved*' *quasi-local mass-angular momentum object*

$$A_{\alpha\beta}Z^{\alpha}Z^{\beta} = \frac{-i}{4\pi G}\oint \nu\Big\{(\Psi_1 - \Phi_{10})\omega^0\omega^0 + 2(\Psi_2 - \Phi_{11} - \Lambda)\omega^0\omega^1 + (\Psi_3 - \Phi_{21})\omega^1\omega^1\Big\}\,\underline{\mathcal{S}} \qquad \text{(I)}$$

where $\underline{\mathcal{S}}$ is the surface area element. If ν is indeed of unit modulus, in accordance with K.P.Tod's conjecture, then (I) seems well-enough defined for its necessary purposes. Otherwise some procedure would be needed to fix $|\nu|$, e.g. normalizing so that the average $|\nu|$ over $\mathcal{S}$ is unity.

Now, by the methods of standard flat-space twistor theory (cf. Penrose & MacCallum 1973) we can evaluate (A), and find

$$A_{\alpha\beta} \doteq \frac{K^{(-1)}{}_{\alpha\beta}}{\sqrt{detK^{\rho\sigma}}} \qquad \text{or} \qquad 0 \qquad \text{(J)}$$

according as: $\mathcal{S}$ surrounds the source (once); or else can be shrunk to a point without crossing the source region. To compute the mass from (J) when the source *is* linked we still have some problems. Perhaps Tod's method of forming a determinant and then taking $|\cdots\cdots|^{1/4}$ may be the best. But this seems to need Tod's conjecture.

Part (2). (Outline of my original attempt to prove property K for Schwarzchild.) We need a usable method of distinguishing Schwarschild case from C-metric (since we *know* $A_{\alpha\beta} \neq 0$ in case when $\mathcal{S}$ lies on $\mathcal{I}^+$ above the 'bullet holes' of the C-metric) and also from Kerr case (since K.P.Tod and W.T.Shaw have shown that property K is *false* for Kerr). Now $\mathcal{M}_m$ has 3-rotational Killing vectors x^a, y^a, z^a, with

$$X_{AB}\tilde{X}_{A'B'} = x_{AA'}x_{BB'} + y_{AA'}y_{BB'} + z_{AA'}z_{BB'} + \lambda\varepsilon_{AB}\varepsilon_{A'B'} \tag{K}$$

(Here $\tilde{X}_{A'B'}$ is the complex conjugate of X_{AB}, but will need to be 'freed' from it shortly.) The tangential parts of

$$\nabla_{A'(A}X_{BC)} = 0, \quad \nabla_{A(A'}\tilde{X}_{B'C')} = 0, \quad \nabla_{(A|(A'}x_{B')|B)} = 0, \ \cdots y, \ \cdots z \cdots$$

give

$$\eth X_{00} + 2\sigma X_{10} = 0, \ \eth' X_{11} + 2\sigma' X_{01} = 0, \ \cdots\tilde{X}\cdots, \ \cdots, \ \eth' x_{01'} + \sigma x_{11'} + \bar{\sigma}' x_{00'} = 0, \ \cdots, \ y, \cdots z \cdots,$$

which provide 10 differential equations relating derivatives of the 10 'nice' components

$$X_{00}, \ X_{11}, \ \tilde{X}_{0'0'}, \ \tilde{X}_{1'1'}, \ x_{01'}, \ x_{10'}, \ y_{01'}, \ y_{10'}, \ z_{01'}, \ z_{10'}$$

to the 9 'nasty' components

$$X_{01}, \ \tilde{X}_{0'1'}, \ x_{00'}, \ y_{11'}, \ y_{00'}, \ y_{11'}, \ z_{00'}, \ z_{11'}, \ \lambda.$$

From (K) we have 2 equations

$$X_{00}\tilde{X}_{1'1'} = x^2_{01'} + y^2_{01'} + z^2_{01'},$$
$$X_{11}\tilde{X}_{0'0'} = x^2_{10'} + y^2_{10'} + z^2_{10'}$$

relating only nice components and 6 equations

$$X_{00}\tilde{X}_{0'1'} = x_{01'}x_{00'} + y_{01'}y_{00'} + z_{01'}z_{10'} \quad \text{etc.}$$

which relate nasty ones to nice ones. Two more equations on nasty components are provided by

$$X_{01}{}^2 - X_{00}X_{11} = \lambda = \tilde{X}_{0'1'} - \tilde{X}_{0'0'}\tilde{X}_{1'1'} \tag{L}$$

and we also have, from the skew parts of (K),

$$X_{01}\tilde{X}_{0'1'} = x_{01'}x_{10'} + y_{01'}y_{10'} + z_{01'}z_{10'} - \lambda = x_{00'}x_{11'} + y_{00'}y_{11'} + z_{00'}z_{11'} + \lambda \tag{M}$$

though only *one* of these is independent of (L), in effect. In principle, we could use these $6+2+1$ equations to express all the nasty components in terms of nice ones and substitute back into the 10 differential equations. The question is: how big is the solution space? The idea was to try to show that it is no bigger than in the flat case, where we know that *all* such X's, $\tilde{X}$'s, x's, y's, z's do belong to their proper $\mathbb{T}\odot\mathbb{T}$, $\mathbb{T}^*\odot\mathbb{T}^*$, $\mathbb{T}\otimes\mathbb{T}^*$, $\mathbb{T}\otimes\mathbb{T}^*$, $\mathbb{T}\otimes\mathbb{T}^*$. Then since, in $\mathcal{M}_m$, all of X_{AB}, $\tilde{X}_{A'B'}$, $x_{AA'}$, $y_{AA'}$, $z_{AA'}$ actually *do* satisfy all the equations, it must follow that they belong to the required spaces $\mathbb{T}(\mathcal{S})\odot\mathbb{T}(\mathcal{S})$, etc. To see how big the solution space actually is, the idea was to consider the 'canonical' case first, when $\mathcal{S}$ is spherically symmetrically situated, so $\sigma=0$, $\sigma'=0$. The differential equations decouple and can be solved explicitly for the nice components. From (L) and (M) we can find X_{01} and $\tilde{X}_{0'1'}$ in terms of nice components and then solve for the remaining nasty ones. The hope had been that a general argument could be invoked to show that no new solutions appear on perturbation. Howver we know now that this does not work.

Thanks to K.P.Tod, R.M.Kelly, W.T.Shaw and N.M.J.Woodhouse.

References

Penrose, R. (1982) Proc. Roy. Soc. Lond. **A381**, 53-63.

Penrose, R. & MacCallum, M.A.H. (1973) Phys. Repts. **6C**, no. 4, 241.

Shaw, W.T. (1984) Proc. Roy. Soc. Lond., **A390**, 191-215.

Tod, K.P. (1983) Some examples of Penrose's quasi-local mass construction, Proc. Roy. Soc. Lond. **A388** 457.

Kelly, R., Tod, K.P. and Woodhouse, N.M.J. (1986) Quasi-local mass for small surfaces, Class. & Quant. Grav. **3**, 1151-67).

Walker, M. & Penrose, R. (1970) Comm. Math. Phys., **18**, 265-274

Woodhouse, N.M.J. (1987) Ambiguities in the definition of quasi-local mass, Class. Quant. Grav., **4**, L121-123

§II.4.16 **Quasi-local mass** *by N.M.J.Woodhouse* (TN 19, Jan. 1985)

In §II.4.14, K.P.Tod gives the results of some calculations that he and Ron Kelly had done on the quasi-local mass inside a small sphere in vacuum. I have checked their results by another method that can also be applied to small surfaces of other shapes.

Let Σ be a space-like hypersurface in a vacuum space-time, with unit normal t^a. Suppose that we have a family of 2-surfaces in Σ given by $(f = \text{const.})$ and that we solve the 2-surface twistor equation on each surface. Then, provided that the solutions vary smoothly from surface to surface, we have

$$\nabla_{AA'}\omega^B = -i\varepsilon_A{}^B\pi_{A'} + (\nabla_{AA'}f)\Omega^B \quad \text{mod } t_{AA'} \tag{1}$$

for some Ω^B.

For the small sphere calculation, we can take $2f$ to be the square of the geodesic distance from a point $0 \in \Sigma$. Then

$$\nabla_{AA'}f = 0 \quad \text{and} \quad \nabla_a\nabla_b f = -g_{ab} + t_a t_b \tag{2}$$

at 0. Of course, there is no obvious justification for the assumption that Ω^B is nonsingular at O, although some thought shows that it is not implausible (and, in fact, may not be necessary).

Successive differentiation of (1) leads to a sequence of restrictions on the derivatives of Ω^B, in the same way that differentiating the 3-surface twistor equation leads to curvature obstructions. By taking two derivatives, one finds that the values of Ω^A and $\nabla_{AA'}\Omega^A$ at 0 may be chosen arbitrarily (this freedom arises because one can make independent linear transformations in the 2-surface twistor space on each $f = \text{const.}$), but that if

$$\alpha_{ABC} = -\sqrt{2}\,(\nabla^{A'}_{(A}\Omega_B)\,t_{C)A'}\Big|_0 , \tag{3}$$

then

$$\sqrt{2}\,t^A_{A'}\alpha_{ABC} = \tfrac{1}{3}H_{BCA'B'}\pi^{B'} - \tfrac{1}{3}F_{BCA'D}\omega^D\Big|_0 \tag{4}$$

where, with $V_{ABA'B'} = 2\,t_{A'}{}^C t_{B'}{}^D \Psi_{ABCD}$,

$$H_{ABA'B'} = i(\overline{V}_{ABA'B'} - V_{ABA'B'})$$

$$F_{ABA'C} = \nabla_{B'(A} V_{B)CA'}{}^{B'} + 2t_{(A}{}^{E'} t_{B)}{}^{F'} \nabla_{BE'} V^{B}{}_{CA'F'} \tag{5}$$

The choices made for $\Omega^A(0)$ and $\nabla_{AA'}\Omega^A(0)$ do not affect the value of $A_{\alpha\beta}Z^\alpha Z^\beta$. One finds that

$$A_{\alpha\beta}Z^\alpha Z^\beta = \frac{i}{2\pi}\int_{f \leq \epsilon^2} \Psi_{ABCD}\omega^A(\nabla^C_{C'}\omega^B)t^{C'D}d\tau$$

$$= -\frac{\sqrt{2}iV\epsilon^2}{20\pi}\Psi_{ABCD}\omega^A\alpha^{BCD}\Big|_0 + O(r^6) \tag{6}$$

for the surface $f = \epsilon^2$, which yields the (corrected version) of Tod's expression for $P_{AA'}$. Here $V = \text{vol}\{f \leq \epsilon^2\}$.

This can be repeated for a family of small ellipsoids, by replacing f by f' where

$$2\, t^{A'}_C t^{B'}_D \nabla_{AA'}\nabla_{BB'}(f' - f) = \beta_{ABCD} = \beta_{(ABCD)}. \tag{7}$$

In this case, (4) becomes

$$\sqrt{2}\, t^A_{A'}(\alpha_{EF(A}\beta_{BC)}{}^{EF} + \alpha_{ABC}) = \frac{1}{3}H_{BCA'B'}\pi^{B'} - \frac{1}{3}F_{BCA'D}\omega^D, \tag{8}$$

which determines α_{ABC}; and (6) becomes

$$A_{\alpha\beta}Z^\alpha Z^\beta = -\frac{\sqrt{2}iV'\epsilon^2}{20\pi}\Psi_{ABCD}\omega^A\alpha^{BCD} + O(r^6); \qquad V' = \text{Vol}\{f' \leq \epsilon^2\}. \tag{9}$$

Comment (1994). What about the 'improved' construction? It is shown by the same method in Kelly, Tod & Woodhouse (1986) that the momentum vanishes for small ellipsoids at $O(r^5)$ when the determinant factor is inserted in the integral. [The original calculation reported in Twistor Newsletter contained a mistake and seemed to indicate that this was not so.]

Reference

Kelly, R.M., Tod, K.P. & Woodhouse, N.M.J. (1986) Quasi-local mass for small surfaces, Class. Quant. Grav., **3** 1151-67.

§II.4.17 **Two-surface twistors and Killing vectors** *by B.P.Jeffryes* (TN 19, January 1985)

So far little has been published on how the presence of Killing vectors affects two-surface twistor space, the only example being Paul Tod's work on spherically symmetric two-surfaces in Tod (1983). If the Killing vectors are neither tangential nor orthogonal to the two surface we still know nothing of their influence, however for a tangential (rotational) Killing vector or an orthogonal *and* hyper-surface orthogonal Killing vector simplifications result. Regrettably it appears that no similar simplifications hold if there is a two-surface orthogonal Killing vector which is not hypersurface orthogonal.

Before discussing Killing vectors some words on dual two-surface twistor space ($\mathbb{T}^*(\mathcal{S})$) are in order (see §II.4.12). Two-surface twistor space $\mathbb{T}$ is the four dimensional (we hope) space of solutions to the equations

$$\eth'\omega^0 = \sigma'\omega^1 \qquad \eth\omega^1 = \sigma\omega^0 \tag{1,2}$$

where ω^0 is of type $\{-1,0\}$, ω^1 of type $\{1,0\}$. The corresponding π's are then defined by

$$-i\pi_{0'} = \eth'\omega^1 - \rho'\omega^0 \qquad -i\pi_{1'} = \eth\omega^0 - \rho\omega^1 \tag{3,4}$$

We may find expressions for $\eth\pi_{0'}$ and $\eth'\pi_{1'}$ by applying the commutator $[\eth,\eth']$ to ω^0 and ω^1, but all we know of $\eth'\pi_{0'}$ and $\eth\pi_{1'}$ is that expressions of the form

$$\begin{aligned} -i(\eth\pi_{1'} + \overline{s}'\pi_{0'} - B'\pi_{1'}) &= C'\omega^1 + D'\omega^0 \\ -i(\eth'\pi_{0'} + \overline{s}\pi_{1'} - B\pi_{0'}) &= C\omega^0 + D\omega^1 \end{aligned} \tag{5,6}$$

must exist for some $\overline{s},\overline{s}'$, B, B', C, C', D, D' because $\mathbb{T}$ is four dimensional. In a conformally flat space we would have $\overline{s} = \overline{\sigma}$, $\overline{s}' = \overline{\sigma}'$, $B = B' = 0$. Note that $B = \eth'$ ln Det $\mathbb{T}$, $B = \eth$ ln Det $\mathbb{T}$. $\mathbb{T}^*$ is defined simply as the space algebraically dual to $\mathbb{T}$, i.e. $(\tilde{\omega}^{A'}, \tilde{\pi}_A)$ s.t.

$$\Sigma = \omega^A\tilde{\pi}_A + \tilde{\omega}^{A'}\pi_{A'}$$

is constant. Clearly in conformally flat space $\tilde{\omega}^{A'} = \overline{\omega}^{A'}$, $\tilde{\pi}_A = \overline{\pi}_A$, and more generally we find

$$\eth\tilde{\omega}^{0'} = \overline{s}'\overline{\omega}^{1'} \qquad \eth'\tilde{\omega}^{1'} = \overline{s}\tilde{\omega}^{0'} \tag{7,8}$$

and also

$$\eth\eth' \ln \mathrm{Det}\ \mathbb{T} = \Psi_2 - \overline{\Psi}_2 + \overline{\sigma}\overline{\sigma}' - \overline{s}\overline{s}'. \tag{9}$$

Note that the r.h.s. of (9) is in general complex.

Killing vectors. Consider a Killing vector of the form

$$k_a = \xi' l_a + \xi n_a - \eta' m_a - \eta \overline{m}_a. \tag{10}$$

The tangential components of the Killing vector equation $\nabla_{(a}k_{b)} = 0$ are

$$\eth\eta = \sigma\xi' + \overline{\sigma}'\xi \tag{11}$$
$$\eth'\eta' = \overline{\sigma}\xi' + \sigma'\xi \tag{12}$$
$$\eth\eta' + \eth'\eta = 2\rho\xi' + 2\rho'\xi. \tag{13}$$

If the Killing vector is tangential to $\mathcal{S}$ then the right hand sides of (11-13) vanish automatically, if orthogonal the left hand sides vanish.

Tangential Killing vectors. If there is a Killing vector tangential to $\mathcal{S}$ then we may make a choice of phase (except at the poles), such that $m_a - \overline{m}_a$ is aligned along the Killing vector direction, and thus regarding ϕ as a Killing vector coordinate and θ as another suitably chosen coordinate we have

$$m^a\nabla_a = \frac{\partial}{\partial\theta} + \frac{i}{f(\theta)}\frac{\partial}{\partial\phi}\,. \tag{14}$$

Putting this into (11-13) we see that

$$\beta + \overline{\beta}' = \overline{\beta} + \beta' \tag{15}$$

It is convenient also to choose a boost such that

$$\overline{\beta} - \beta' = i\frac{d\gamma}{d\theta} = \overline{\beta}' - \beta \qquad (\gamma \in \mathbb{R}) \tag{16}$$

where

$$-2i\eth\eth'\gamma = \Psi_2 - \overline{\Psi}_2 + \overline{\sigma\sigma}' - \sigma\sigma'. \tag{17}$$

We write solutions to $(1,2)$ as

$$\omega^0 = e^{\pm\frac{i}{2}\phi} W^0_{\pm}(\theta) \qquad \omega^1 = e^{\pm\frac{i}{2}\phi} W^1_{\pm}(\theta) \tag{18}$$

and thus obtain equations for $\frac{dW^{0,1}_{\pm}}{d\theta}$ in terms of $W^0_{\pm}$ and $W^1_{\pm}$ and thus write $\pi_{0'}$, $\pi_{1'}$ in terms of $W^0_{\pm}$, $W^1_{\pm}$. These expressions may be substituted into the two sides of $(5,6)$ and coefficients compared. One finds

$$\sigma = \overline{s}, \quad \sigma' = \overline{s}', \quad \beta = -2(\overline{\beta}' - \beta), \quad \beta' = -2(\overline{\beta} - \beta'). \tag{19,20}$$

Firstly we see that $\ln \operatorname{Det} \mathbb{T} \in i\mathbb{R}$, and thus $\operatorname{Det} \mathbb{T}$ has constant modulus. Comparing $(1,2)$ with $(7,8)$ we also find a direct correspondence between ω^A and $\tilde{\omega}^{A'}$. We find that the solutions $\tilde{\omega}^{0'}$, $\tilde{\omega}^{1'}$ are

$$\tilde{\omega}^{0'} = e^{\mp\frac{i}{2}\phi} e^{-i\gamma} W^0_{\pm}, \qquad \tilde{\omega}^{1'} = e^{\mp\frac{i}{2}\phi} e^{-i\gamma} W^1_{\pm}. \tag{21,22}$$

Since $\overline{\omega}^{0'} = e^{\mp\frac{i}{2}\phi}\overline{W}^0_{\pm}$, etc. this gives an isomorphism from the complete conjugate to the dual, and thus a norm on axi-symmetric two-surfaces. If $\mathcal{S}$ is not only axisymmetric but spatial reflection symmetric (i.e. invariance under $\phi \to -\phi$) then the shears must be aligned along the chosen basis and $\Psi_2 \in \mathbb{R}$, and hence $\sigma = \overline{\sigma} = \overline{s}$, $\sigma' = \overline{\sigma}' = \overline{s}'$, $\operatorname{Det} \mathbb{T}$ is constant and $\mathcal{S}$ is uncontorted.

If we consider $A_{\alpha\beta}$ for axi-symmetric $\mathcal{S}$ then clearly only four (complex) components survive. If reflection symmetry is added then these components are real and $A_{\alpha\beta}$ consists only of a real 4-momentum.

Hypersurface orthogonal Killing vectors orthogonal to $\mathcal{S}$. Let us assume there is a hypersurface orthogonal Killing vector orthogonal to $\mathcal{S}$, i.e. in addition to $\nabla_{(a}k_{b)} = 0$ we have

$$k^a\,{}^*(\nabla_a k_b) = 0 \tag{23}$$

and thus

$$k^a k^b\,{}^*R_{cadb} = 0. \tag{24}$$

Let us assume k^a is timelike (the spacelike case involves a few sign changes), and fix the boost so that

$$k_a = \frac{1}{\sqrt{2}}\,|k|(l_a + n_a)$$

and thus from (11-13) & (24) we find

$$\Psi_2 - \overline{\Psi}_2 = \Psi_1 + \overline{\Psi}_3 = \phi_{01} + \phi_{12} = 0 \qquad (25)$$

$$\sigma + \overline{\sigma}' = \rho + \rho' = 0, \qquad (26)$$

and using (23) we find

$$\beta - \overline{\beta}' = 0. \qquad (27)$$

Now compare the equations for $\mathbb{T}$ and $\overline{\mathbb{T}}$.

$$\begin{aligned} \eth'\omega^0 &= -\overline{\sigma}\omega^1 & \qquad \eth\overline{\omega}^{0'} &= -\sigma\overline{\omega}^{1'} \\ \eth\omega^1 &= \sigma\omega^0 & \qquad \eth'\overline{\omega}^{1'} &= \overline{\sigma}\overline{\omega}^{0'} \end{aligned} \qquad (28\text{-}31)$$

since $\beta = \overline{\beta}'$, ($\eth'$ acting on ω^0) $=$ ($\eth'$ acting on $\overline{\omega}^{1'}$) and thus if $(\omega^0, \omega^1, \pi_{0'}, \pi_{1'}) \in \mathbb{T}$, we have $(-\overline{\omega}^{1'}, \overline{\omega}^{0'}, \overline{\pi}_{1'}, -\overline{\pi}_{0'}) \in \mathbb{T}$. Since complex conjugates are both elements of $\mathbb{T}$ the same must be true for $\mathbb{T}^*$ and thus

$$s + \overline{s}' = 0, \qquad \text{and } \eth\eth' \ln \mathrm{Det}\ \mathbb{T} = s\overline{s} - \sigma\overline{\sigma} \in \mathbb{R} \qquad (32)$$

hence Det $\mathbb{T}$ is real. In addition for either the modified or original version of $A_{\alpha\beta}$ (see Penrose's article §II.4.15) we see that if $\underset{3}{W}^\alpha = \overline{\underset{1}{Z}}^\alpha$, $\underset{4}{W}^\alpha = \overline{\underset{2}{Z}}^\alpha$ then $A_{\alpha\beta}\underset{1}{Z}^\alpha\underset{2}{Z}^\beta = A_{\alpha\beta}\underset{3}{W}^\alpha\underset{4}{W}^\beta$, and thus $A_{\alpha\beta}$ has four complex and two real components.

If the hypersurface orthogonality condition is relaxed then we still have $\sigma\sigma' = \overline{\sigma}\overline{\sigma}'$ (using (11, 12)), but we do not have the algebraic restriction on $\nabla_a k_b$ which enabled us to find that $\sigma + \overline{\sigma}' = 0 \Leftrightarrow \beta - \overline{\beta}' = 0$. More generally we may choose a boost so that $\sigma + \overline{\sigma}' = 0$, *or* we may choose another boost so that $\beta - \overline{\beta}' = i\eth\gamma$, $\gamma \in \mathbb{R}$, in order that we may find a similar relation between $\mathbb{T}$ and $\overline{\mathbb{T}}$ the two boosts must coincide. Writing $\nabla_a k_b = F_{ab}$ we find this condition is equivalent to

$$\gamma_a = \frac{k^b\ {}^*F_{ab}}{k^e k_e} \quad \text{being a gradient, or} \quad \nabla_{[a}\gamma_{b]} = 0$$

implying

$$0 = |k|^2 k^e k^f\ {}^*R_{e[ab]f} + 2k^e F_{e[a}{}^*F_{b]f}k^f \tag{33}$$

The term involving ${}^*R_{eabf}$ vanishes if $k^a R_{ab} \propto k_b$, hence if this holds (e.g. if $R_{ab} = 0$) then a very stringent condition on $\nabla^a|k|$ remains, and most Killing vectors orthogonal to $\mathcal{S}$ do not satisfy this condition.

Thanks to K.P.Tod for discussions.

Comment (1994). A development of this work was published in Jeffryes (1987).

Reference

Jeffryes, B.P. (1987) 2-surface twistors, embeddings and symmetries, Proc. Roy. Soc. Lond. **A411**, 59-83.

Tod, K.P. (1983) Proc. Roy. Soc. Lond. **A 388**, 457.

§II.4.18 **Two-surface twistors for large spheres** *by W.T.Shaw* (TN 19, January 1985)

In §II.4.14 K.P.Tod discusses the behaviour of 2-surface twistors on small spheres, including the effects of the contorsion factor, also described in §II.4.15 by R.Penrose. In this note some results on the structure of 2-surface twistors on large spheres will be described. By a large sphere I mean the following: Let S_∞ be a cross-section of future null infinity $\mathscr{I}^+$, and let $\mathcal{N}$ be the associated outgoing null surface. Choose an affine parameter r for $\mathcal{N}$ asymptotic to an area coordinate for the 2-surface cross-sections (of constant r) S_r of $\mathcal{N}$. The picture to bear in mind is the following:

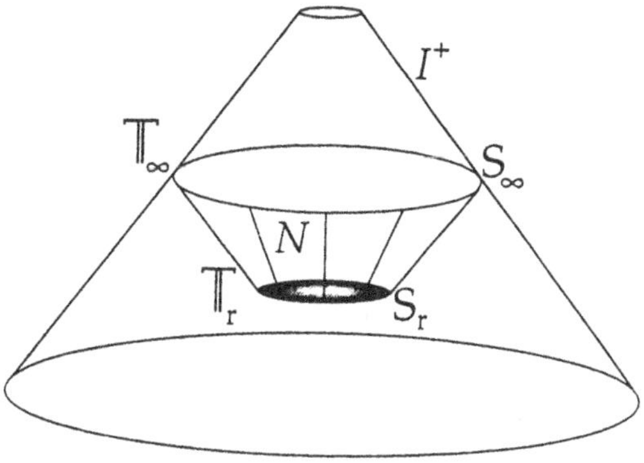

If one is given solutions to the 2-surface twistor equations on S_∞ then one can seek a power series expansion in r for the solutions on S_r. The contorsion factor, the twistor norm and the angular momentum twistor can then all be given in powers of r. The equations are difficult to solve in general, but a detailed analysis of all stationary Einstein-Maxwell space-times has been made. In this case, the ease with which one can solve the equations depends on the shear of S_∞. So far, I have found solutions to first and second order in r^{-1} for S_∞ a general cross-section of infinity, but third order solutions have only been found for the case when S_∞ is a good cut.

The degree of contorsion depends on the magnetic multipole structure of the space-time. If $\underline{J}$ is the angular momentum then a scalar function J is given by $J = \underline{J} \cdot \underline{e}$, where $\underline{e}$ is a triplet of basis functions given by

$$e_1 = \sin\theta\cos\phi,\ e_2 = \sin\theta\sin\phi,\ e_3 = \cos\theta.$$

The contorsion factor C is then given by the following expression:

$$C = 1 + 6iJr^{-2} - 2i(B_M + 2qm)r^{-3} + O(r^{-4}),$$

where B_M is a function describing the total magnetic gravitational quadrupole, q is the charge and m (similarly to J) describes the magnetic dipole moment of the Maxwell field. Like all formulae in this note, it is correct to second order for arbitrary S_∞ and correct to third order provided S_∞ is a good cut. The norm structure presents considerable problems. Firstly suppose one computes the standard candidate

$$H = H_{\alpha\alpha'} Z^\alpha \bar{Z}^{\alpha'} = \omega^A \bar{\pi}_A + \bar{\omega}^{A'} \pi_{A'}.$$

Then one can expand this in a sequence of matrices as:

$$H = H^\infty_{\alpha\alpha'} \overset{\infty}{Z}{}^\alpha \overset{\infty}{\bar{Z}}{}^{\alpha'} + r^{-2} H^2_{\alpha\alpha'} \overset{\infty}{Z}{}^\alpha \overset{\infty}{\bar{Z}}{}^{\alpha'} + r^{-3} H^3_{\alpha\alpha'} \overset{\infty}{Z}{}^\alpha \overset{\infty}{\bar{Z}}{}^{\alpha'}$$

in terms of suitable coordinates for the 2-surface twistor space on S_∞. The first term (the norm at infinity) is constant since the shear of S_∞ is necessarily electric. However, the higher order terms are not constant on the sphere. In §II.4.14 K.P.Tod suggests applying a complex conformal transformation to remove the non-constant terms. Here, if one chooses a rescaling given by:

$$\hat{\varepsilon}_{AB} = \Omega \varepsilon_{AB}, \qquad \hat{\varepsilon}_{A'B'} = \bar{\Omega} \varepsilon_{A'B'},$$

where

$$\Omega = 1 - i[3Jr^{-2} - (B_M + 2qm)r^{-3}],$$

then the contorsion is removed and several of the terms in the norm are eliminated. However, to obtain a constant norm one must go further. In fact, in general one must apply (i) a complex supertranslation and (ii) a complex coordinate transformation. The implications of (ii) are not fully understood, but at least one obtains a constant norm, which can be described as follows. Let S' be the unique shear-free cross-section of $\mathscr{I}^+$ nearest to S_∞ (with 'nearest' defined in the conformal scale for $\mathscr{I}^+$ induced by the Killing vector). One can regard S' as a cross-section of the $\mathscr{I}^+$ of flat space-time, and is given as the intersection of the future light-cone of some point O with $\mathscr{I}^+$. One can extend the 2-surface twistors on S_∞ to O using the twistor equation for flat space-time. This gives natural coordinates

$$\overset{\infty}{Z}{}^\alpha = (\omega^A(O), \pi_{A'})$$

The norm is then given by

$$g_{\alpha\alpha'} Z^\alpha \bar{Z}^{\alpha'} = G_{\alpha\alpha'} \overset{\infty}{Z}{}^\alpha \overset{\infty}{\bar{Z}}{}^{\alpha'}, \qquad G_{\alpha\beta'} = \begin{bmatrix} -3J_{AB'} r^{-3} & \delta_A{}^B \\ \delta_{B'}{}^{A'} & 0 \end{bmatrix}.$$

Now consider the angular momentum twistor. To first order, the only contributions are from the electromagnetic stress tensor, and with O as above one finds:

$$A_{\alpha\beta} = \begin{bmatrix} 0 & t_A{}^{B'}(M - \frac{q^2}{2r}) \\ t^{A'}{}_B(M - \frac{q^2}{2r}) & 2i[\mu^{A'B'}(O) - \frac{3q}{2r}\underline{\mu}^{A'B'}(O)] \end{bmatrix}$$

where $Mt^{AA'}$ is the total energy-momentum, $\mu^{A'B'}(O)$ is the total angular momentum and $\underline{\mu}^{A'B'}(O)$ is the total electromagnetic dipole moment. Now let $\underline{D}(O)$ and $\underline{d}(O)$ denote the total mass and electric dipole moment respectively. Then the orbital and dipole angular momenta at radius r are:

$$\underline{J}(r) = \underline{J} - \frac{2q}{3r}\underline{m}; \qquad \underline{D}(r) = \underline{D}(O) - \frac{2q}{3r}\underline{d}(O).$$

The angular momentum twistor has 10 real components and the Penrose mass is just

$$m_p = M - \frac{1}{2r}q^2.$$

At second order gravitational non-linearities become important. The detailed structure of the angular momentum twistor depends on where the original or 'new improved' definition is applied. Suppose the integrand for the original construction is multiplied (cf. §II.4.14 and §II.4.15) by a factor

$$\nu = 1 - n + nC$$

so that $n = 0, 1$ correpsonds to the original and modified construction respectively. One finds that the angular momentum twistor still has a zero in the top left corner, but the 'momentum' component is now

$$P^a = t^a(M - \frac{1}{2r}q^2) + i(2n+1)MJ^a r^{-2}\,.$$

However, the imaginary term here can be removed by a transformation of the twistor coordinates for S_∞. In any case, since $J_a t^a = 0$, the mass is still

$$m_p = M - \frac{1}{2r}q^2 + O(r^{-3}),$$

and this holds for all S_∞.

The angular momentum components can also be evaluated. The results simplify considerably if one takes S_∞ to be shear-free. The second order contributions to the orbital and dipole terms are just:

$$\Delta \underline{J} = q(q\underline{J} - M\underline{m})r^{-2},$$
$$\Delta \underline{D} = q(q\underline{D} - M\underline{d})r^{-2} + [\tfrac{1}{3}\underline{m}_\wedge \underline{d} + 3(n+1)\underline{J}_\wedge \underline{D}]r^{-2}.$$

In the absence of an infinity twistor the identification of these terms as 'angular momenta' is tentative, but the first terms in each come directly from the electromagnetic stress tensor, and are absent in flat space-time. The combinations $(q\underline{J} - M\underline{m})$ and $(q\underline{D} - M\underline{d})$ are the 6 'Newman-Penrose' constants for the Maxwell field of a stationary space-time (Exton, Newman & Penrose 1969). Now we see that these quantities have a physical interpretation as part of the field's angular momentum.

At third order (S_∞ shear-free) the angular momentum twistor has 20 real components, but they are not algebraically independent. The details of a suitable untwisting procedure and the structure of the infinity twistor have yet to be worked out. At third order the NP constants also contribute to the quasi-local momentum terms. The mass can be computed, using the norm given above, and one finds

$$m_p = M - \frac{1}{2r}q^2 - \tfrac{1}{3}[\underline{m}^2 + \underline{d}^2]r^{-3} - 8n\underline{J}^2 r^{-3}.$$

The electromagnetic terms are those found in flat space-time, but the modified construction has a further contribution from the rotation of the space-time. A physical interpretation of the term $8\underline{J}^2 r^{-3}$ has yet to be found although numerical factors other than 8 can be justified—see Shaw (1986).

An encouraging feature of these results is that when it can be computed, the mass formula displays the same dependence on S_∞ as one would expect in flat space-time. This compares very favourably with the results from some other definitions, where the mass is sensitive to the choice of S_∞. For example, Ludvigsen & Vickers (1983) have given a definition of a quasi-local momentum for corss-sections of outgoing null hypersurfaces. From this one can compute a mass (it makes no difference whether one computes the energy component or the momentum norm to these orders), and one finds,

$$m_{LV} = M - \frac{1}{2r}q^2 - [\tfrac{1}{3}(\underline{m}^2 + \underline{d}^2) + \tfrac{1}{4}(\underline{J}^2 + \underline{D}^2)]r^{-3},$$

where the third order terms are given for S_∞ shear-free. For example, in the Schwarzschild space-time the mass to third order is

$$m_{LV} = M - \tfrac{1}{4}\underline{D}^2 r^{-3}$$

which *is* sensitive to the choice of S_∞. In the absence of a monopole energy density (as there is for the electromagnetic field) the presence of such a cross-section dependent term is unphysical. For the Hawking (1968) definition the sensitivity is worse. In this case on recovers the first order result $m_H = M - \frac{1}{2r}q^2$ only if S_∞ is a good cut. Clearly the twistor analysis requires some more work. In particular, the structure of the infinity twistor requires elucidation.

Comment (1994). Further progress and details of this analysis was reported in Shaw (1986).

Many thanks to Roger Penrose, Paul Tod, Ben Jeffryes and Ron Kelly for use discussions.

References

Exton, A.R., Newman, E.T. and Penrose R., (1969) J. Math. Phys. **10**, 1566-1570.

Ludvigsen, M. and Vickers, J.A.G., (1983) J. Phys. A. **16**, 1155-1168

Hawking, S.W. (1968) J. Math. Phys. **9**, 598.

Shaw, W.T. (1986) Class. Quant. Grav. **3**, 1069.

§II.4.19 **An example of a two-surface twistor space with complex determinant** *by B.P.Jeffryes* (TN 19, January 1985)

For terminology regarding the dual space see my articles §II.4.17 and §II.4.12. If we consider a two-surface near $\mathscr{I}$ then we find that, to leading order in r

$$\sigma = \frac{\sigma^0}{r^2}, \qquad \sigma^1 = \frac{-\dot{\overline{\sigma}}^0}{r}, \qquad \Psi_2 = \frac{\Psi_2^0}{r^3}$$

$$\Psi_2^0 - \overline{\Psi_2^0} = \overline{\sigma}^0\dot{\sigma}^0 - \sigma^0\dot{\overline{\sigma}}^0 + \eth'^2\sigma^0 - \eth^2\overline{\sigma}^0$$

where r is an affine parameter on the null ray approaching $\mathscr{I}$, and dot denotes time derivative, and $\eth$ and $\eth'$ are those for a metric sphere i.e. $[\eth, \eth'] = q - p$. On examining the equations obeyed by the dual shears $\overline{s}, \overline{s}'$ we find that

$$\overline{s} = \frac{\overline{s}^0}{r^2}, \qquad \overline{s}' = \frac{\overline{s}'^0}{r}, \qquad \sigma^0 = \eth^2\alpha\,, \qquad \overline{s}^0 = \eth'^2\alpha$$

$$\dot{\overline{\sigma}}^0 = \eth'^2\dot{\overline{\alpha}}\,, \qquad \overline{s}'^0 = -\eth^2\dot{\overline{\alpha}}$$

Substituing into the equation for $\mathrm{Det}\,\mathbb{T}$ we find

$$r^3\eth\eth' \ln \mathrm{Det}\,\mathbb{T} \;= \eth^2\eth'^2(\alpha - \overline{\alpha}) - (\eth^2\alpha)(\eth'^2\dot{\overline{\alpha}}) + (\eth'^2\alpha)(\eth^2\dot{\overline{\alpha}}).$$

In general this is complex, but there are a number of circumstances where it is imaginary (i.e. $\mathrm{Det}\,\mathbb{T}$ has constant modulus):

i) non radiating $\quad \dot{\alpha} = 0$

ii) axi-symmetry $\quad \dfrac{\eth^2\alpha}{\eth'^2\alpha} = \dfrac{\eth^2\dot{\overline{\alpha}}}{\eth'^2\dot{\overline{\alpha}}}$

iii) $\dfrac{\alpha}{\overline{\alpha}} = \dfrac{\dot{\alpha}}{\dot{\overline{\alpha}}} = e^{2ik}, \quad k \in \mathbb{R}$, constant.

Case iii) is interesting, since a norm naturally arises which coincides with the usual norm if $\alpha = \overline{\alpha}$ and $\Psi_2^0 = \overline{\Psi}_2^0$. If $(\overline{\omega}^{0'}, \overline{\omega}^{1'})$ is the principal part of a complex conjugate twistor then $(e^{ik}\overline{\omega}^{0'},\ e^{-ik}\overline{\omega}^{1'})$ is the principal part of a dual twistor, and

$$\Sigma = e^{-ik}(\omega^0\bar{\pi}_0 + \pi_{1'}\bar{\omega}^{1'}) + e^{ik}(\bar{\omega}^{0'}\pi_{0'} + \omega^1\bar{\pi}_1) + ie^{ik}B\omega^1\bar{\omega}^{0'} + ie^{-ik}B'\omega^0\bar{\omega}^{1'}$$

is constant, where $B = \eth \ \ln \mathrm{Det}\mathbb{T}$, $B' = \eth' \ \ln \mathrm{Det}\mathbb{T}$. Thanks to W.T.Shaw for the spin-coefficients.

§II.4.20 **A suggested further modification to the quasi-local formula** *by R.Penrose* (TN 20, September 1985)

In view of the anomalous expression for the quasi-local mass for small contorted ellipsoids found by N.M.J.Woodhouse (§II.4.16), it seems that one must seek a further modification beyond that given by the determinant factor as suggested in §II.4.15. An appealing feature of the *original* form had always been the existence of the Tod expression $-\frac{i}{2\pi G}\oint \pi_{0'}\pi_{1'}\underline{\mathcal{S}}$ ($\underline{\mathcal{S}}$ being the surface-area 2-form). The determinant modification

$$-\frac{i}{4\pi G}\oint \{(\Psi_1 - \Phi_{01})(\omega^0)^2 + (\Psi_2 - \Phi_{11} - \Lambda)2\omega^0\omega^1 + (\Psi_3 - \Phi_{21})(\omega^1)^2\}\eta\underline{\mathcal{S}}$$

with $\eta \propto \begin{vmatrix} \underset{1}{\omega^0} & \underset{2}{\omega^0} & \underset{3}{\omega^0} & \underset{4}{\omega^0} \\ \underset{1}{\omega^1} & \cdots & \cdots & \cdots \\ \eth'\underset{1}{\omega^0} & \cdots & \cdots & \cdots \\ \eth'\underset{1}{\omega^1} & \cdots & \cdots & \cdots \end{vmatrix}$ leads, by integration by parts, to a 'Tod form'

$$-\frac{i}{2\pi G}\oint (\hat{\pi}_{0'}\hat{\pi}_{1'} + \omega^0\omega^1\eth\nu\eth'\nu)\underline{\mathcal{S}}$$

where $\hat{\pi}_{0'} = \nu\pi_{0'} + i\omega^1\eth'\nu$, $\hat{\pi}_{1'} = \nu\pi_{1'} + i\omega^0\eth\nu$ and $\nu^2 = \eta$. These modifications to $\pi_{0'}$, $\pi_{1'}$ seem natural enough as 'complex conformal factor' terms, but the extra term $\omega^0\omega^1\eth\nu\eth'\nu$ has an awkward appearance. Though the motivation is a little flimsy, it would seem to be worth trying

$$-\frac{i}{2\pi G}\oint \hat{\pi}_{0'}\hat{\pi}_{1'}\mathcal{S}$$

as a possible modified definition for $A_{\alpha\beta}Z^{\alpha}Z^{\beta}$. The *good* results to date seem to be unaffected by this change. Thanks to K.P.Tod, N.M.J.Woodhouse for comments.

Reference

Penrose, R. & Rindler, W. (1986) Spinors and Space-Time Vol. 2, C.U.P.

§II.4.21 **Higher-dimensional two-surface twistors** *by R.Penrose* (TN 20, September 1985)

Generalizations of the 2-surface twistor concept to higher dimensions can be achieved in two different directions. In the first place we can keep the surface $\mathcal{S}$ two-dimensional, but allow the ambient space to become higher-dimensional. In the second place we can allow $\mathcal{S}$ itself to become higher-dimensional. In any case, the spinors and twistors need to be those for a higher-dimensional ambient space-time $\mathcal{M}$. For simplicity, I shall consider $\mathcal{M}$ to be 6-dimensional and $\mathcal{S}$ to be either 2-dimensional (Case *A*) or 4-dimensional (Case *B*), but higher generalizations can also be readily achieved.

Here the reduced spinors for $\mathcal{M}$ are 4-dimensional and the twistors for $\mathcal{M}$, if $\mathcal{M}$ happens to be conformally flat, would be 8-dimensional (i.e. reduced spinors for the pseudo-orthogonal group for which the points of $\mathcal{M}$ correspond to generators of the null cone; see the Appendix to Penrose & Rindler (1986) for the relevant n-dimensional spinor and twistor concepts). For the moment, at least, everything is complex, so the signature is irrelevant.

We need to see how to describe, in spinor terms, the projection operator E which takes us down from the 6-space to the 2-space (A) tangential to $\mathcal{S}$ or (B) normal to $\mathcal{S}$ and its orthogonal complement $F = I - E$ which takes us from the 6-space to the 4-space normal to $\mathcal{S}$ (case A) tangential to $\mathscr{S}$ (case B).

We use a 'twistor-type' description of vectors in $\mathcal{M}$, where lower case Greek indices are 4-dimensional ('twistor-like') spinor indices, skew pairs of which represent tangent vectors in $\mathcal{M}$. The tangent space (Case A) or normal space (Case B) to $\mathscr{S}$ at a point p ($\mathscr{S}$ assumed non-null) contains two (complex) null directions U, V. Now null vectors in $\mathcal{M}$ correspond to simple bivectors in the (twistor-like) spin space of $\mathcal{M}$:

$$U^{\alpha\beta} = a^{[\alpha}b^{\beta]}, \qquad V^{\alpha\beta} = c^{[\alpha}d^{\beta]}$$
$$U_{\alpha\beta} = \frac{1}{2}\varepsilon_{\alpha\beta\gamma\delta}a^{\gamma}b^{\delta}, \qquad V_{\alpha\beta} = \frac{1}{2}\varepsilon_{\alpha\beta\gamma\delta}c^{\gamma}d^{\delta}$$

and take $a^{\alpha}b^{\beta}c^{\gamma}d^{\delta}\varepsilon_{\alpha\beta\gamma\delta} = 4$. The projector E is then given by

$$E^{\alpha\beta}_{\gamma\delta} = \frac{1}{2}U^{\alpha\beta}V_{\gamma\delta} + \frac{1}{2}V^{\alpha\beta}U_{\gamma\delta}$$

and

$$F^{\alpha\beta}_{\gamma\delta} = \delta^{[\alpha}_{\gamma}\delta^{\beta]}_{\delta} - E^{\alpha\beta}_{\gamma\delta}.$$

It is convenient to consider corresponding projectors in the spin-space:

$$e^{\alpha}_{\beta} = U^{\alpha\beta}V_{\beta\gamma}, \qquad f^{\alpha}_{\beta} = V^{\alpha\gamma}U_{\beta\gamma}$$

which decompose the spin-space:

$$\delta^{\alpha}_{\beta} = e^{\alpha}_{\beta} + f^{\alpha}_{\beta} \tag{A}$$

where

$$e^{\alpha}_{\beta}e^{\beta}_{\gamma} = e^{\alpha}_{\gamma}, \quad f^{\alpha}_{\beta}f^{\beta}_{\gamma} = f^{\alpha}_{\gamma}, \quad e^{\alpha}_{\beta}f^{\beta}_{\gamma} = 0 = f^{\alpha}_{\beta}e^{\beta}_{\gamma}.$$

Note also

$$e^{[\alpha}_{\gamma}e^{\beta]}_{\delta} = \frac{1}{2}U^{\alpha\beta}V_{\gamma\delta}, \quad f^{[\alpha}_{\gamma}f^{\beta]}_{\delta} = \frac{1}{2}V^{\alpha\beta}U_{\gamma\delta}. \tag{B}$$

The corresponding decomposition of the vector space is given by

$$\begin{array}{lcccc} \boxed{\text{Case A}}:\longrightarrow \text{normal} & & & & \text{tangential} \\ \delta^{[\alpha}_{\gamma}\delta^{\beta]}_{\delta} & = & \overbrace{2e^{[\alpha}_{[\gamma}f^{\beta]}_{\delta]}} & + & \overbrace{e^{[\alpha}_{\gamma}e^{\beta]}_{\delta} + f^{[\alpha}_{\gamma}f^{\beta]}_{\delta}} \\ \boxed{\text{Case B}}:\longrightarrow \text{tangential} & & & & \text{normal} \end{array} \tag{C}$$

Case A. Let $\mathscr{S}$ be 'spacelike' with S^2 topology. The twistor equation in (conformally flat) 6-space can be written

$$\nabla_{\alpha\beta}\omega^\gamma = \delta^\gamma_{[\alpha}\pi_{\beta]} \tag{D}$$

By (C), $e^{[\cdot}_{\cdot}e^{\cdot]}_{\cdot}$ and $f^{[\cdot}_{\cdot}f^{\cdot]}_{\cdot}$ project tangentially to $\mathscr{S}$ and we can extract the tangential parts of (D) as

$$e^\lambda_\gamma f^\alpha_\mu f^\beta_\nu \nabla_{\alpha\beta}\omega^\gamma = 0, \qquad f^\lambda_\gamma e^\alpha_\mu e^\beta_\nu \nabla_{\alpha\beta}\omega^\gamma = 0$$

i.e. (by B)

$$e^\lambda_\gamma V^{\alpha\beta}\nabla_{\alpha\beta}\omega^\gamma = 0, \qquad f^\lambda_\gamma U^{\alpha\beta}\nabla_{\alpha\beta}\omega^\gamma = 0$$

i.e.

$$\underbrace{e^\mu_\lambda V^{\alpha\beta}\nabla_{\alpha\beta}(\omega^\gamma e^\lambda_\gamma)}_{} = \underbrace{\omega^\gamma f^\lambda_\gamma}_{}\ \underbrace{e^\mu_\nu V^{\alpha\beta}\nabla_{\alpha\beta}e^\nu_\lambda}_{} \tag{E}$$

analogue: $\eth'$ ω^0 ω^1 σ'

$$\underbrace{f^\mu_\lambda U^{\alpha\beta}\nabla_{\alpha\beta}(\omega^\gamma f^\lambda_\gamma)}_{} = \underbrace{\omega^\gamma e^\lambda_\gamma}_{}\ \underbrace{f^\mu_\nu U^{\alpha\beta}\nabla_{\alpha\beta}f^\nu_\gamma}_{}$$

analogue: $\eth$ ω^1 ω^0 σ

Similar use of the Atiayh-Singer theorem to that used for the original S^2 in 4-space applies here also. (Look at a canonical case, with its adjoint, and then deform.) In general we get an 8-dimensional space of solutions of (E).

Case B. Let $\mathscr{S}$ become a 4-dimensional space-time $\mathscr{M}$ of embedding class 2 (or 1), $\mathscr{M}$ being now a conformally flat 6-space. For example $\mathscr{S}$ could be the Schwarzschild solution $\mathscr{M}_S$. The restriction of (D) to $\mathscr{M}$ turns out to take the form

$$\tilde\nabla^{A'}_{(A}\omega_{B)} = \Gamma^{A'B'}_{AB}\tilde\omega_{B'}, \qquad \tilde\nabla'^{A}_{(A'}\tilde\omega_{B')} = \tilde\Gamma^{AB}_{A'B'}\omega_B$$

where $\tilde\nabla_{AA'}$ and $\tilde\nabla'_{AA'}$ are analogues of $\eth$ and $\eth'$, which differ from $\nabla_{AA'}$ by 'gauge' dependent terms - like α, β, γ, ϵ of the spin-coefficient-formalism. Strictly ω_A and $\tilde\omega_{A'}$ are 'boost-weighted' spinors on $\mathscr{M}$, the boost weight referring to the perpendicular frame transformations. Thanks to Toby Bailey, Lane Hughston, Paul Tod and Ian Gatenby.

References

Penrose, R. & Rindler, W. (1986) Spinors and space-time. vol. 2, C.U.P.

§II.4.22 **Embedding 2-Surfaces in CM** *by R.Penrose* ($\mathbb{TN}$ 20, September 1985)

A result of 2-surface twistor theory—apparently of the status of a 'folk-lore theorem'—is that any spacelike 2-surface $\mathcal{S} \subset \mathcal{M}$ for which the space of 2-surface twistors is 4-dimensional (e.g. a generic S^2) can be embedded in complex conformally flat 4-space in such a way that σ and σ' remain unchanged (except for the standard scalings) whereas $\overline{\sigma}$ and $\overline{\sigma}'$ get replaced by new quantities $\tilde{\sigma}$ and $\tilde{\sigma}'$. This embedding (or perhaps immersion) is to be achieved as follows: We have ('natural' immersion)

$$\mathcal{S} \mapsto \mathbb{CM}^{\#}(\mathcal{S}) \tag{1}$$

where the points of $\mathbb{CM}^{\#}(\mathcal{S})$ are the 2-dimensional linear subspaces of $\mathbb{T}^*(\mathcal{S})$ (or, equivalently, of $\mathbb{T}(\mathcal{S})$). To obtain (1), first identify each pair $(p,\ \lambda_A)$—for which $p \in \mathcal{S}$ and λ_A is a spin-covector at p in $\mathcal{M}$—with the element of $\mathbb{T}^*(\mathcal{S})$ which carries the 2-surface twistor $\{\omega^A\}$ to $\omega^A\lambda_A$ (at p). Then p itself gets identified with the linear span of all the λ_A at p.

Now it turns out that the 2-surface twistors on $\mathcal{S}$ can be identified with the restrictions to $\mathcal{S}$ of the solutions of the twistor equation

$$\nabla^{(A}_{A'}\omega^{B)} = 0 \tag{2}$$

in $\mathbb{CM}^{\#}(\mathcal{S})$—which is a (conformally flat) complex manifold, so (2) is meaningful. To see this, note first that the solutions of (2) are precisely the spin-vector fields in $\mathbb{CM}^{\#}(\mathcal{S})$ for which $\omega^A\lambda_A$ is constant over each β-plane in $\mathbb{CM}^{\#}(\mathcal{S})$—where λ_A is such that it is covariantly constant along the β-plane and $\lambda^A\xi^{A'}$ is tangent to it for all $\xi^{A'}$. This follows from the vanishing of

$$\lambda^A\xi^{A'}\nabla_{AA'}(\lambda_B\omega^B) = \xi^{A'}\lambda^A\lambda_B\nabla_{AA'}\omega^B = -\xi^{A'}\lambda_A\lambda_B\nabla^{(A}_{A'}\omega^{B)}. \tag{3}$$

Second, note that the same equation holds at $\mathcal{S}$ *in* $\mathcal{M}$ for those directions for which $\lambda^A\xi^{A'}$ is *tangent* to $\mathcal{S}$ by virtue of the 2-surface twistor equation on $\mathcal{S}$. Third, note that two neighbouring points of $\mathcal{S}$ are null separated if and only if their images in $\mathbb{CM}^{\#}(\mathcal{S})$ are null separated; for if neighbouring points $p, p' \in \mathcal{S}$ satisfy $(pp')^{AA'}\lambda_A = 0$, then (p, λ_A) and (p', λ_A) both annihilate the same $\{\omega^A\}$, namely one given by $\omega^A\lambda_A = 0$ at p and therefore at p' since $\lambda^A\xi^{A'}\nabla_{AA'}(\omega^B\lambda_B) = 0$. Thus, the *complex conformal metric of $\mathcal{S}$ is preserved* by its embedding in $\mathbb{CM}^{\#}(\mathcal{S})$. Now the spin-vector fields ω^A in $\mathbb{CM}^{\#}(\mathcal{S})$ for which $\omega^A\lambda_A$ is constant along the $\lambda^A\xi^{A'}$ directions in

$\mathbb{CM}^{\#}(\mathscr{S})$ induce spin-vector fields on $\mathscr{S}$ which are constant along those same null directions at $\mathscr{S}$ when they happen to be tangent to $\mathscr{S}$ in $\mathscr{M}$, which, by the above, gives the 2-surface twistor equation.

In $\mathbb{CM}^{\#}(\mathscr{S})$, the 2-surface twistor equation takes the standard form

$$ð'\omega^0 = \sigma'\omega^1, \qquad ð\omega^1 = \sigma\omega^0 \tag{4}$$

with respect to a spinor basis associated with the tangent 2-plane element, and with respect to some complex flat metric for $\mathbb{CM}(\mathscr{S})$. These σ, σ' must scale in the standard way under conformal, and dyad, rescalings—which amounts to:

$$\sigma \mapsto \Gamma^2 \tilde{\Gamma}' \Gamma'^{-1} \sigma, \qquad \sigma' \mapsto \Gamma'^2 \tilde{\Gamma} \Gamma^{-1} \sigma' \tag{5}$$

under

$$o^A \mapsto \Gamma o^A, \qquad \iota^A \mapsto \Gamma' \iota^A, \qquad \tilde{o}^{A'} \mapsto \tilde{\Gamma} \tilde{o}^{A'}, \qquad \tilde{\iota}^{A'} \mapsto \tilde{\Gamma}' \tilde{\iota}^{A'}$$

so with normalized dyads,

$$\varepsilon_{AB} \mapsto (\Gamma\Gamma')^{-1} \varepsilon_{AB}, \qquad \varepsilon_{A'B'} \mapsto (\tilde{\Gamma}\tilde{\Gamma}')^{-1} \varepsilon_{A'B'}.$$

Comparing (4) in $\mathbb{CM}^{\#}(\mathscr{S})$ with (4) in $\mathscr{M}$, we find that the corresponding σ, σ' must scale, as in (5), between the two as required. More can be done using these ideas—later!

§II.4.23 **Asymptotically anti-de Sitter space-times** *by R.Kelly*[1] (TN 20, September 1985)

1. The Angular Momentum Twistor. I will first briefly review the work of Ashtekar & Magnon (1984). A space time $(\hat{M}, \hat{g}_{ab})$ is said to be asymptotically anti-de Sitter if there exists a manifold M with a metric g_{ab} and a diffeomorphism from $\hat{M}$ to $M - \partial M$ such that

(i) There exists Ω on M such that $g_{ab} = \Omega^2 \hat{g}_{ab}$ on $\hat{M}$;

(ii) $\mathscr{I} = \partial M$ is topologically $S^2 \times \mathbb{R}$ and $\Omega = 0$ on $\mathscr{I}$;

(iii) $\hat{g}_{ab}$ satisfies $\hat{R}_{ab} - \frac{1}{2}\hat{R}\hat{g}_{ab} + \lambda \hat{g}_{ab} = -8\pi G \hat{T}_{ab}$ with $\lambda < 0$ and where $\Omega^{-4}\hat{T}_a{}^b$ admits a smooth limit as $\Omega \to 0$;

(iv) $B^1_{ab} \equiv \Omega^{-1} B_{ab} \doteq 0$, where B_{ab} is the magnetic part of the Weyl tensor C_{abcd} of M and the symbol '$\doteq$' means 'in the limit as $\Omega \to 0$'.

Notes: (a) In [1] they assumed the rather weaker condition that $\Omega^{-3}\hat{T}_a{}^b$ admits a smooth limit as $\Omega \to 0$. The scaling chosen in (iii) leaves the conservation equation $\nabla^a T_{ab} = 0$ conformally invariant, see Penrose & Rindler (1986) page 371.

(b) Condition (iii) and the equations for the change in the spinor decomposition of the curvature under conformal scalings imply that (1) If $S_a = \nabla_a \Omega$ then $S^a S_a \doteq \frac{\lambda}{3}$ and (2) $C_{abcd} \doteq 0$. Thus $\mathscr{I}$ is timelike. We may further choose Ω so that $\nabla_a S_b \doteq 0$.

(c) Condition (iv) is equivalent to the Cotton-York tensor of $\mathscr{I}$ vanishing and thus $\mathscr{I}$ is conformally flat. It is also equivalent to the condition ${}^{\mathscr{I}}D_{[a}V_{b]c} = 0$ where $V_{ab} = (\Phi_{ab} - \Lambda g_{ab}) - E_{ab}$ and ${}^{\mathscr{I}}D_a$ denotes the intrinsic covariant derivative of $\mathscr{I}$. This, together with the condition $E_{ab} \doteq 0$ (which is trivially satisfied from (b2)), implies that $\mathscr{I}$ is embeddable in conformally flat space-time with the same first and second fundamental forms (see reference [3]), and thus three-surface twistors exist on $\mathscr{I}$. We may therefore think of $\mathscr{I}$ as the conformal infinity of anti-de Sitter space-time which can be embedded in the Einstein static universe as the time axis multiplied by the two-sphere equator of the three-sphere cross-sections of constant time.

It follows that the conformal group of $\mathscr{I}$ is the anti-de Sitter group $O(2,3)$ and there are 10 conformal Killing vectors on $\mathscr{I}$ (with respect to the intrinsic metric). Hawking (1983) has shown that (iv) is equivalent to gravitational radiation obeying a reflective boundary condition at $\mathscr{I}$.

Given a cross section C of $\mathscr{I}$ and a conformal Killing vector ξ^a on $\mathscr{I}$ define a conserved quantity by

[1]Editorial note: Although every effort has been made, we are unable to trace Dr R. Kelly. We would appreciate any information which would enable us to do so.

$$Q_\xi[C] = -\frac{1}{8\pi G}\oint E^1_{ab}\xi^a dS^b$$

(see Ashtekar & Magnon 1984). This expression is conformally invariant. We may show that the flux $F_\xi(\Delta)$ across any region Δ of $\mathscr{I}$ bounded by two cross-sections is given by

$$F_\xi(\Delta) = \int_\Delta [\lim_{\Omega\to 0}\Omega^{-4}\hat{T}_a{}^c]S^a\xi_c d\Sigma.$$

Note that if there is no matter near $\mathscr{I}$ the flux vanishes and there is no Bondi leakage.

Schwarzschild anti-de Sitter space-time has metric

$$dS^2 = (1 - \frac{2-MG}{r} + a^2r^2)dt^2 - (1 - \frac{2MG}{r} + a^2r^2)^{-1}dr^2 - r^2d\Omega^2$$

where $d\Omega^2 = d\theta^2 + \sin^2\theta d\phi^2$ and $a = \sqrt{-\lambda/3}$. If $\xi^a = \frac{1}{a}\frac{\partial}{\partial t}$ (which with $\Omega = 1/r$ is a unit vector at $\mathscr{I}$ with respect to the rescaled metric) we obtain $Q_\xi[C] = M$ for any cross-section C of $\mathscr{I}$. To show that the above expression for conserved quantities is really Penrose's expression for quasi-local mass and angular momentum (modified for asymptotically anti-de Sitter space-times by the subtraction of the cosmological constant term) we first need the following lemma.

Lemma. If Σ is a hypersurface in a conformally flat space-time with normal S^a and $\nabla_a S_b = 0$ then $\xi^a = \omega^A S^{A'}_B \omega^B$ is a (null) conformal Killing vector on $\mathscr{I}$ if $\nabla_{A'(A}\omega_{B)} = 0$.

The proof is simply substitution of the solution of the twistor equation in a conformally flat space-time. For a general conformal Killing vector ξ^a, $\xi_a = F_{ab}S^b$, where

$$F_{ab} = \sum_{i=1}^{2}\epsilon_{A'B'}\underset{i}{\omega}_{(A}\underset{i}{\tilde{\omega}}_{B)}$$

and $\underset{i}{\omega}_A$, $\underset{i}{\tilde{\omega}}_A$ are solutions to the twistor equation. Then if we write $\xi^a = 2i\omega^A\overline{S}^{A'}_B\omega^B$ where $\overline{S}^b = S^b/a$ and let t^a be the normal to the cross-section C of $\mathscr{I}$ that lies in $\mathscr{I}$,

$$\frac{-1}{8\pi G}\oint_C E^1_{ab}\xi^a dS^b = \frac{-1}{8\pi G}\oint_C 2\phi_{ABCD}\overline{S}^C_{A'}\overline{S}^D_{B'}2i\omega^A\overline{S}^{A'}_E\omega^E t^{BB'}dS$$
$$= \frac{-i}{4\pi G}\oint_C \phi_{ABCD}\omega^A\omega^C o^B\iota^D dS$$

where $\phi_{ABCD} = \Omega^{-1}\Psi_{ABCD}$. This is Penrose's expression, Penrose (1982).

We shall choose our momentum and angular momentum conformal Killing vectors by embedding $\mathscr{I}$ as the boundary of conformally anti-de Sitter space-time on the Einstein cylinder and then restricting solutions of the twistor equation in conformally flat space-time to it. This depends on how we embed Minkowski space-time on the Einstein cylinder with respect to anti-de Sitter space-time. We choose to do this symmetrically.

With respect to the constant spinor basis (α^A, β^A) in Minkowski space the solution to the twistor equation is given by $\omega^A = \Omega^A - iX^{AA'}\pi_{A'}$ where $\Omega^A = \Omega^0\alpha^A + \Omega^1\beta^A$, $\pi_{A'} = \pi^{0'}\overline{\alpha}_{A'} + \pi^{1'}\overline{\beta}_{A'}$ and Ω^0, Ω^1, $\pi^{0'}$, $\pi^{1'}$ are constants. Then with respect to the null tetrad:

$$l^a = \frac{1}{\sqrt{2}}\Big(\frac{\partial}{\partial t'} + \frac{\partial}{\partial r'}\Big), \qquad n^a = \frac{1}{\sqrt{2}}\Big(\frac{\partial}{\partial t'} - \frac{\partial}{\partial r'}\Big),$$

$$m^a = \frac{1}{\sqrt{2}\sin r'}\Big(\frac{\partial}{\partial \theta} - \frac{i}{\sin\theta}\frac{\partial}{\partial \phi}\Big)$$

on the Einstein cylinder we find that

$$\omega^A = \omega^0 o^A + \omega^1 \iota^A$$

where

$$\omega^0 = \sqrt{2}\cos\tfrac{1}{2}(t'+r')\Big(\Omega^0 e^{-i\phi/2}\cos\tfrac{\theta}{2} + \Omega^1 e^{i\phi/2}\sin\tfrac{\theta}{2}\Big) + i\sin\tfrac{1}{2}(t'+r')\Big(-\pi^{0'}e^{i\phi/2}\sin\tfrac{\theta}{2} + \pi^{1'}e^{-i\phi/2}\cos\tfrac{\theta}{2}\Big)$$

$$\omega^1 = \sqrt{2}\cos\tfrac{1}{2}(t'-r')\Big(-\Omega^0 e^{-i\phi/2}\sin\tfrac{\theta}{2} + \Omega^1 e^{i\phi/2}\cos\tfrac{\theta}{2}\Big) + i\sin\tfrac{1}{2}(t'-r')\Big(\pi^{0'}e^{i\phi/2}\cos\tfrac{\theta}{2} + \pi^{1'}e^{-i\phi/2}\sin\tfrac{\theta}{2}\Big)$$

Since the twistor equation is conformally invariant this is also a solution of the twistor equation on the Einstein cylinder. From these expressions we may calculate the conformal Killing vector $\xi^a = 2i\omega^{(A}\overline{S}_{B}^{|A'|}\tilde{\omega}^{B)}$ which lies in $r' = \frac{\pi}{2}$ and we find that

$$-\frac{i}{4\pi G}\oint_C \phi_{ABCD}\omega^A\tilde{\omega}^B dS^{CD} = A_{\alpha\beta}Z^\alpha Z^\beta$$

where $Z^\alpha = (\Omega^A, \pi_{A'})$ and $A_{\alpha\beta}$ is of the form $A_{\alpha\beta} = \begin{bmatrix} 2\Phi_{AB} & P_A{}^{B'} \\ P^{A'}{}_B & \Phi^{A'B'} \end{bmatrix}$. In particular

$$P^{\underline{a}} = -\frac{1}{8\pi G}\oint_C dS^b E_{cb}\Big(\gamma^c, \frac{\eta^c + \overline{\eta^c}}{2}, \frac{\eta^c - \overline{\eta^c}}{2i}, \beta^c\Big)$$

where

$$\gamma^a = \frac{\partial}{\partial t'}, \quad \eta^a = e^{i\phi}\Big(\sin t' \sin\theta \frac{\partial}{\partial t'} - \cos t' \cos\theta \frac{\partial}{\partial \theta} - i\cos t' \frac{\partial}{\partial \phi}\Big),$$

$$\beta^a = \cos\theta \sin t' \frac{\partial}{\partial t'} + \sin\theta \cos t' \frac{\partial}{\partial \theta}.$$

For Schwarzschild–anti–de Sitter space-time in the coordinates described earlier we find that $t' = at$ and thus $P^a = (M, 0, 0, 0)$, $\Phi_{AB} = 0$ and we recover M as the mass.

The angular momentum twistor $A_{\alpha\beta}$ obeys the hermiticity property

$$A_{\alpha\beta} I^{\beta\gamma} = \overline{A_{\gamma\beta} I^{\beta\alpha}}$$

with respect to the infinity twistor

$$I^{\alpha\beta} = \begin{bmatrix} -\frac{1}{2}\varepsilon^{AB} & 0 \\ 0 & \varepsilon_{A'B'} \end{bmatrix}.$$

2. A positive energy theorem. We first need to look at the work of Gibbons et al (1983). Consider a spacelike hypersurface Σ in an asymptotically anti-de Sitter space-time (with possibly an inner boundary on a $\begin{Bmatrix}\text{past} \\ \text{future}\end{Bmatrix}$ apparent horizon H), which asymptotically approaches the $t' = 0$ cut of $\mathscr{I}$. Define a 'supercovariant' derivative on the 4-spinor $(\alpha^A, \beta^{A'})$ by

$$\hat\nabla_{MM'}\alpha_A = \nabla_{MM'}\alpha_A + \tfrac{a}{\sqrt{2}}\varepsilon_{MA}\beta_{M'} \text{ and } \hat\nabla_{MM'}\beta_A = \nabla_{MM'}\beta_{A'} + \tfrac{a}{\sqrt{2}}\varepsilon_{M'A}\alpha_M \,.$$

Then if D_a is the projection of the 4-dimensional connection into Σ and α_A, $\beta_{A'}$ satisfy the 'supercovariant' Witten equation on Σ, i.e.

$$\hat D_{AA'}\alpha^A \equiv D_{AA'}\alpha^A + \tfrac{3a}{2\sqrt{2}}\beta_{A'} = 0 \text{ and } \hat D_{AA'}\beta^{A'} \equiv D_{AA'}\beta^{A'} + \tfrac{3a}{2\sqrt{2}}\alpha_A = 0,$$

we may show that

$$-D_m[t^{AA'}(\overline{\alpha}_{A'}\hat D^m\alpha_A + \overline{\beta}_A\hat D^m\beta_{A'})] = -t^{AB'}[\hat D_m\alpha_A\overline{(\hat D^m\alpha_{B'})} + \hat D_m\beta_{B'}\overline{(\hat D^m\beta_A)}] + 4\pi G T_{ab}t^a\xi^b$$

where $\xi^a = \alpha^A\overline{\alpha}^{A'} + \beta^{A'}\overline{\beta}^A$. Thus if T_{ab} satisfies the dominant energy condition then the right hand side is non-negative. We may choose our spinors α^A, $\beta^{A'}$ to obey boundary conditions on an inner apparent horizon H such that, at H:

$$t^{AA'}(\overline{\alpha}_{A'}\hat{D}^m\alpha_A + \overline{\beta}_A\hat{D}^m\beta_{A'})N_m = 0,$$

where N^m is the normal to H lying in Σ. If we use Green's theorem on the above identity we therefore obtain

$$-\oint_{t'=0} t^{AA'}(\overline{\alpha}_A\hat{D}^m\alpha_A + \overline{\beta}_A\hat{D}^m\beta_{A'})\overline{S}_m dS \quad \geq \quad 0 \qquad (*)$$

This integral is finite, providing that, as our spinors α_A, $\beta_{A'}$ tend to $\mathcal{I}$,

$$\hat{D}_{MM'}\alpha_A \to 0 \text{ and } \hat{D}_{MM'}\beta_{A'} \to 0.$$

In the original paper it was proposed that the boundary term could be written as

$$4\pi G[P^{S\,AA'}(\alpha^0_A\overline{\alpha}^0_{A'} + \beta^0_{A'}\overline{\beta}^0_A) - X^{AB}\alpha^0_A\overline{\beta}^0_B - \overline{X}^{A'B'}\overline{\alpha}_{A'}\beta_{B'}]$$

(credited as a private communication from D.Z.Freedman) where α^0_A, $\beta^0_{A'}$ are the supercovariantly constant spinors that are the limits of α_A, $\beta_{A'}$ as they tend to $\mathcal{I}$, (when divided by some suitable power of the conformal factor). For Schwarzschild anti-de Sitter space-time the component

$$P^{S\,0} \equiv \frac{P^{S\,00'} + P^{S\,11'}}{\sqrt{2}}$$

is proportional (via a positive constant of proportionality!) to the mass parameter M. In general this inequality obviously implies $P^{S\,0} \geq 0$ and so if we identify this term with the mass, as was done by Gibbons et al (1983) we have shown that the mass is positive.

But, as shall be described, if we evaluate the above integral explicitly we obtain Penrose's expression and we find information about $A_{\alpha\beta}$. We first write the metric in terms of coordinates (u, s, θ, ϕ) as

$$dS^2 = \frac{1}{\sin^2 s}\left(du^2 - 2\frac{du\,ds}{a} - \cos^2 s\, d\Omega^2\right) + O(s^{-1}) \times \begin{pmatrix} \text{products of two} \\ \text{differentials except } ds^2 \end{pmatrix}$$

The $u =$ constant surfaces are outgoing null hypersurfaces from $\mathscr{I}$, (with conformal factor $\Omega = \sin s$, $\mathscr{I}$ is the hypersurface $s = 0$) and s is a (non-affine) parameter along each null geodesic generator. The null geodesic generators are labelled by $x^2 = \theta$ and $x^3 = \phi$. We shall choose a null tetrad by defining

$$l_a = \frac{du}{\sin s}, \qquad l^a = -a \sin s \frac{\partial}{\partial s},$$

n^a is chosen to be the ingoing null vector orthogonal to the two-spheres $s =$ constant and m^a, $\overline{m}^a$ lie in these two-spheres. m^a may be defined so that $\varepsilon \cdot \overline{\varepsilon} = 0$. We thus have

$$n^a = \sin s\Big(\frac{\partial}{\partial u} + a\, U\frac{\partial}{\partial s} + X^k\frac{\partial}{\partial x^k}\Big),$$

$$m^a = \frac{a}{\sqrt{2}}\,\xi^k\frac{\partial}{\partial x^k}.$$

For anti-de Sitter space-time $U = \frac{1}{2}$, $X^K = 0$, $\xi^2 = \tan s$ and $\xi^3 = -\,i\tan s/\sin\theta$ and these are therefore our boundary conditions on the functions defining our tetrad. Also since $\Omega^{-4}\hat{T}_a{}^b$ tends to a finite limit as $\Omega \to 0$ we have $\Phi_{ij} = O(s^4)$, $\Lambda - \frac{\lambda}{6} = O(s^4)$, $\Psi_i = O(s^3)$. We may therefore solve for the spin coefficients as power series expansions in s and then we may calculate α_A, $\beta_{A'}$ similarly. We find that

$$\alpha^A = s^{-\frac{1}{2}}\sum_{i=0}^{\infty}\alpha_i^A s^i \qquad\qquad \beta^{A'} = s^{-\frac{1}{2}}\sum_{i=0}^{\infty}\beta_i^{A'} s^i$$

where $\beta_0^{A'} = \sqrt{2}\alpha_0^A \overline{S}_A^{A'}$ (this is really an identity in the rescaled space-time with, on $S = 0$,

$$l^a = -a\frac{\partial}{\partial s}, \qquad n^a = \frac{\partial}{\partial u} + \frac{a}{2}\frac{\partial}{\partial s} \text{ and } \overline{S}^a = \frac{1}{2}\, l^a - n^a\ ).$$

The integral $(*)$ becomes

$$-\frac{\sqrt{2}}{a^3}\oint_{\substack{t'=0\\ s=0}}[\psi_1^{(3)}\alpha_0^0\overline{\beta_0^0} + \psi_2^{(3)}[\alpha_0^1\overline{\beta_0^0} + \alpha_0^0\overline{\beta_0^1}] + \psi_3^{(3)}\alpha_0^1\overline{\beta_0^1}]dS_0$$

$$= -\frac{\mathrm{i}}{a^3}\oint_{\substack{t'=0\\ s=0}}[\psi_1^{(3)}\omega^0\tilde{\omega}^0 + \psi_2^{(3)}(\omega^1\tilde{\omega}^0 + \omega^0\tilde{\omega}^1) + \psi_3^{(3)}\omega^1\tilde{\omega}^1]dS_0$$

$$= \frac{4\pi G}{a^3} A_{\alpha\beta} Z^\alpha \tilde{Z}^\beta$$

where $\omega^A = \alpha_0^A$, $\tilde{\omega}^A = -\sqrt{2}i\bar{\beta}_0^A$ since ω^A, $\tilde{\omega}^A$ are found to be two-surface twistors on the $t' = 0$ cross-section of $\mathscr{I}$. But from earlier work we have expressions for ω^0, ω^1 at $t' = 0$ in terms of our twistor coordinates $(\Omega^A, \pi_{A'})$ and the relationship between α_0^A and $\beta_0^{A'}$ gives us that $\tilde{Z}^\alpha = 2I^{\alpha\beta}\bar{Z}_\beta$. We have therefore proved that $A_{\alpha\beta} Z^\alpha I^{\beta\gamma} \bar{Z}_\gamma \geq 0$. In particular this implies that P^a is timelike and future-pointing. From the angular momentum twistor $A_{\alpha\beta}$ we may calculate the associated mass m_P as

$$m_P^2 = -\tfrac{1}{2} A_{\alpha\beta} \bar{A}^{\alpha\beta} = P_a P^a - \Phi_{AB} \Phi^{AB} - \Phi_{A'B'} \bar{\Phi}^{A'B'}.$$

The above inequality implies that m_P^2 is non-negative and also provides a further inequality relating components of $P_{AA'}$ and Φ_{AB}. An alternative definition of mass is

$$m_D^4 = 4 \, Det \, A_{\alpha\beta} = 4\varepsilon^{\alpha\beta\gamma\delta} A_{\alpha 1} A_{\beta 2} A_{\gamma 3} A_{\delta 4}$$

and the above inequality also implies that m_D^4 is non-negative. $P^a P_a = m_P^2 = m_D^2$ when $\Phi_{AB} = 0$ but in general it is not possible to decide which mass is larger.

I would like to thank K.P.Tod for considerable assistance in this work, and R.Penrose and W.T.Shaw for useful discussions.

References

Ashtekar, A. & Magnon, A. (1984) Class. Quant. Grav. 1, L39. See also Henneaux, M. & Teitelboim, C., Comm. math. Phys. **98**, 391.

Gibbons, G.W., Hawking, S.W., Horowitz & Perry, M.J. (1983) Comm. Math. Phys. **88**, 295.

Hawking, S.W. (1983) Phys. Lett. **126B**, 175.

Penrose, R. & Rindler, W. (1986) *Spinors & space-time*, Vol. 2, C.U.P.

Penrose, R. (1982) Proc. Roy. Soc. **A381**, 53.

§II.4.24 **Two-surface pseudo-twistors** *by B.P.Jeffryes* (TN 21, February 1986)

Two-surface twistor space can be defined as the solution space to the equations (written in GHP spin coefficient notation)

$$ð\omega^1 = \sigma\omega^0, \qquad ð'\omega^0 = \sigma'\omega^1, \tag{1}$$

where ω^1 is type $(\frac{1}{2}, 0)$ and ω^0 of type $(-\frac{1}{2}, 0)$ and the complex null direction m^a of the null tetrad is tangential to a spacelike 2-surface (normally taken to be of spherical topology). In order to understand more about the generic properties of such solution spaces it seemed interesting to look at the solution space of a family of equations which includes the 2-surface twistor equations. Namely

$$ðX = \sigma Y \qquad ð'Y = \sigma'X \tag{2}$$

where X is of type $\left(\frac{b+s+1}{2}, \frac{b-s}{2}\right)$ and Y is of type $\left(\frac{b+s-1}{2}, \frac{b-s}{2}\right)$. Henceforth the 2-surface will be assumed spherical. If $b = s = 0$ then these are the 2-surface twistor equations.

The dimension of the solution space of these equations will be independent of b and it is easily calculated (using a metric sphere with vanishing shears) that the index of each system of equations is 4. For such a sphere we see that the solution space for $|s| = 0$, $\frac{1}{2}$ or 1 is 4 dimensional, but for $|s| > 1$ the solution space is $2(|s| + 1)$ dimensional.

The name 2-surface pseudo-twistors seemed appropriate since for this simple example just as 2-surface twistor space splits naturally into two 2-dimensional subspaces ($\omega^1 = 0$ and $\omega^0 = 0$ respectively), the $|s| = \frac{1}{2}$ solution spaces each split into a 3-dimensional and a 1-dimensional subspace.

The solution spaces for $|s| > 1$ are interesting because they have 'extra' solutions for the sphere with vanishing shears. It can be shown very simply that some of these 'extra' solutions vanish with the addition of arbitrarily small shear. First a case where the extra solutions persist. Consider the equations with $s \geq 1$ and $\sigma' = 0$

$$ðX = \sigma Y, \qquad ð'Y = 0. \tag{3,4}$$

There are no solutions to equation (4) other than $Y = 0$ and hence all solutions have

$$\eth X = 0 \tag{5}$$

which has $2s+2$ linearly independent solutions. Now we consider a case where some of the extra solutions disappear. Examine the case $\sigma = 0$. We have

$$\eth X = 0, \qquad\qquad \eth' Y = \sigma' X. \tag{6,7}$$

Equation (6) on its own again has $2s+2$ solutions. Without loss of generality we assume that $\eth$ and $\eth'$ are those for a metric sphere in Minkowski space, and hence the solutions to (6) are the spin weight $s+\frac{1}{2}$ spherical harmonics. These also obey $\eth'^{2s+2}X = 0$, and thus if (7) has $2s+2$ solutions we can write the solutions as

$$Y = \sum_{i=0}^{2s+1} (-1)^i a_i \eth'^i X \tag{8}$$

for some spin-weighted functions a_i. Substituting into (7) we find

$$\begin{aligned} &\eth' a_{i+1} = a_i \qquad i < 2s+2 \\ &\eth' a_{2s+2} = \sigma' \end{aligned} \tag{9}$$

so that there exists a potential α such that

$$\eth'^{2s+2}\alpha = \sigma' \tag{10}$$

and thus if the lowest term in the expansion of σ' in spin-weighted spherical harmonics is of $l = N$ then there will not be $2s+2$ solutions to equation (7) for $s > \frac{N}{2}$. Since the disappearance of the extra solutions depends not on the 'size' of σ' in some norm but its expansion in spin-weighted spherical harmonics we may find spheres arbitrarily close to the shear-free case with extra solutions.

I do not know if this is of any use. A lot of the utility of 2-surface twistors comes because they are solutions to the restriction to a 2-surface of a 4-dimensional equation. For the pseudo-twistors there is no such 4-dimensional equation in general, and so even for a 2-surface in Minkowski space very little can be said about the solutions. I have been unable for instance to find any norm except for shear-free 2-surfaces. Their use may well be pedagogical, providing a simple example of extra solutions and demonstrating that the shear-free 2-surface can give non-generic results.

§II.4.25 **Two-surface twistors and hypersurface twistors** *by R.Penrose* (TN 21, February 1986)

Let $\mathcal{S}$ be a real analytic 2-surface (topology S^2) in an analytic spacelike hypersurface $\mathcal{H}$ in an analytic space-time, and let $\mathbb{C}\mathcal{S}$, $\mathbb{C}\mathcal{H}$ be respective complex 'thickenings' of $\mathcal{S}$, $\mathcal{H}$. Let ω^A be a solution of the 2-surface twistor equation on $\mathcal{S}$. Then

$$\xi_A \xi^B \eta^{C'} \nabla_{BC'} \omega^A = 0 \tag{1}$$

whenever $\xi^A \eta^{A'}$ is tangent to $\mathcal{S}$. Let ζ be a β-curve in $\mathbb{C}\mathcal{H}$ with associated (tangent) spinor λ_A, so that

$$N^{BA'} \lambda_B \lambda^A \nabla_{AA'} \lambda_{C'} = 0 \tag{2}$$

along ζ, where N^a is normal to $\mathcal{H}$. (Note that $N^{BA'}\lambda_B\lambda^A$ is tangent to ζ.) Let $\mathcal{T}^*$ be the dual hypersurface twistor space for $\mathcal{H}$. Consider the function Ω, on a region $\mathcal{R}$ of $\mathcal{T}^*$, defined as the value of

$$\omega^A \lambda_A$$

at the intersection point of ζ with $\mathcal{S}$—assuming for the moment, that this point is unique. It is clear from the holomorphic nature of this construction that Ω is holomorphic on $\mathcal{R}$. Now consider the situations where ζ is *tangent* to $\mathcal{S}$. If Ω is to remain holomorphic we require that the derivative of $\omega^A \lambda_A$ along the β-curve ζ vanishes. Indeed we have

$$N^{BA'} \lambda_B \lambda^A \nabla_{AA'} (\omega^C \lambda_C) = N^{BA'} \lambda_B \lambda^A \lambda_C \nabla_{AA'} \omega^C + \omega^C N^{BA'} \lambda_B \lambda^A \nabla_{AA'} \lambda_C = 0$$

since the first term vanishes by (1) and the second by (2).

Though necessary, the vanishing of this derivative is not sufficient for Ω to be holomorphic at the points of $\mathcal{T}^*$ where ζ touches $\mathcal{S}$. For near such places ζ will have 2 intersections with $\mathcal{S}$ and so the function Ω is liable to be 2-valued there. The condition just established ensures that to *first order* these 2 values coincide. But they need not coincide to higher orders.

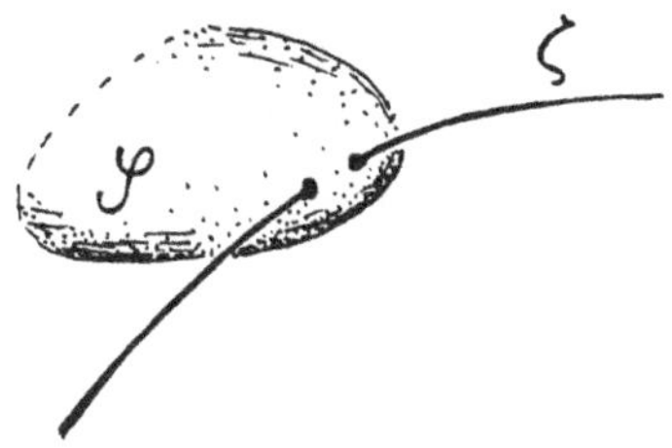

The (linearly independent) elements of $\mathbb{T}(\mathcal{S})$ (i.e. 4 independent ω^A's) provide *coordinates* ('linear' functions) on $\mathcal{R}$ via their corresponding functions Ω. Such coordinates are characterized by the fact that the branching vanishes to 1st order—and this fact serves to *define* the 2-surface twistor space $\mathbb{T}(\mathcal{S})$ in terms of $\mathcal{T}^*$. Only when $\mathcal{T}^*$ is flat (so 3-surface twistors exist on $\mathcal{H}$) can holomorphic coordinates extend single-valuedly to the *whole* of $\mathcal{T}^*$, so only in this case is the branching completely absent.

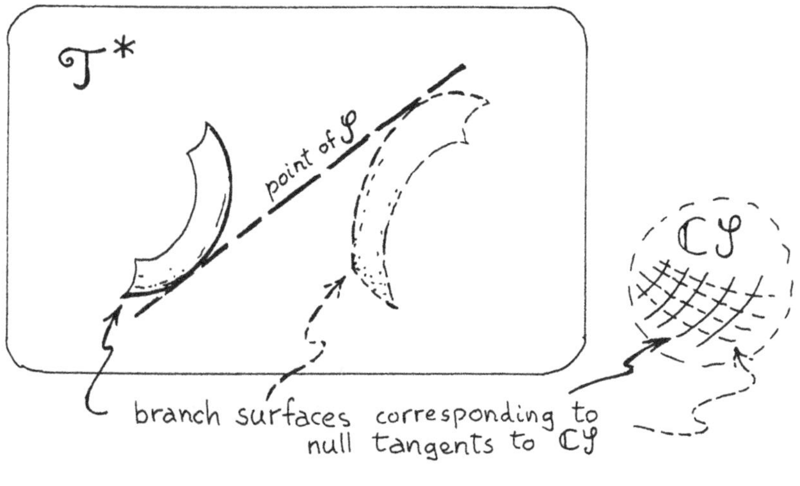

Similar arguments can be applied to β-planes (in place of β-curves) in the complexified space-time, when these exist.

§II.4.26 **A quasi-local mass construction with positive energy** *A.Dougan and L.J.Mason* (TN 30, June 1990)

In this note we propose a pair of modifications to Penrose's quasi-local mass construction that not only lead to a definition of a real 4-momentum and mass of the gravitational and matter fields within a two surface $\mathcal{S}$, but also have the property that the momentum can be proved to be future pointing when the 2-surface can be spanned by a three surface on which the data satisfies the dominant energy condition (the proof also requires that the 2-surface be convex). The new definition reproduces the good properties of the quasi-local mass construction—it gives zero in flat space, and the correct results in linearized theory and at infinity.

Motivation: In Mason (1989) (see also Mason & Frauendiener 1990) the components of the angular momentum twistor associated to a 2-surface $\mathcal{S}$ are interpreted as the values of the Hamiltonians that generate motions of a spanning 3-surface $\mathcal{H}$ whose boundary value on $\mathcal{S}$ are 'quasi-Killing vectors' constructed out of solutions to the 2-surface twistor equations. (The value of the Hamiltonians that generate motions of $\mathcal{H}$ in space-time depends only on the boundary value of the deformation 4-vector field on $\mathcal{S}$.)

For Penrose's quasi-local mass construction (Penrose 1982) the quasi-Killing vectors are constructed out of the four linearly independent solutions of the twistor equation $\omega^A_\alpha = (\omega^A_0, \ldots, \omega^A_3)$. They are given by $K^{AA'} = K^{\alpha\beta}\omega^A_\alpha\pi^{A'}_\beta$ where $K^{\alpha\beta} = K^{(\alpha\beta)}$ is a matrix of constants and $\pi_{A'\alpha}$ are the π-parts of the ω^A_α defined by $d\omega^A|_{\mathcal{S}} = -i\pi_{A'}dx^{AA'}|_{\mathcal{S}}$. The value of the Hamiltonian that generates deformations of $\mathcal{H}$ with boundary value $K^{AA'}$ on $\mathcal{S}$ is obtained by inserting this decomposition of $K^{AA'}$ into the Witten-Nestor integral:

$$H(K^{AA'}) = A_{\alpha\beta}K^{\alpha\beta} = -i\oint_{\mathcal{S}} K^{\alpha\beta}\omega^A_\alpha d\pi^{A'}_\beta \wedge dx_{AA'}.$$

This expression depends on $K^{AA'}$ *and* its decomposition into spinors. By use of $d(-i\pi_{A'}dx^{AA'}) = d^2\omega^A = R_B{}^A\omega^B$ it can be seen that this is equivalent to Penrose's original definition.

The new momentum definition: In order to define a real 4-momentum we must have a definition of real 'quasi-translations' at $\mathcal{S}$. Two definitions follow. The equation $\bar{\eth}\pi_{A'} = 0$, resp. $\eth\pi_{A'} = 0$ (where $\bar{\eth} = \bar{m}^a\nabla_a$, $\eth = m^a\nabla_a$, $\bar{m}^a = \iota^A o^{A'}$ and $m^a = o^A\iota^{A'}$ with $o^{A'}o^A$, $\iota^A\iota^{A'}$ the outward resp.

inward null normal etc.) in general has just 2 linearly independent solutions on $\mathcal{S}$, since this equation can be thought of as the condition that $\pi_{A'}$ is a holomorphic (resp. anti-holomorphic) section of the spin bundle $\mathbf{S}_{A'}$ on the sphere $\mathcal{S}$ where the complex structure on $\mathcal{S}$ is that induced from the space-time metric, and that on $\mathbf{S}_{A'}$ arises from the space-time spin connection. Generically $\mathbf{S}_{A'}$ is trivial as a holomorphic vector bundle on $\mathcal{S}$ and so there exists precisely two solutions $(\pi^{0'}_{A'}, \pi^{1'}_{A'}) = \pi^{\underline{A}'}_{A'}$. (This type of idea is used in the definition of quasi-local charges for Yang-Mills in Tod 1983.)

We can now define a 'quasi-translation' to be a 4-vector field on $\mathcal{S}$ of the form

$$K_{AA'} = K_{\underline{A}\underline{A}'}\pi^{\underline{A}'}_{A'}\bar{\pi}^{\underline{A}}_{A}$$

where the $K_{\underline{A}\underline{A}'}$ are constants. This can now be inserted into the Witten-Nestor form to obtain the corresponding values of the momenta. The quasi-local momentum can thus be defined (modulo irrelevant overall real constants) as:

$$P^{\underline{A}\underline{A}'} = i\oint_{\mathcal{S}} \pi^{\underline{A}'}_{A'} d\bar{\pi}^{\underline{A}}_{A} \wedge dx^{AA'}$$

The mass. In order to define a mass, we must be able to define a constant $\varepsilon_{\underline{A}\underline{B}}$ so that we can define:

$$m^2 = P^{\underline{A}\underline{A}'}P^{\underline{B}\underline{B}'}\varepsilon_{\underline{A}\underline{B}}\varepsilon_{\underline{A}'\underline{B}'}.$$

The natural definition is $\varepsilon^{\underline{A}'\underline{B}'} = \pi^{\underline{A}'}_{A'}\pi^{\underline{B}'}_{B'}\varepsilon^{A'B'}$. It follows from $\bar{\eth}\pi^{\underline{A}'}_{A'} = 0$ that $\bar{\eth}\varepsilon^{\underline{A}'\underline{B}'} = 0$, so that the $\varepsilon^{\underline{A}'\underline{B}'}$ are holomorphic and global functions on the sphere and hence, by Liouville's theorem, constant.

Flat space, linearized theory and infinity. In flat space, the $\pi_{A'}$ are guaranteed to be the restriction to $\mathcal{S}$ of the constant spinors, since they certainly satisfy the equation, and the solutions are unique. The integrand therefore vanishes giving the correct answer $P_{\underline{A}\underline{A}'} = 0$. In linearized theory one can again, with a little work, see that the right answer is obtained (one needs to integrate potentially awkward terms by parts in order to see that they vanish). Asymptotically at space-like infinity, the $\pi_{A'}$'s are the asymptotically constant spinors (again because the asymptotically constant spinors satisfy $\bar{\eth}\pi_{A'} = 0$ and therefore span the solution space) and the expression reduces to the Witten-Nestor expression for the ADM energy. At null infinity there is the subtlety that only one of the definitions gives the correct asymptotic spin space depending on whether one is at future or past null infinity.

Positivity. It is essential for a good definition of momentum that it should be future pointing. The following argument is an adaptation of ideas in Ludvigsen & Vickers (1983) based on Witten (1981). In the following we show that $P^{00'}$ is positive, and write, for simplicity, $\pi_{A'} = \pi^{0'}_{A'}$.

> ***Theorem.*** The quasi-local momentum $P^{00'}$ defined by $\eth\pi_{A'} = 0$ (resp. $\bar{\eth}\pi_{A'} = 0$) is positive whenever $\rho < 0$ (resp. $\rho' > 0$).

Proof. Let $\lambda_{A'}$ be some field defined on a 2-surface $\mathcal{S}$ spanned by some non singular space-like 3-surface $\mathcal{H}$. Let $I_\lambda(\mathcal{S})$ be the integral of the Witten-Nestor 2-form $\Lambda = -i\bar{\lambda}_A d\lambda_{A'} \wedge dx^{AA'}$ over $\mathcal{S}$. In spin coefficients and the GHP formalism this may be written:

$$I_\lambda(\mathcal{S}) = \oint_{\mathcal{S}} \{\bar{\lambda}_1(\eth\lambda_{0'} + \rho\lambda_{1'}) - \bar{\lambda}_0(\bar{\eth}\lambda_{1'} + \rho'\lambda_{0'})\}dS \qquad (1).$$

Consider first the system of equations $\eth\pi_{A'} = 0$:

$$\eth\pi_{0'} + \rho\pi_{1'} = 0, \qquad \eth\pi_{1'} + \bar{\sigma}'\pi_{0'} = 0 \qquad (2,3).$$

Then using (2) and integrating by parts we get:

$$I_\pi(\mathcal{S}) = -\oint_{\mathcal{S}} (\rho'\bar{\pi}_0\pi_{0'} + \rho\bar{\pi}_1\pi_{1'})dS \qquad (4)$$

Since the Sen-Witten equation on $\mathcal{H}$ consists of an elliptic system of two first order P.D.E's, we may find a solution $\tilde{\pi}_{A'}$ satisfying the boundary condition

$$\tilde{\pi}_{0'} = \pi_{0'} \qquad (5)$$

on $\mathcal{S}$. In general $\tilde{\pi}_{1'}$ will differ from $\pi_{1'}$ on $\mathcal{S}$. Denote this difference by:

$$Y = \tilde{\pi}_{1'} - \pi_{1'} \qquad (6)$$

We now relate $I_\pi(\mathcal{S})$ to $I_{\tilde{\pi}}(\mathcal{S})$:

$$\begin{aligned}
I_{\tilde{\pi}}(\mathcal{S}) &= \oint_{\mathcal{S}} \Big\{\bar{\tilde{\pi}}_1(\eth\tilde{\pi}_{0'} + \rho\tilde{\pi}_{1'}) - \bar{\tilde{\pi}}_0(\bar{\eth}\tilde{\pi}_{1'} + \rho'\tilde{\pi}_{0'})\Big\} dS \\
&= \oint_{\mathcal{S}} \Big\{\bar{\tilde{\pi}}_1(\eth\pi_{0'} + \rho\tilde{\pi}_{1'}) - \bar{\pi}_0(\bar{\eth}\tilde{\pi}_{1'} + \rho'\pi_{0'})\Big\} dS \\
&= \oint_{\mathcal{S}} \Big\{\bar{\tilde{\pi}}_1(-\rho\pi_{1'} + \rho\tilde{\pi}_{1'}) - \rho\bar{\pi}_1\tilde{\pi}_{1'} - \rho'\bar{\pi}_0\pi_{0'})\Big\} dS \\
&= \oint_{\mathcal{S}} \Big\{-\rho'\bar{\pi}_0\pi_{0'} - \rho\bar{\pi}_1\pi_{1'} + \rho(\bar{\tilde{\pi}}_1 - \bar{\pi}_1)(\tilde{\pi}_{1'} - \pi_{1'})\Big\} dS \\
&= I_\pi(\mathcal{S}) + \oint_{\mathcal{S}} \rho Y\bar{Y} dS
\end{aligned}$$

Where we have used equations (2), (4), (5) and (6) and an integration by parts. As is well known (Witten 1980) for matter satisfying the dominant energy condition $I_{\tilde{\pi}}(\mathcal{S}) \geq 0$ so that whenever $\rho \leq 0$, $I_\pi(\mathcal{S}) \geq 0$. This implies that $P^{\underline{A}\underline{A}'}$ is future pointing as required.

Considering next the equation $\bar{\eth}\pi_{A'} = 0$, an analogous argument to the one above but now with $\tilde{\pi}_{1'} = \pi_{1'}$ as boundary conditions for the Sen-Witten equation will show positivity whenever $\rho' \geq 0$. □

The conditions $\rho \leq 0$ or $\rho' \geq 0$ are the condition that the two surface is convex, i.e. that there should be no indentations. This will be satisfied by a wide class of 2-surfaces in a generic space-time.

Angular momentum: One can define more general quasi-Killing vectors using local twistors, $(\omega^A, \pi_{A'})$ restricted to $\mathcal{S}$ satifying either $\eth(\omega^A, \pi_{A'}) = 0$ or $\bar{\eth}(\omega^A, \pi_{A'}) = 0$ where $\eth$ and $\bar{\eth}$ act according to the local twistor connection. These equations are guaranteed to have just four independent solutions generically since as before these are $\bar{\partial}$ equations whose solutions are holomorphic sections of a holomorphic vector bundle on the sphere $\mathcal{S}$. Generically the holomorphic vector bundle will be trivial and so there will be just four linearly independent solutions. These can be used to define quasi-Killing vectors, and quasi-conformal Killing vectors as in Mason & Frauendiener which then give rise to 'conserved' quantities by substitution into the Witten-Nestor form. (When $R_{ab} = 0$ on $\mathcal{S}$, it makes consistent sense to set $\pi_{A'} = 0$ in such a local twistor and then we can retrieve the quasi-local momentum above within the scheme.)

Many thanks to K.P.Tod for input and useful discussions.

Comment (1994). This article became Dougan & Mason (1991). Further calculations and developments appeared in Dougan (1992).

References

Dougan, A.J. (1992) Class. Quant. Grav..

Dougan, A.J. & Mason, L.J. (1991) Phys. Rev. Lett., **67**, 16, 2119-2122.

Ludvigsen, M. & Vickers, J. (1983) Momentum, angular momentum and their quasi-local null surface extensions, J. Phys. A: Math. Gen., **16**, 1155-1168.

Mason, L.J. (1989) A Hamiltonian interpretation of Penrose's quasi-local mass, Class. Quant. Grav. **6** , L7-L13.

Mason, L.J. & Frauendiener, J. (1990) The Sparling 3-form, Ashtekar variables and quasi-local mass, in *Twistors in mathematics and Physics*, (R.J.Baston, & T.N.Bailey eds), LMS lecture note series, CUP.

Penrose, R. (1982) Quasi-local mass and angular momentum in general relativity, Proc. R. Soc. London, A **381**, 53-63.

Penrose, R. & Rindler, W. (1986) Spinors and space-time, Vol. 2, C.U.P.

Tod, K.P. (1983) Quasi-local charges in Yang-Mills theory, Proc. R. Soc. Lond. A **389**, 369-377.

Tod, K.P. (1990) Penrose's quasi-local mass, in *Twistors in mathematics and Physics*, (R.J.Baston & T.N.Bailey, LMS lecture note series, CUP.

Witten, E. (1981) A new proof of the positive energy theorem, Comm. Math. Phys. **80**, 381-402

Index

Printed and bound by CPI Group (UK) Ltd, Croydon, CR0 4YY
01/11/2024
01782610-0018